Active Plasmonic Nanomaterials

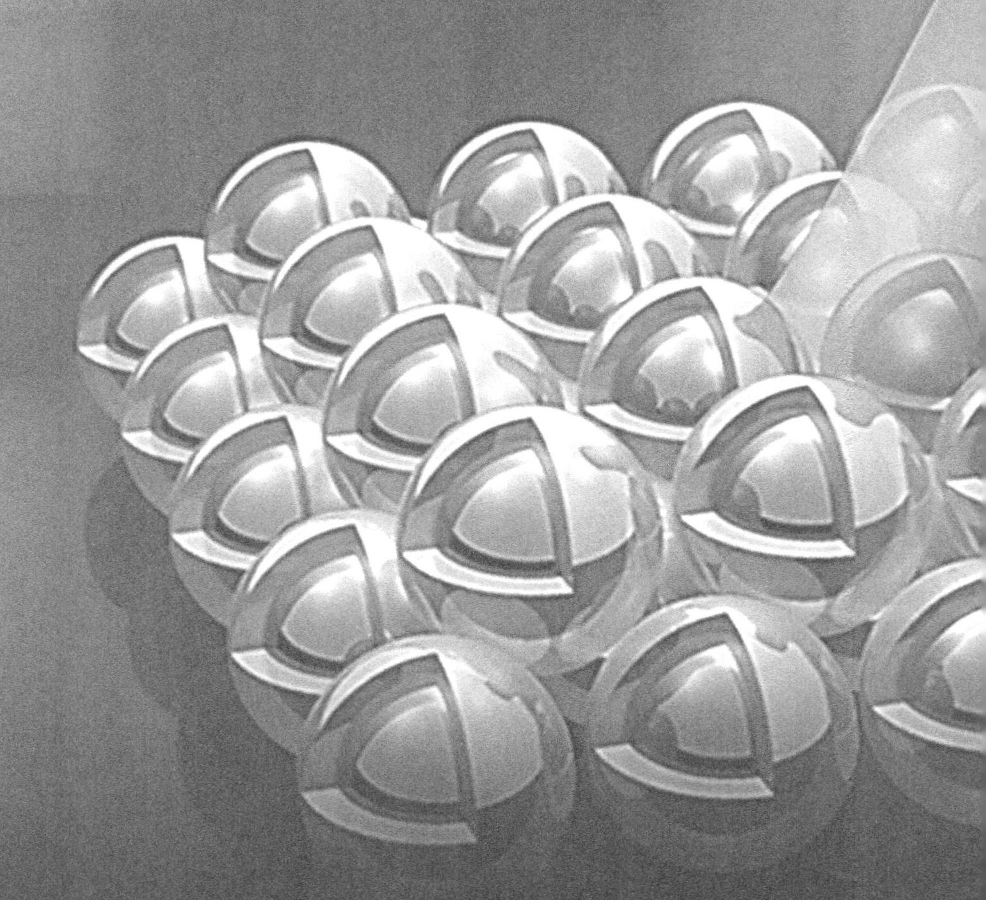

Active Plasmonic Nanomaterials

edited by

Luciano De Sio

Published by

Pan Stanford Publishing Pte. Ltd.
Penthouse Level, Suntec Tower 3
8 Temasek Boulevard
Singapore 038988

Email: editorial@panstanford.com
Web: www.panstanford.com

British Library Cataloguing-in-Publication Data
A catalogue record for this book is available from the British Library.

Active Plasmonic Nanomaterials

ISBN 978-981-4613-00-2 (Hardcover)
ISBN 978-981-4613-01-9 (eBook)

Printed in the USA

The editor and those among the authors who had the fortune to meet her, want to dedicate this book to Daniela Pucci, a dear friend and a clever scientist. She was a pioneer member of the Metallomesogens Group at the University of Calabria and, recently, she started to be enthusiastically involved in researches on metal-containing chromonic systems, of interest for plasmonics.

Ciao Daniela!

Contents

Preface

Science and technology seldom progress separately; more often they go hand in hand, in a combination that yields to a number of intellectual and practical paths. Nanotechnology is part of one of such vibrant paths, leading to a number of new findings that are maturing and thriving fast in both academia and industry, while nanophotonics gathers together all that can be utilized to manipulate light at the nanoscale, for a number of photonic applications. As a matter of fact, nanophotonic devices are of great interest for integrated optics, plasmonic circuits, biosensing and quantum information processing. Plasmonics is a subfield of nanophotonics, which involves exploitation of surface plasmons to realize the control of light at the nanoscale. Indeed, one promising way to localize the optical radiation into a nanometer-sized volume has been realized by using the unique properties of plasmonic metallic nanoparticles (NPs), which, represent an effective bridge between bulk materials and atomic or molecular structures. NPs can exhibit a highly vibrant color, which is absent both in the bulk material and in individual atoms; the physics behind this phenomenon can be understood by considering the collective oscillation of the conducting free electrons of the metallic NP, an effect that is referred as the "localized plasmonic resonance (LPR)". Such a resonance is strongly dependent on particle size and shape, and on the dielectric function of the medium surrounding the NPs; as an example, these parameters enable a "static" design of the resonance frequency throughout the visible and into the near-infrared spectra when keeping the particle size below 100 nm. A more effective method to influence the LPR is to vary the refractive index of the medium surrounding the NPs; the use of reconfigurable matter as the surrounding medium could hence provide an "active way" for controlling the plasmonic resonance.

This book is aimed at reviewing some recent efforts committed to utilize NPs in a number of research fields that include, but are not particularly limited to, photonics, optics, chemistry, material science, or metamaterials. In this framework, a particular interest is devoted to "active plasmonics", a concept related to NPs that play an 'active' role, and includes realization of gain-assisted means, utilization of NPs embedded in liquid crystalline materials, and exploitation of NPs for solar energy or even flexible plasmonics. Moreover, the book is designed to provide a powerful tool to people that are enthusiastic of technology, but have no, or minimal, scientific background. The book, indeed, begins with a theoretical background on the physical interaction between light and plasmonic materials. Further on, the most important techniques enabling size and or shape-controlled growth of NPs are highlighted, with an additional focus on their surface functionalization.

In editing and organizing the book, I have made all possible attempts to cover the growing field of plasmonic nanomaterials and related technologies. With this aim, the publication gathers various contributions of outstanding research groups all over the world, which are related to the field of plasmonic nanomaterials and provide both the basics and the necessary advanced knowledge in the field of plasmonics, photonics, and optics. I express my deep and sincere thanks to all the authors, who contributed their expertise and research findings for the success of this publication. Special thanks to Antonio De Luca for providing the cover image of the book and to Roberto Bartolino, Timothy Bunning Nelson Tabiryan, and Cesare Umeton for their nurturing and long-standing support. I thank Stanford Chong for his never-ending help in organizing and completing the book and to Hari M. Atkuri for talking over its content while participating with me in official and unofficial 10 km runs. I would like to express my gratitude to my family without whose support I would not have been able to published this work. I owe thanks to Francesca Petronella for encouraging me to materialize and finalize the book.

Luciano De Sio
Summer 2015

Chapter 1

Plasmonics: A Theoretical Background

Luigia Pezzi, Giovanna Palermo, and Cesare Umeton
Department of Physics and Centre of Excellence for the Study of Innovative Functional Materials, University of Calabria, 87036 Arcavacata di Rende, Italy
luigia.pezzi@fis.unical.it

1.1 Introduction

The paper that best embodies the genesis of processes involved in plasmonics of nanostructured materials is the one by the Russian physicist Veselago (1968). In his visionary work, he lays the foundation of many topics that have been clarified in recent years. Plasmonics is a branch of optics that investigates the behavior of electromagnetic (EM) waves in the visible range in nanostructured materials. In order to study the propagation of EM waves in matter, the dispersion equation represents the starting point:

$$\left| \frac{\omega^2}{c^2} \varepsilon_{il} \mu_{lj} - k^2 \delta_{ij} + k_i k_j \right| = 0 \quad (1.1)$$

In this equation, the only involved parameters of the medium are ε_{il} and μ_{lj}, which are the components of the tensors dielectric permittivity ε and magnetic permeability μ, respectively; δ_{ij} is the

Active Plasmonic Nanomaterials
Edited by Luciano De Sio

ISBN 978-981-4613-00-2 (Hardcover), 978-981-4613-01-9 (eBook)
www.panstanford.com

Kronecker delta; ω is the frequency of a monochromatic impinging wave and k its wave vector. If we consider an isotropic material, Eq. (1.1) takes the simpler form:

$$k^2 = \frac{\omega^2}{c^2} n^2 \tag{1.2}$$

where n^2 is the square of the refractive index of the material and is given by

$$n^2 = \varepsilon\mu. \tag{1.3}$$

Veselago's analysis starts by assuming that a material with negative values of both dielectric permittivity ε and magnetic permeability μ might exist. This hypothesis does not change Eq. (1.3), but by considering the case $\varepsilon < 0$ and $\mu < 0$ from a purely formal point of view in Maxwell's equation, Veselago comes to rewrite some important laws of physics and to presume that new and interesting phenomena could occur.

The first case that he presents is the so called "left-handed" medium. For a monochromatic plane wave, in which all fields contain the phase factor $e^{i(kz-\omega t)}$, Maxwell's equations that involve the curl of the EM field become

$$\begin{aligned} \mathbf{k} \times \mathbf{E} &= \omega\mu\mathbf{H} \\ \mathbf{k} \times \mathbf{H} &= -\omega\varepsilon\mathbf{E}. \end{aligned} \tag{1.4}$$

It is evident that if $\varepsilon > 0$ and $\mu > 0$, then **E**, **H**, and **k** form a right-handed triplet of vectors, but if $\varepsilon < 0$ and $\mu < 0$, then the vectors form a left-handed set. This means that the energy flux carried by the wave, which is determined by the Poynting vector **S**:

$$\mathbf{S} = \mathbf{E} \times \mathbf{H} \tag{1.5}$$

always forms a right-handed set with **E** and **H**. Now for right-handed substances, **k** (related to phase velocity) and **S** (related to energy flux and group velocity) are in the same direction, while for left-handed materials, **k** and **S** have opposite directions, with the consequence that left-handed materials exhibit a negative group velocity. Introducing the direction cosines for vectors **E**, **H**, and **k**, the determinant of the resulting matrix p is always $+1$ for right-handed set and -1 for left-handed set; thus, p characterizes the "right-ness" of the given medium, and with this difference in mind,

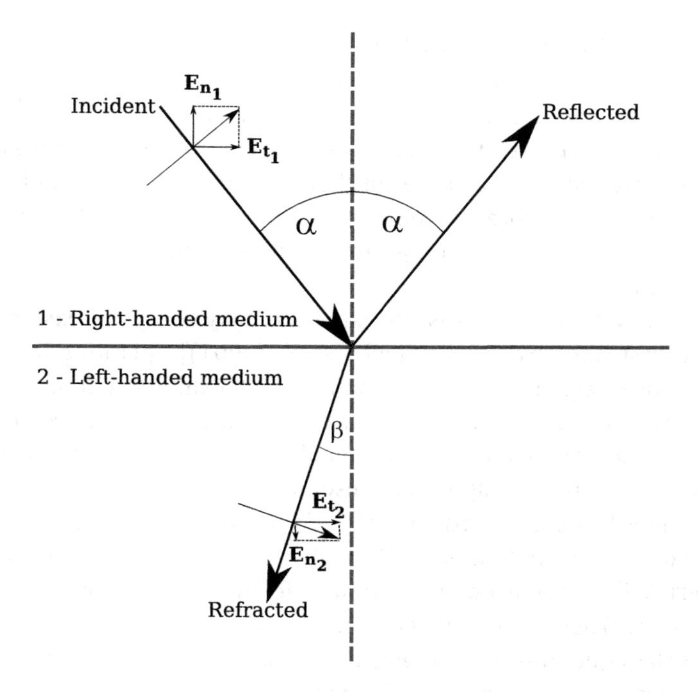

Figure 1.1 Passage of a light beam through the boundary between two media: from a right-handed medium to a left-handed one.

Veselago rewrites the laws of a lot of physical effects, like Doppler shift, Vavilov-Cerenkov effect, and Snell's law. In particular, for the refraction of a light beam from one medium into another, boundary conditions must be satisfied independently of the rightness of the media:

$$E_{t_1} = E_{t_2} \qquad H_{t_1} = H_{t_2} \tag{1.6}$$

$$\varepsilon_1 E_{n_1} = \varepsilon_2 E_{n_2} \qquad \mu_1 H_{n_1} = \mu_2 H_{n_2} \tag{1.7}$$

where subscripts 1,2 indicate the two media, while n and t stay for "normal" and "tangential" to the separation surface, respectively. It follows that the path of the resulting refracted beam (Fig. 1.1) lies on the opposite side with respect to the normal axis to the separation surface only in case of a right-handed second medium. Indeed, according to the above considerations, with notations of

Fig. 1.1, Snell's law is now written as

$$\frac{\sin\alpha}{\sin\beta} = n_{1,2} = \frac{p_2}{p_1}\left|\sqrt{\frac{\varepsilon_2\mu_2}{\varepsilon_1\mu_1}}\right| \tag{1.8}$$

where p_1 and p_2 are the "right-ness" of the two media and $\varepsilon_{1,2}$ and $\mu_{1,2}$ are their dielectric permittivity and magnetic permeability, respectively. Therefore, by going from a right-handed medium to a left-handed one, the angle of refraction turns out to be negative (the β angle in Fig. 1.1).

This law allowed scientists to experimentally verify the existence of the first *"metamaterial"* [Shelby et al. (2001)] operating in the microwave region, that is, a material that exhibits a frequency range where the effective index of refraction (n) is negative. The material consists of a two-dimensional array of repeated unit cells of copper strips and split ring resonators on interlocking strips of standard circuit board material. By measuring the scattering angle of the transmitted beam through a prism fabricated with this material, Shelby et al. determined the effective n value that satisfies Snell's law. These experiments directly confirmed the prediction of Maxwell's equations that n is given by the negative square root of $\varepsilon\mu$ for the frequencies where both the permittivity and the permeability are negative. An interesting case, pointed out by Veselago, is that of a beam going from a medium with $\varepsilon_1 > 0$, $\mu_1 > 0$ into another one with $\varepsilon_2 = -\varepsilon_1$, $\mu_2 = -\mu_1$. In this case, the beam undergoes refraction at the interface between the two media, but there is no reflected beam, since in Fresnel's formulas [see Born and Wolf (1999)] only the absolute values of quantities ε, μ, n, α, β have to be used; then the reflection coefficient turns out to be zero. The use of left-handed media would allow the design of very unusual refracting systems. An example (Fig. 1.2) is a simple plate of thickness d made of a left-handed medium with $n = -1$ and put in the vacuum (where $n = 1$). Such a plate can focus in a given point the radiation from a point source located at a distance $l < d$ from the plate, but it is not a lens in the usual sense of the word since it will not focus in a given point a bundle of rays coming from infinity. Above considerations have been exposed by Pendry (2000) in his letter in which he assumed also the possibility to build a perfect lens. In addition, Pendry envisaged that the left-handed medium can prevent the decay of evanescent waves. Indeed, such waves decay in

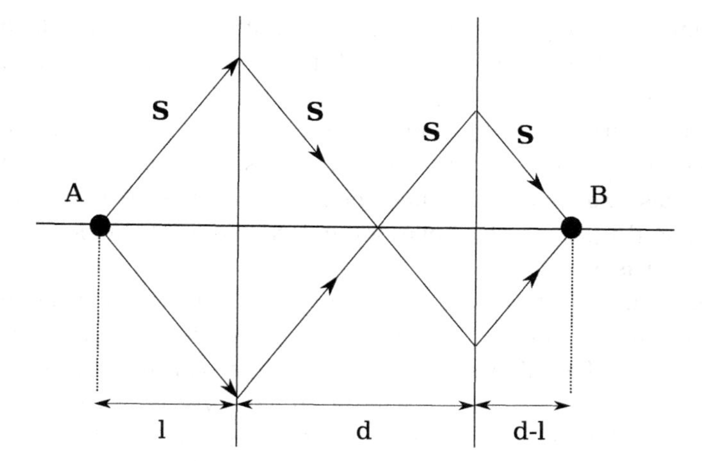

Figure 1.2 Passage of rays of light through a plate of thickness d made of left-handed substance. A is a source of radiation, B is the detector of radiation.

amplitude, but not in phase, as they propagate away from the object plane; therefore, in order to be focused, they need to be amplified rather than corrected in their phase. Pendry showed that in above materials, evanescent waves emerge from the far side of the sample enhanced in amplitude by the transmission process. This does not violate the energy conservation, since evanescent waves do not transport energy, but by using this new lens, a distinctive feature can be exploited, represented by the circumstance that both propagating and evanescent waves contribute to the resolution of the image. As a consequence, there is no physical obstacle to realize a perfect reconstruction of the image beyond those practical limitations that are, in general, introduced by the numerical aperture value and by defects in the lens surface.

As a matter of fact, in order to get all the fantastic properties listed so far (and much more), it is, however, necessary to exploit a material that simultaneously exhibits negative values of both ε and μ. As this feature is not found in natural substances known to date, it is necessary to obtain it artificially. Today, it seems that the best way to obtain both $\varepsilon < 0$ and $\mu < 0$ is the fabrication of nanostructured (nonmagnetic) metallic materials which would naturally exhibit this characteristic. In the microwave region, this is already a reality [Pendry et al. (1999); Smith et al. (2000); Shelby et al. (2001)], while, for the optical frequency range, there is still much to do.

As a starting point, in order to study these nanostructured materials, it is necessary to know in details the electromagnetism of metals and associated phenomena, which involve the volume plasmon (VP), the surface plasmon polaritons (SPP), and localized surface plasmon (LSP).

As well known from everyday experience, for frequencies up to the infrared part of the spectrum, metals are highly reflective and do not allow EM waves to propagate through them. At higher frequencies, toward the near-infrared and visible part of the spectrum, the EM field penetration increases significantly, leading to an increased dissipation. Finally, at ultraviolet frequencies, metals acquire dielectric character and allow the propagation of EM waves, albeit with varying degrees of attenuation, depending on the details of the electronic band structure of the medium. These dispersive properties can be described by a complex dielectric function $\varepsilon(\omega)$, which provides the basis of all previously listed phenomena (PV, SPP, LSP) associated with EM of metals.

The interaction of metals with EM fields can be investigated in a classical framework based on Maxwell's equations. Even metallic nanostructures down to sizes of the order of a few nanometres can be described without a need to resort to quantum physics. Indeed, this has been introduced as a solution to the failure of the classical model for physical systems at the atomic and molecular scale. The "unit" that is necessary to take into account for a particular physical system to be assigned a macroscopic or microscopic character is represented by Planck's constant $\hbar$: In the limit where $\hbar \to 0$, the formalism must reduce to the classical one. In fact, in a black body, assuming that the energy is discretely packed in energy packets called "quanta" with energy $h\nu = \hbar\omega$, by making use of the kinetic theory, Plank's Formula for the average energy gives $\overline{E} = h\nu/(e^{\frac{h\nu}{k_B T}} - 1)$; in the limit where Plank's constant tends to zero (i.e., when $k_B T \gg h\nu$), the classical result is obtained:

$$\lim_{\frac{h\nu}{k_B T}\to 0} \overline{E} \simeq \frac{h\nu}{1 + \frac{h\nu}{k_B T} + \cdots - 1} \simeq k_B T \tag{1.9}$$

Optics of metals was developed in the framework of Drude's theory, which considers (as we will see in the next section) only the effects of the electrons that are in the conduction band. Since the high density of free carriers results in a quite minute spacing between the electron energy levels if compared to the thermal energy $k_B T$ at room temperature, the argument falls within the realms of the classical theory.

1.2 Electromagnetism of Metals

For a long time the most known property of metals was the high electrical conductivity. After three years from Thompson's discovery of the electron, scientists became more interested in studying mechanisms of interaction between metals and electromagnetic fields. Around 1900, Paul Drude, a German physicist, used new concepts to postulate a classical model that well explained several phenomena related to the interaction between radiation and metals. This model links optical and electric properties of a metal through the behavior of electrons. The assumptions of Drude's model are as follows:

- Metals are made of heavy, static, positively charged ions immersed in a cloud of light, negatively charged, easily mobile electrons, which form an *electron gas* that follows the Maxwell–Boltzmann statistics.
- The electron–electron interactions can be neglected.
- The only considered interaction are the electron–ion collisions.

By following the kinetic theory of gases, electrons in the gas move in straight lines and make collisions only with the ion cores. The probability for an electron to make a collision in a short time dt is dt/τ, where τ is the mean time between collisions, called *relaxation time*. This quantity, which is typically of the order of 10^{-14} s at room temperature, is related to another important quantity $\gamma = \frac{1}{\tau}$, which represents the collision frequency and thus has values of the order of 100 THz.

Drude's model successfully determined the form of Ohm's law in terms of free electrons and the relation between electrical and thermal conduction [Drude (1900)], but it failed to explain electron heat capacity and the magnetic susceptibility of conduction electrons. Failures of the model are the result of the limitations of the classical model (and Maxwell–Boltzmann statistics in particular) [Ashcroft and Mermin (1976)].

In microscopic physics, it is common to express Ohm's law in terms of a dimension-independent conductivity, which is intrinsic to the substance that the wire is made of. In this framework, Ohm's law writes as:

$$\mathbf{J} = \sigma \mathbf{E} \tag{1.10}$$

where $\mathbf{E}$ represents the electric field, $\mathbf{J}$ the current density, and σ the conductivity of the material. We consider a wire of cross-sectional area A, where an electrical current flows, which consists of N electrons per volume unit, all moving in the same direction with velocity v. The number of electrons flowing through the area A in time dt is given by dN=NAvdt, while the charge crossing A in dt is dQ= −edN= −NevAdt, so that $\mathbf{J} = -Ne\mathbf{v}$. In the absence of electric fields, electrons move randomly inside the conductor due to their thermal energy, but when an electric field is applied, electrons are affected by the force $\mathbf{F} = -e\mathbf{E}$ that pushes them to move all in the same direction, with an average speed given by:

$$\mathbf{v} = -\frac{e\tau}{m}\mathbf{E} \tag{1.11}$$

Thus, substituting in $\mathbf{J} = -Ne\mathbf{v}$, Eq. (1.11) yields

$$\mathbf{J} = -\frac{Ne^2\tau}{m}\mathbf{E} \tag{1.12}$$

Comparison with Eq. (1.10) gives the DC-Drude conductivity:

$$\sigma_0 = -\frac{Ne^2\tau}{m} \tag{1.13}$$

Drude's model can also predict a current as a response to an oscillating electric field with angular frequency ω. This can be achieved by considering that the equation of motion for an electron of the electron gas subjected to an external electric field $\mathbf{E}$, is:

$$m\ddot{\mathbf{x}} + m\gamma\dot{\mathbf{x}} = -e\mathbf{E} \tag{1.14}$$

where m is the effective mass and γ is the already mentioned collision frequency that produces the damping. This expression can be rewritten as:

$$\dot{\mathbf{p}} = -\frac{\mathbf{p}}{\tau} - e\mathbf{E} \tag{1.15}$$

where $\mathbf{p} = m\dot{\mathbf{x}}$ is the momentum of an individual free electron. If $\mathbf{E}$ assumes the form $\mathbf{E} = \mathbf{E}_0 e^{-i\omega t}$, we consider as a solution to Eq. (1.15) the expression $\mathbf{p}(t) = \mathbf{p}_0 e^{-i\omega t}$, by substituting we obtain:

$$-i\omega\mathbf{p}_0 = -\frac{\mathbf{p}_0}{\tau} - e\mathbf{E_0} \tag{1.16}$$

$$\mathbf{J} = -\frac{Ne\mathbf{p}}{m} = \frac{\sigma_0}{1 - i\omega\tau}\mathbf{E} = \sigma(\omega)\mathbf{E} \tag{1.17}$$

Thus, the AC-Drude conductivity is given by:

$$\sigma(\omega) = \frac{\sigma_0}{1 - i\omega\tau} \tag{1.18}$$

A useful application of the Drude model is the description of the propagation of electromagnetic waves in metals by considering a complex dielectric function $\varepsilon(\omega)$, which shows the dispersive properties of the substance. In order to derive the expression of $\varepsilon(\omega)$, we consider again Eq. (1.14), which takes into account the oscillations of the free electron gas induced by the electric field $\mathbf{E}(t)$. A solution to Eq. (1.14) is given by $\mathbf{x}(t) = \mathbf{x}_0 e^{-i\omega t}$, which, when replaced in Eq. (1.14), yields:

$$\mathbf{x}(t) = \frac{e}{m(\omega^2 + i\gamma\omega)}\mathbf{E}(t). \tag{1.19}$$

Since the electric displacement $\mathbf{D}$ and the macroscopic polarization $\mathbf{P}$ are given by:

$$\mathbf{D} = \varepsilon_0\mathbf{E} + \mathbf{P} = \varepsilon_0\varepsilon\mathbf{E} \tag{1.20}$$

$$\mathbf{P} = -Ne\mathbf{x} \tag{1.21}$$

respectively, where N is the number of electrons per unit volume. Thus,

$$\mathbf{D} = \varepsilon_0\mathbf{E} - \frac{Ne^2}{m(\omega^2 + i\gamma\omega)}\mathbf{E} \tag{1.22}$$

and

$$\varepsilon(\omega) = 1 - \frac{\omega_p^2}{\omega^2 + i\gamma\omega} \tag{1.23}$$

that represents the dielectric function in the Drude model, where ω_p is the plasma frequency of the free electron gas, is defined by:

$$\omega_p = \left(\frac{Ne^2}{\varepsilon_0 m}\right)^{1/2} \tag{1.24}$$

From Eq. (1.23) it can be easily derived that the real and imaginary components of this complex dielectric function $\varepsilon(\omega) = \varepsilon_1(\omega) + i\varepsilon_2(\omega)$ are given by:

$$\varepsilon_1(\omega) = 1 - \frac{\omega_p^2 \tau^2}{1 + \omega^2 \tau^2} \tag{1.25}$$

$$\varepsilon_2(\omega) = \frac{\omega_p^2 \tau}{\omega(1 + \omega^2 \tau^2)} \tag{1.26}$$

The complex dielectric function $\varepsilon(\omega)$ is related to the complex refractive index of the medium $\tilde{n} = n(\omega) + i\kappa(\omega)$ through the relation $\tilde{n} = \sqrt{\varepsilon_r}$. Explicitly, this yields:

$$\varepsilon_1 = n^2 - \kappa^2 \tag{1.27}$$

$$\varepsilon_2 = 2n\kappa \tag{1.28}$$

$$n^2 = \frac{\varepsilon_1}{2} + \frac{1}{2}\sqrt{\varepsilon_1^2 + \varepsilon_2^2} \tag{1.29}$$

$$\kappa = \frac{\varepsilon_2}{2n} \tag{1.30}$$

Here, κ is called *extinction coefficient* and determines the absorption of optical EM waves propagating through the medium; it is linked to the absorption coefficient α of Beer's law, which describes the exponential attenuation of a beam intensity $I(x)$ propagating through the medium via $I(x) = I_0 e^{-\alpha x}$. Indeed, since $E \propto \exp(i\frac{\omega}{c}\tilde{n}x)$ and then $I \propto E^2 \propto \exp(2i\frac{\omega}{c}(n + i\kappa)x)$, we have:

$$\alpha(\omega) = \frac{2\kappa(\omega)\omega}{c} \tag{1.31}$$

Equations (1.25) and (1.26) allow as to study the EM response of metals (related to the plasma frequency ω_p) distinguishing the three cases $\omega > \omega_p$, $\omega < \omega_p$, and $\omega = \omega_p$).

In the case $\omega > \omega_p$, $\tilde{n}$ is positive because $\varepsilon(\omega)$ is real and positive, and $\varepsilon(\omega) \to 1$, which implies that the electromagnetic wave propagates through the metal that appears transparent. For noble metals, it is necessary to take into account that the response of the material in this region is dominated by free s-electrons, since the filled d-band close to the Fermi surface causes a highly polarized environment [Maier (2007)]. This contribution to the polarization related to the ion cores can be considered by adding the term $\mathbf{P}_\infty = \varepsilon_0(\varepsilon_\infty - 1)\mathbf{E}$ to Eq. (1.20). This effect is, therefore, described by a dielectric constant ε_∞ (usually $1 \leq \varepsilon_\infty \leq 10$), and we can write:

$$\varepsilon(\omega) = \varepsilon_\infty - \frac{\omega_p^2}{\omega^2 + i\gamma\omega} \tag{1.32}$$

The validity limits of the free-electron description (1.32) are illustrated for the case of gold in Fig. (1.3). It shows the real and imaginary components ε_1 and ε_2 of the dielectric function of a free electron gas, fitted with experimental data of the dielectric function of gold [Johnson and Christy (1972)]. In the visible frequency range, the applicability of the free-electron model clearly breaks down due to the occurrence of interband transitions, leading to an increase in ε_2.

For frequencies $\omega < \omega_p$, we distinguish two subcases: $\omega\tau \gg 1$ and $\omega\tau \ll 1$.

In the case $\omega\tau \gg 1$, we are in the condition of frequency very close to ω_p and, as we have seen at the beginning of this discussion, metals totally reflect the EM waves; in this range the real and imaginary parts of the dielectric function become:

$$\varepsilon_1(\omega) = 1 - \frac{\omega_p^2}{\omega^2} \tag{1.33}$$

$$\varepsilon_2(\omega) \approx 0 \tag{1.34}$$

respectively. As we can see from these equations, the permittivity is real, which implies that there is no absorption; metals retain their metallic character of *perfect conductor*. This behavior is common among different metals but not for noble metals, in which the response is again affected by the interband transitions.

In the case $\omega\tau \ll 1$, we are in the condition of frequency very far from ω_p and the real and imaginary parts of the dielectric function become:

$$\varepsilon_1(\omega) = 1 - \omega_p^2\tau^2 \tag{1.35}$$

$$\varepsilon_2(\omega) \approx \frac{\omega_p^2}{\omega} \tag{1.36}$$

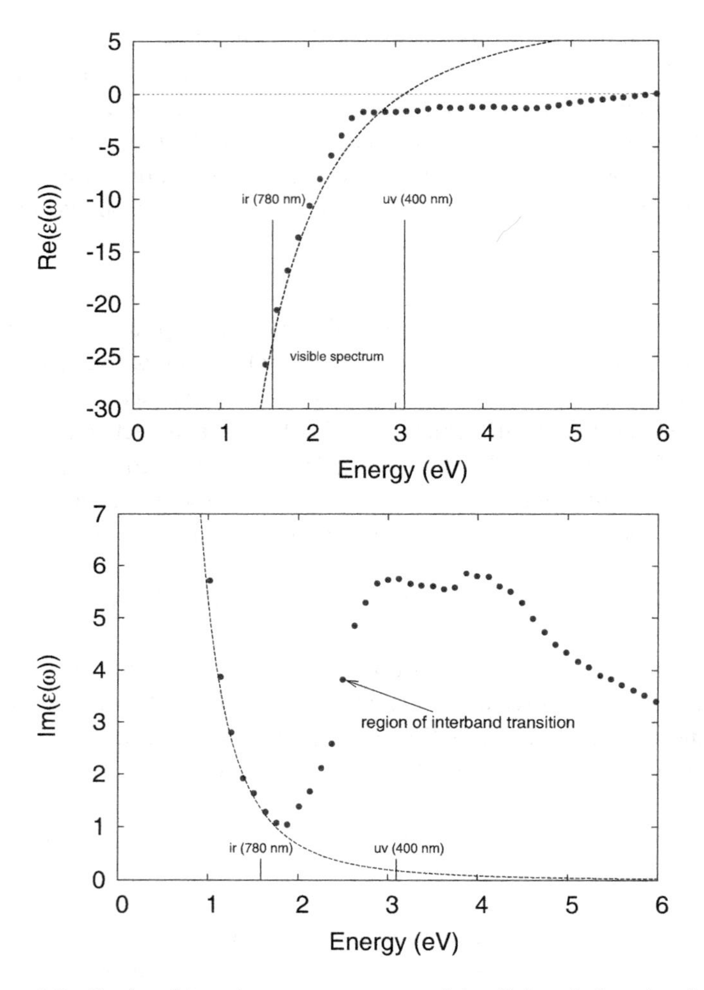

Figure 1.3 Real and imaginary components of the dielectric function for a free electron gas (dash) fitted to experimental data of the dielectric function of gold (dot).

respectively. In this case, $\varepsilon_2 \gg \varepsilon_1$, and the real and imaginary parts of the refractive index have a comparable magnitude:

$$n \approx \kappa = \sqrt{\frac{\varepsilon_2}{2}} = \sqrt{\frac{\tau\omega_p^2}{2\omega}} \tag{1.37}$$

In this region, metals are mainly absorbing, with an absorption coefficient given by

$$\alpha(\omega) = \frac{2\kappa(\omega)\omega}{c} = \left(\frac{2\omega_p^2\tau\omega}{c^2}\right)^{1/2}. \tag{1.38}$$

Remembering the DC-Drude conductivity (1.13) and the expression for the plasma frequency ω_p, the expression of σ becomes $\sigma_0 = \omega_p^2\tau\varepsilon_0$ and then

$$\alpha = \sqrt{2\sigma_0\omega\mu_0} \tag{1.39}$$

This coefficient is closely related to the skin depth, which represents the depth of penetration of the wave in the metal:

$$\delta = \frac{2}{\alpha} = \frac{c}{\kappa\omega} = \sqrt{\frac{2}{\sigma_0\omega\mu_0}} \tag{1.40}$$

In order to complete the study of the response of metals to an EM field, it is necessary to consider the particular case $\omega = \omega_p$. To understand what happens in this case, it is necessary to introduce the fundamental relation that links conductivity and dielectric function in the Fourier domain. The relation can be derived by starting from Maxwell's equations [details of calculations in Maier (2007)] and can be written as

$$\varepsilon(\mathbf{K}, \omega) = 1 + \frac{i\sigma(\mathbf{K}, \omega)}{\varepsilon_0\omega} \tag{1.41}$$

We consider the traveling-wave solution to Maxwell's equations in the absence of external stimuli. Combining the curl equations:

$$\nabla \times \mathbf{E} = -\frac{\partial \mathbf{B}}{\partial t} \tag{1.42}$$

$$\nabla \times \mathbf{H} = \mathbf{J}_{\text{ext}} + \frac{\partial \mathbf{D}}{\partial t} \tag{1.43}$$

leads to the *wave equation*, which in the Fourier domain becomes:

$$\mathbf{K}(\mathbf{K} \cdot \mathbf{E}) - K^2\mathbf{E} = -\varepsilon(\mathbf{K}, \omega)\frac{\omega^2}{c^2}\mathbf{E} \tag{1.44}$$

where $c = \frac{1}{\sqrt{\varepsilon_0\mu_0}}$ is the speed of light in vacuum. Looking at this equation, two cases have to be distinguished: the case of transverse waves, $\mathbf{K} \cdot \mathbf{E} = 0$, yielding the generic dispersion relation:

$$K^2 = \varepsilon(\mathbf{K}, \omega)\frac{\omega^2}{c^2} \tag{1.45}$$

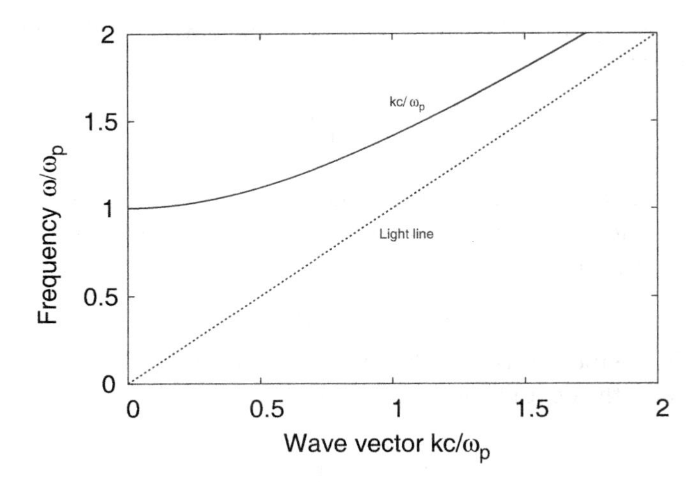

Figure 1.4 Dispersion relation of free electron gas.

where $\varepsilon(\mathbf{K}, \omega)$ is given by Eq. (1.41), and the case of longitudinal waves, for which Eq. (1.44) implies:

$$\varepsilon(\mathbf{K}, \omega) = 0 \tag{1.46}$$

indicating that longitudinal collective oscillations can only occur at frequencies corresponding to zeros of $\varepsilon(\omega)$. The meaning of this oscillation can be elucidated by considering the dispersion relation of the traveling wave obtained by using Eq. (1.33) in Eq. (1.45):

$$\omega^2 = \omega_p^2 + K^2 c^2 \tag{1.47}$$

Figure 1.4 shows the plot of the dispersion relation for the traveling wave given by Eq. (1.47): there is clearly no propagation of EM waves below the plasmon frequency ($\omega < \omega_p$), while for $\omega > \omega_p$ waves propagate with a group velocity $v_g = \frac{d\omega}{dK} < c$; the special case $\omega = \omega_p$ can be interpreted in the following way. In the small damping limit, $\mathbf{K} = 0$ and $\varepsilon(\omega_p) = 0$; this implies that $\mathbf{D} = 0$ and that the electric field becomes a pure depolarization field ($\mathbf{E} = \frac{-\mathbf{P}}{\varepsilon_0}$). This leads to a collective longitudinal oscillation of the conduction electron gas with respect to the fixed background of positive ion cores in a plasma slab (Fig. 1.5). A collective displacement u of the electron cloud leads to a surface charge density $\sigma = \pm Neu$ at the slab boundaries and yields a homogeneous electric field $\mathbf{E} = \frac{Neu}{\varepsilon_0}$

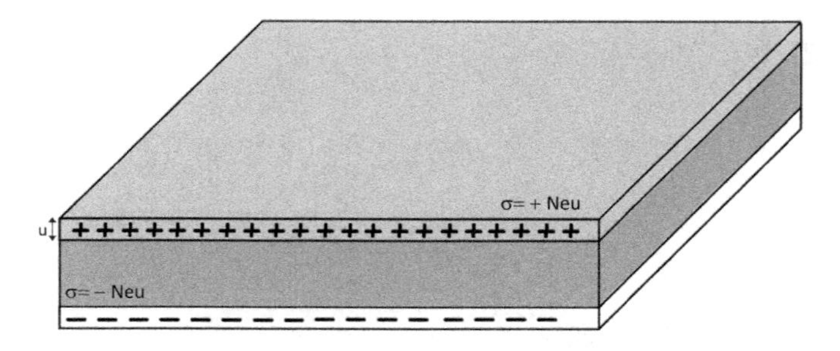

Figure 1.5 Longitudinal collective oscillations of the conduction electrons of a metal:volume plasmons.

inside the slab. Thus, the displaced electrons experience a restoring force, and their movement can be described by the equation of motion $Nm\ddot{u} = -Ne\mathbf{E}$. Inserting the expression for the electric field leads to

$$Nm\ddot{u} = -\frac{N^2e^2u}{\varepsilon_0} \tag{1.48}$$

$$\ddot{u} + \omega_p^2 u = 0 \tag{1.49}$$

Thus, the plasma frequency ω_p represents the natural frequency of a free oscillation of the electron sea and the quanta of these charge oscillations are called *plasmons or volume plasmons* (VPs). Due to the longitudinal nature of the excitation, VPs do not couple to transverse EM waves and can only be excited by particle impacts.

1.3 Surface Plasmon Polariton

In the framework of a classical approach, not only the VP, but also the other two fundamental excitations of plasmonics can be described: surface plasmon polariton (SPP) and localized surface plasmon (LSP). In this section, the SPP will be treated. It occurs when an EM radiation enters sliding at the interface of separation between a dielectric and a conductor and represents an EM excitation propagating at the interface, evanescently confined in the perpendicular direction. In order to investigate the physical properties of SPPs, it is convenient to start from the *Helmholtz*

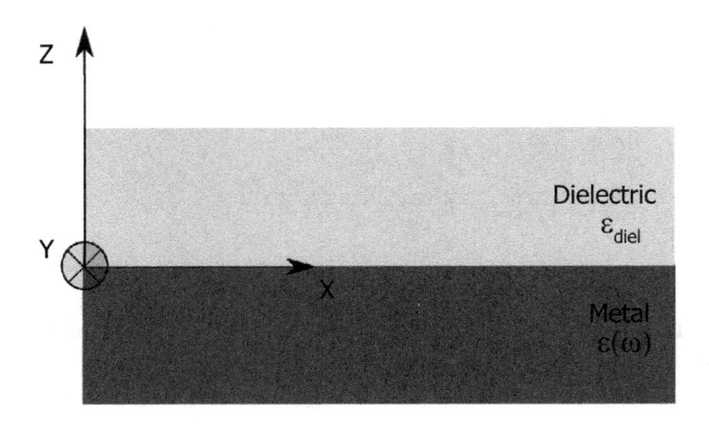

Figure 1.6 Sketch of the geometry of the system.

equation [Riley and Bence (2002)]:

$$\nabla^2 \mathbf{E} + k_0^2 \varepsilon \mathbf{E} = 0 \tag{1.50}$$

where $k_0 = \omega/c$ is the wave vector of the EM wave propagating in vacuum. This equation is obtained from Maxwell's equations under the conditions and assumptions:

- Absence of external stimuli: $\nabla \cdot \mathbf{D} = 0$
- Negligible variation of the profile of the dielectric susceptivity $\varepsilon(r)$ over distances of the order of one optical wavelength: $\nabla\varepsilon/\varepsilon \simeq 0$
- Harmonic time dependence of the EM field: $\mathbf{E}(\mathbf{r}, t) = \mathbf{E}_0(\mathbf{r})e^{-i\omega t}$

The propagation geometry is defined as follows (Fig. 1.6):

- Assumption of a one-dimensional problem, that is, ε depends on one spatial coordinate only: $\varepsilon = \varepsilon(z)$.
- Waves propagate along the x-direction of the Cartesian coordinate system.
- Waves show no spatial variation along the y-direction.
- The plane $z = 0$ coincides with the interface.

With the hypothesis that the wave propagates in the x-direction, while slow variations of the electric field amplitude can occur only in the z-direction, we look for a solution that can be written as $\mathbf{E}(x, y, z) = \mathbf{E}(z)e^{i\beta x}$; thus, Eq. (1.50) assumes the form

$$\frac{\partial^2 \mathbf{E}(z)}{\partial z^2} + (k_0^2 \varepsilon - \beta^2)\mathbf{E}(z) = 0 \tag{1.51}$$

Of course, the same relation exists for the magnetic field **H**. As a matter of fact, Eq. (1.51) has to be solved separately in regions of different, constant ε values, and the obtained solutions have to be matched by exploiting suitable boundary conditions.

In order to use Eq. (1.51) for determining the spatial field profile and the dispersion of propagating waves, we need to find explicit expressions for the different field components of **E** and **H**. This can be achieved by using the curl of Maxwell's equations (1.42) and (1.43) in the specific case of harmonic time dependence ($\frac{\partial}{\partial t} = -i\omega$), propagation along the x-direction ($\frac{\partial}{\partial x} = i\beta$) and homogeneity in the y-direction ($\frac{\partial}{\partial y} = 0$). The obtained system of equations is

$$\frac{\partial E_y}{\partial z} = -i\omega\mu_0 H_x \tag{1.52a}$$

$$\frac{\partial E_x}{\partial z} - i\beta E_z = i\omega\mu_0 H_y \tag{1.52b}$$

$$i\beta E_y = i\omega\mu_0 H_z \tag{1.52c}$$

$$\frac{\partial H_y}{\partial z} = i\omega\varepsilon_0\varepsilon E_x \tag{1.52d}$$

$$\frac{\partial H_x}{\partial z} - i\beta H_z = -i\omega\varepsilon_0\varepsilon E_y \tag{1.52e}$$

$$i\beta H_y = -i\omega\varepsilon_0\varepsilon E_z \tag{1.52f}$$

which provides two sets of self-consistent solutions with different polarization characteristics of the propagating waves: the transverse magnetic (TM) mode, where only the field components E_x, E_z, and H_y are nonzero, and the transverse electric (TE) mode, with only H_x, H_z, and E_y being nonzero. For TM modes, by starting from Eq. (1.52d) and Eq. (1.52f), we obtain the expression of E_x and E_z as functions of H_y:

$$E_x = -i\frac{1}{\omega\varepsilon_0\varepsilon}\frac{\partial H_y}{\partial z} \tag{1.53a}$$

$$E_z = -\frac{\beta}{\omega\varepsilon_0\varepsilon}H_y \tag{1.53b}$$

Here, H_y has to be obtained from the solution to the TM wave equation

$$\frac{\partial^2 H_y}{\partial z^2} + (k_0^2 \varepsilon - \beta^2) H_y = 0 \quad (1.53c)$$

For TE modes, the analogous set is

$$H_x = i \frac{1}{\omega \mu_0} \frac{\partial E_y}{\partial z} \quad (1.54a)$$

$$H_z = -\frac{\beta}{\omega \mu_0} E_y \quad (1.54b)$$

where E_y has to be obtained from the TE wave equation

$$\frac{\partial^2 E_y}{\partial z^2} + (k_0^2 \varepsilon - \beta^2) E_y = 0 \quad (1.54c)$$

By utilizing the above equations, we are able to describe SPPs. The simplest geometry sustaining SPPs is the one of a single, flat interface (Fig. 1.6) between a dielectric, nonabsorbing half space ($z > 0$) characterized by a positive real dielectric constant ε_{diel}, and a conducting half space ($z < 0$) characterized by a dielectric function $\varepsilon(\omega)$, where the requirement of a metallic character implies that $\Re[\varepsilon(\omega)] < 0$. As shown in the previous section, in metals this condition is fulfilled at frequencies that are below the bulk plasmon frequency ω_p. We look for propagating waves confined at the interface, that is, with evanescent decay in the perpendicular z-direction (separately for the two cases TM and TE).

Let us first look at TM solutions. Using the equation set (1.53) in both half spaces and searching for solutions to (1.53c), which are propagating in the x-direction and exponentially decreasing along the z-direction, we obtain

$$H_y(z) = A_2 e^{i\beta x} e^{-k_2 z} \quad (1.55a)$$

$$E_x(z) = i A_2 \frac{1}{\omega \varepsilon_0 \varepsilon_2} k_2 e^{i\beta x} e^{-k_2 z} \quad (1.55b)$$

$$E_z(z) = -i A_2 \frac{\beta}{\omega \varepsilon_0 \varepsilon_2} e^{i\beta x} e^{-k_2 z} \quad (1.55c)$$

for $z > 0$ and

$$H_y(z) = A_1 e^{i\beta x} e^{k_1 z} \quad (1.56a)$$

$$E_x(z) = -i A_1 \frac{1}{\omega \varepsilon_0 \varepsilon_1} k_1 e^{i\beta x} e^{k_1 z} \quad (1.56b)$$

$$E_z(z) = -A_1 \frac{\beta}{\omega \varepsilon_0 \varepsilon_1} e^{i\beta x} e^{k_1 z} \quad (1.56c)$$

for $z < 0$.

Here A_1, A_2 are magnetic field amplitudes, $k_{z,i}$ ($i = 1, 2$) is the component of the wave vector perpendicular to the interface in the two media; its reciprocal value, $\hat{z} = \frac{1}{|k_z|}$, defines the *evanescent decay length* of fields perpendicular to the interface, which quantifies the confinement of the wave. Continuity of H_y, E_x, and $\varepsilon_i E_z$ at the interface ($z = 0$) requires that $A_1 = A_2$ and

$$\frac{k_2}{k_1} = -\frac{\varepsilon_{diel}}{\varepsilon(\omega)} \tag{1.57}$$

According to the convention assumed for signs in the exponents in Eqs. (1.55, 1.56), confinement to the surface demands that $\Re[\varepsilon(\omega)] < 0$ if $\varepsilon_{diel} > 0$, thus surface waves can exist only at interfaces between materials with opposite signs of the real part of their dielectric permittivity, that is, between a conductor and an insulator. In addition, the expression for H_y has to fulfill the wave equation (1.53c), yielding

$$k_1^2 = \beta^2 - k_0^2 \varepsilon(\omega) \tag{1.58a}$$

$$k_2^2 = \beta^2 - k_0^2 \varepsilon_{diel} \tag{1.58b}$$

By combining Eqs. (1.57) and (1.58), we obtain the main result concerning the argument of this section, that is, the dispersion relation of SPPs propagating at the interface between a conductor half space and an insulator one:

$$\beta = k_0 \sqrt{\frac{\varepsilon(\omega)\varepsilon_{diel}}{\varepsilon(\omega) + \varepsilon_{diel}}} \tag{1.59}$$

This expression is valid both for real and complex ε, that is, for conductors without and with attenuation.

Where TE solutions are concerned, expressions for the field components can be obtained by using Eq. (1.54) and are

$$E_y(z) = A_2 e^{i\beta x} e^{-k_2 z} \tag{1.60a}$$

$$H_x(z) = -i A_2 \frac{1}{\omega \mu_0} k_2 e^{i\beta x} e^{-k_2 z} \tag{1.60b}$$

$$H_z(z) = A_2 \frac{\beta}{\omega \mu_0} e^{i\beta x} e^{-k_2 z} \tag{1.60c}$$

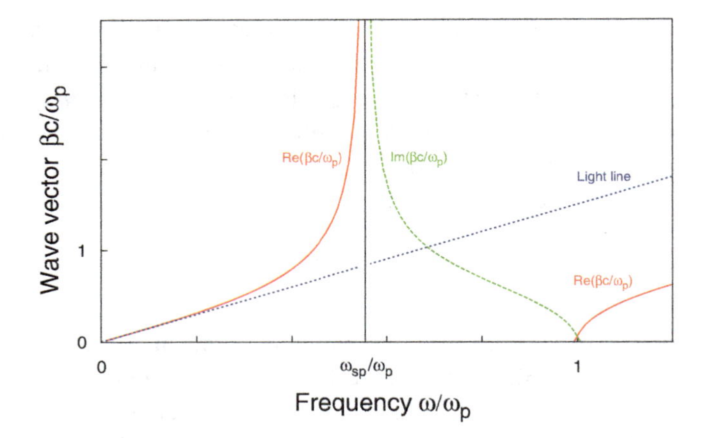

Figure 1.7 Dispersion relation of SPPs real part (red curve) and imaginary part (green curve), at the interface between a metal with negligible collision frequency and silica.

for $z > 0$, and

$$E_y(z) = A_1 e^{i\beta x} e^{k_1 z} \tag{1.61a}$$

$$H_x(z) = i A_1 \frac{1}{\omega \mu_0} k_1 e^{i\beta x} e^{k_1 z} \tag{1.61b}$$

$$H_z(z) = A_1 \frac{\beta}{\omega \mu_0} e^{i\beta x} e^{k_1 z} \tag{1.61c}$$

for $z < 0$.

Here, A_1, A_2 are now electric field amplitudes. Continuity of E_y and H_x at the interface leads to the condition $A_1 = A_2$ and

$$A_1(k_1 + k_2) = 0. \tag{1.62}$$

Since confinement at the surface requires $\Re[k_1] > 0$ and $\Re[k_2] > 0$, condition (1.62) is only fulfilled if $A_1 = 0$, that means also $A_2 = 0$; thus, no surface modes can exist with TE polarization.

Figure 1.7 shows its plots for a metal, with negligible damping described by the real (Drude) dielectric function (1.33) interfaced with fused silica ($\varepsilon_{diel} = 2.25$). In the plot, both the real (red curves) and the imaginary parts (green curves) of the wave vector β are shown (frequency ω and wave vector β are normalized to the plasma frequency ω_p). The light line represents plane wave propagation in the dielectric, while ω_{sp} is the characteristic *surface plasmon frequency*, whose physical meaning will be discussed later on.

- For $\omega > \omega_p$, we are in the transparency regime of metals: the propagation constant is only real and radiation propagates through the metal.
- For $\omega < \omega_p$, different behaviors occur:
  - For very low frequencies, the SPP propagation constant is close to k_0 at the light line: waves extend over many wavelengths into the dielectric space because the metal does not allow to be crossed.
  - For higher frequencies, the SPPs approach the characteristic surface plasmon frequency ω_{sp}, which is the frequency value that makes the denominator in Eq. (1.59) (with $\varepsilon = 1 - \omega_p^2/\omega^2$) vanishing:

$$\omega_{sp} = \frac{\omega_p}{\sqrt{1 + \varepsilon_{diel}}} \tag{1.63}$$

    In the limit of negligible damping of the conduction electron oscillation ($\Im[\varepsilon(\omega)] = 0$), the wave vector β goes to infinity as the frequency approaches ω_{sp}, and the group velocity $v_g = d\omega/dk \to 0$. The mode thus acquires an electrostatic character and is known as the *surface plasmon*. The electrostatic character is confirmed by the circumstance that the condition $\varepsilon(\omega) + \varepsilon_{diel} = 0$ is obtained also via straightforward solution to the Laplace equation $\nabla^2\Phi = 0$ (imposing the continuity of Φ and $\varepsilon\partial\Phi/\partial z$ in order to ensure the continuity of tangential and normal field components).
  - In the region between ω_{sp} and ω_p, the propagation constant is only imaginary and no SPP propagation occurs.

So far we have assumed an ideal conductor that presents $\Im[\beta] = 0$; however, real metals present a complex $\varepsilon(\omega)$ and also an SPP propagation constant β. Since SPPs travel according to the term $e^{i\beta x}$ ($e^{2i\beta x}$ for the intensity), if β is complex, its imaginary part $\Im[\beta] \neq 0$ establishes an *energy attenuation length* $L = (2\Im[\beta])^{-1}$, which has a maximum finite value in correspondence of the plasmon frequency ω_{sp}. In Fig. 1.8, the actual behavior of silver obtained from data by Johnson and Christy (1972) is shown.

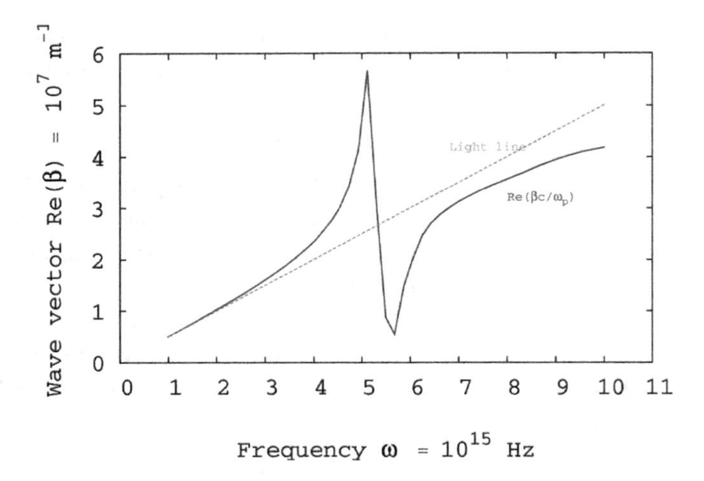

Figure 1.8 Dispersion relation of SPPs real part (red curve) and imaginary part (green curve), at the interface between silver and air.

If we turn our attention to SPPs in multilayers consisting of alternating conducting and dielectric thin films, we can say that each single interface can sustain bound SPPs. When the separation between adjacent interfaces is comparable to or smaller than the decay length $\hat{z}$ of the interface mode, interactions between SPPs give rise to coupled modes. The general properties of coupled SPPs are elucidated in [Maier (2007)].

1.4 Localized Surface Plasmon

When a light beam impinges on a particle, the optical electric field puts into oscillation the electrical charges of matter (conduction electrons and protons). As a consequence, there is an absorption of part of the impinging radiation and an emission of a secondary radiation, a phenomenon known as scattering. In order to describe both effects, it is necessary to write down the expressions of EM fields **E**, **H** starting from Maxwell's equations. The most famous exact solution to Maxwell's equations, for the case of small particles with arbitrary radius and refractive index, has been obtained in the framework of the Mie theory [Bohren and Huffman (1983);

Mie (1908)], developed by Gustav Mie in 1908 with the aim of explaining the different colors exhibited, in absorption and scattering processes, by small colloidal particles of gold suspended in water.

In a linear, isotropic, homogeneous medium, a time-harmonic EM field (**E**, **H**) must satisfy the wave equation:

$$\nabla^2 \mathbf{E} + k^2 \mathbf{E} = 0 \tag{1.64}$$

$$\nabla^2 \mathbf{H} + k^2 \mathbf{H} = 0 \tag{1.65}$$

where $k^2 = \omega^2 \varepsilon \mu$. Fields (**E**, **H**), which have a null divergence:

$$\nabla \cdot \mathbf{E} = 0 \qquad \nabla \cdot \mathbf{H} = 0 \tag{1.66}$$

are not independent, since they are related by the following relations:

$$\nabla \times \mathbf{E} = i\omega\mu \mathbf{H} \qquad \nabla \times \mathbf{H} = -i\omega\varepsilon \mathbf{E} \tag{1.67}$$

Equations (1.64) and (1.65) are complicated because of their vectorial character. Mie's theory simplifies the problem by reducing their solution to the solution of a single scalar wave equation:

$$\nabla^2 \psi + k^2 \psi = 0 \tag{1.68}$$

whose solution enables to obtain the expressions of fields (**E**, **H**).

The detailed procedure is exhaustively reported in [Bohren and Huffman (1983)], and the main steps can be shortly resumed as follows.

Suppose that the scalar function ψ and an arbitrary constant vector **c** define a vectorial field **M**:

$$\mathbf{M} = \nabla \times (\mathbf{c}\psi) \tag{1.69}$$

By keeping in mind that the divergence of the curl of any vectorial field vanishes ($\nabla \cdot \mathbf{M} = 0$), and using the vectorial identities $\nabla \times (\mathbf{A} \times \mathbf{B}) = \mathbf{A}(\nabla \cdot \mathbf{B}) - \mathbf{B}(\nabla \cdot \mathbf{A}) + (\mathbf{B} \cdot \nabla)\mathbf{A} - (\mathbf{A} \cdot \nabla)\mathbf{B}$ and $\nabla(\mathbf{A} \cdot \mathbf{B}) = \mathbf{A} \times (\nabla \times \mathbf{B}) + \mathbf{B} \times (\nabla \times \mathbf{A}) + (\mathbf{B} \cdot \nabla)\mathbf{A} + (\mathbf{A} \cdot \nabla)\mathbf{B}$, we find that **M** satisfies the vector wave equations (1.64) and (1.65) if ψ is a solution to the scalar wave equation (1.68). We may also write $\mathbf{M} = -\mathbf{c} \times \nabla\psi$, which shows that **M** is perpendicular to **c**. Now, we derive from **M** a second vectorial field:

$$\mathbf{N} = \frac{\nabla \times \mathbf{M}}{k} \tag{1.70}$$

which has a null divergence and also satisfies the vector wave equation. Since $\nabla \times \mathbf{N} = k\mathbf{M}$, we can state that both **M** and **N** exhibit all the properties required to be an EM field: they satisfy the vector wave equation, they are divergence-free, the curl of **M** is proportional to **N**, and the curl of **N** is proportional to **M**. Thus, the problem of finding solutions to the wave equations (1.64), (1.65) reduces to the simpler problem of finding solutions to the scalar wave equation (1.68). The scalar function ψ is named *generating function* for the vector harmonics **M** and **N**, while the vector **c** is sometimes called the *guiding or pilot vector*. In general, the choice of *generating functions* is suggested by whatever symmetry may exist in the problem; thus, being interested in scattering by a sphere, we choose functions ψ that satisfy the wave equation written in spherical coordinates r, θ, φ (Fig. 1.9). The choice of the pilot vector is somewhat less obvious. We could choose some arbitrary vector **c**; however, if we write:

$$\mathbf{M} = \nabla \times (\mathbf{r}\psi) \tag{1.71}$$

where **r** is the radius vector, then **M** is a solution to the vector wave equation in spherical coordinates. In problems involving spherical symmetry, therefore, we assume **M** as given in Eq. (1.71) and the associated **N** as the fundamental solutions to the field equations. Note that **M** is everywhere tangential to any sphere $|\mathbf{r}| = constant$ (i.e., $\mathbf{r} \cdot \mathbf{M} = 0$).

The scalar wave equation in spherical coordinates is

$$\frac{1}{r^2}\frac{\partial}{\partial r}\left(r^2\frac{\partial \psi}{\partial r}\right) + \frac{1}{r^2 \sin\theta}\frac{\partial}{\partial \theta}\left(\sin\theta\frac{\partial \psi}{\partial \theta}\right) + \frac{1}{r^2 \sin^2\theta}\left(\frac{\partial^2 \psi}{\partial \varphi^2}\right) + k^2\psi = 0 \tag{1.72}$$

We seek particular solutions to Eq. (1.72) in the form:

$$\psi(r, \theta, \phi) = R(r)\Theta(\theta)\Phi(\varphi) \tag{1.73}$$

which, when substituted into Eq. (1.72), yields three separated equations linked by two separation constants (m and n), to be determined by subsidiary conditions that have to be satisfied by ψ. The request of single-valued functions and linearly independent

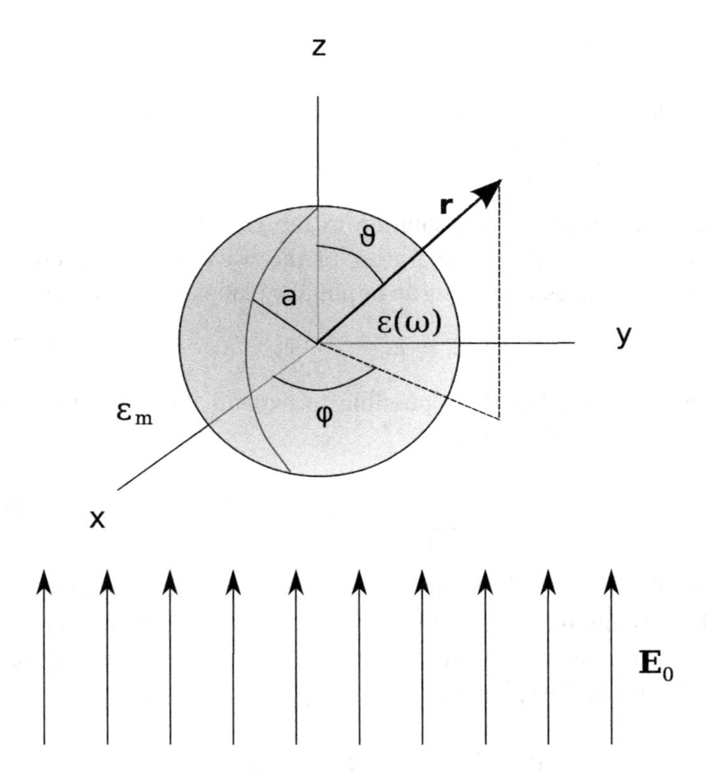

Figure 1.9 Spherical coordinate system centered on a spherical particle of radius a.

solutions produce the following generating functions in spherical coordinates [Bohren and Huffman (1983)]:

$$\psi_{emn} = \cos m\varphi P_n^m(\cos\theta) z_n(kr) \tag{1.74}$$

$$\psi_{omn} = \sin m\varphi P_n^m(\cos\theta) z_n(kr) \tag{1.75}$$

where subscripts e and o denote *even* and *odd*, $P_n^m(\cos\theta)$ are the associated Legendre functions of the first kind and z_n is any of the four spherical Bessel functions j_n, y_n, $h_n^{(1)}$, or $h_n^{(2)}$. Moreover, any function that satisfies the scalar wave equation in spherical coordinates may be expanded as an infinite series in the spherical functions (1.74) and (1.75). The vectorial spherical harmonics generated by ψ_{emn} and ψ_{omn} are:

$$\mathbf{M}_{emn} = \nabla \times (\mathbf{r}\psi_{emn}), \qquad \mathbf{M}_{omn} = \nabla \times (\mathbf{r}\psi_{omn}) \tag{1.76}$$

$$\mathbf{N}_{emn} = \frac{\nabla \times \mathbf{M}_{emn}}{k}, \qquad \mathbf{N}_{omn} = \frac{\nabla \times \mathbf{M}_{omn}}{k} \tag{1.77}$$

In order to obtain the solution to the scattering problem, it is necessary to specify the character of the wave impinging on the spherical particles. Assuming an impinging plane, x-polarized, wave:

$$\mathbf{E}_i = E_0 e^{ikr\cos\theta}\hat{\mathbf{e}}_x \tag{1.78}$$

it can be shown that it is possible to expand the plane wave in vectorial spherical harmonics:

$$\mathbf{E}_i = \sum_{n=1}^{\infty} E_n \left(\mathbf{M}_{o1n} - i\mathbf{N}_{e1n}\right) \tag{1.79}$$

where $E_n = E_0 i^n \frac{2n+1}{n(n+1)}$. The corresponding impinging magnetic field $\mathbf{H}_i$ is obtained from the curl of Eq. (1.79). Of course, it is also possible to expand the scattered EM field $(\mathbf{E}_s \mathbf{H}_s)$ and the field $(\mathbf{E}_1, \mathbf{H}_1)$ inside the sphere in vectorial spherical harmonics:

$$\mathbf{E}_s = \sum_{n=1}^{\infty} \mathbf{E}_n \left(ia_n\mathbf{N}_{e1n} - b_n\mathbf{M}_{o1n}\right) \tag{1.80}$$

$$\mathbf{E}_1 = \sum_{n=1}^{\infty} \mathbf{E}_n \left(c_n\mathbf{M}_{o1n} - id_n\mathbf{N}_{e1n}\right) \tag{1.81}$$

where a_n, b_n, c_n, d_n are the Mie coefficients. The corresponding magnetic field $\mathbf{H}_{s,1}$ is obtained from the curl of $\mathbf{E}_{s,1}$. The Mie coefficients are obtained by satisfying the boundary condition at the interface between the sphere and the surrounding medium:

$$(\mathbf{E}_i + \mathbf{E}_s - \mathbf{E}_1) \times \hat{\mathbf{e}}_r = (\mathbf{H}_i + \mathbf{H}_s - \mathbf{H}_1) \times \hat{\mathbf{e}}_r = 0 \tag{1.82}$$

Mie coefficients are directly involved in the expression of scattering cross section and polarizability of the sphere. In fact, the scattering cross section C_{sca} is the area associated to the sphere that takes into account the amount of scattered energy. C_{sca} is the ratio between the total scattered energy per second and the incident energy per square meter per second. Since $\mathbf{E}_s$ and $\mathbf{H}_s$ are known quantities, by exploiting the Poynting vector we can evaluate the scattered energy and then the scattering cross section:

$$C_{sca} = \frac{W_s}{I_i} = \frac{2\pi}{k^2} \sum_{n=1}^{\infty} (2n+1) \left(|a_n|^2 + |b_n|^2 \right) \tag{1.83}$$

Derivation of the Mie coefficients is exhaustively reported in many books and articles, but the problem of determining the response of sub-wavelength metal particles acted on by a plane wave radiation can be analyzed using the simple quasi-static approximation $a \ll \lambda$. In this case, indeed, since the particle is much smaller than the wavelength of the impinging light, the phase of the harmonically oscillating EM field is, in fact, constant over the particle volume, so that we can calculate the spatial field distribution by considering the simple problem of a particle in an electrostatic field; the harmonic time dependence can then be added to the solution once the field distribution is known. The solution to the problem describes adequately the optical properties of nanoparticles of dimensions below 100 nm in many geometries. We start with the most convenient configuration for an analytical treatment: a homogeneous, isotropic sphere of radius a located at the origin in a uniform, static electric field $\mathbf{E} = E_0\hat{\mathbf{z}}$ (Fig. 1.9); the surrounding medium is isotropic and nonabsorbing with dielectric constant ε_m, and at a sufficient distance from the sphere, the field lines are parallel to the z-direction. The dielectric response of the sphere is described by the dielectric function $\varepsilon(\omega)$, which we take for the moment as a simple complex number ε. In the framework of an electrostatic approach, we are interested in finding a solution to the Laplace equation for the potential $\nabla^2\Phi = 0$, which will enable calculation of the electric field $\mathbf{E} = -\nabla\Phi$. Due to the azimuthal symmetry of the problem, a general solution can be written in the form [Jackson (1999)]:

$$\Phi(r, \theta) = \sum_{l=0}^{\infty} \left[A_l r^l + B_l r^{-(l+1)} \right] P_l(\cos\theta) \tag{1.84}$$

where $P_l(\cos\theta)$ are the Legendre polynomials of order l. Due to the requirement that the potentials remain finite at the origin, the solution for potentials Φ_{in} inside the sphere and Φ_{out} outside the sphere can be written as:

$$\Phi_{in}(r,\theta) = \sum_{l=0}^{\infty} A_l r^l P_l(\cos\theta) \tag{1.85}$$

$$\Phi_{out}(r,\theta) = \sum_{l=0}^{\infty} \left[B_l r^l + C_l r^{-(l+1)}\right] P_l(\cos\theta) \tag{1.86}$$

where coefficients A_l, B_l, and C_l can now be determined by imposing the boundary conditions at $r \to \infty$ and at the sphere surface $r = a$.

By writing:

$$\mathbf{E} = -\nabla\Phi = -\frac{\partial\Phi}{\partial r}\hat{\mathbf{e}}_{\mathbf{r}} - \frac{1}{r}\frac{\partial\Phi}{\partial\theta}\hat{\mathbf{e}}_{\theta} - \frac{1}{r\sin\theta}\frac{\partial\Phi}{\partial\varphi}\hat{\mathbf{e}}_{\varphi} \tag{1.87}$$

the requirements are

- At $r \to \infty$, $\Phi_{out} \to -E_0 z$
- Continuity of the tangential components of the electric field

$$-\frac{1}{a}\frac{\partial\Phi_{in}}{\partial\theta}\bigg|_{r=a} = -\frac{1}{a}\frac{\partial\Phi_{out}}{\partial\theta}\bigg|_{r=a} \tag{1.88}$$

- Continuity of the normal components of the displacement field

$$-\varepsilon_0\varepsilon_m\frac{\partial\Phi_{in}}{\partial r}\bigg|_{r=a} = -\varepsilon_0\varepsilon_m\frac{\partial\Phi_{out}}{\partial r}\bigg|_{r=a} \tag{1.89}$$

In this way, it is straightforward to obtain $B_l = 0$ for $l \neq 1$ and $A_l = C_l = 0$ for $l \neq 1$ and then

$$A_1 = -\frac{3\varepsilon_m}{\varepsilon + 2\varepsilon_m}E_0 \tag{1.90}$$

$$B_1 = -E_0 \tag{1.91}$$

$$C_1 = a^3\frac{\varepsilon - \varepsilon_m}{\varepsilon + 2\varepsilon_m}E_0 \tag{1.92}$$

Therefore

$$\Phi_{in} = -\frac{3\varepsilon_m}{\varepsilon + 2\varepsilon_m}E_0 r\cos\theta \tag{1.93}$$

$$\Phi_{out} = -E_0 r\cos\theta + \frac{\mathbf{p}\cdot\mathbf{r}}{4\pi\varepsilon_0\varepsilon_m r^3} \tag{1.94}$$

Therefore, we can conclude that application of the field induces inside the sphere a dipole moment $\mathbf{p}$ of magnitude proportional to

$|\mathbf{E}_0|$. Introducing the polarizability α defined via $\mathbf{p} = \varepsilon_0\varepsilon_m\alpha\mathbf{E}_0$, we obtain

$$\alpha = 4\pi a^3 \frac{\varepsilon - \varepsilon_m}{\varepsilon + 2\varepsilon_m} \tag{1.95}$$

This function undergoes a resonant enhancement under the condition that $|\varepsilon + 2\varepsilon_m|$ is minimum, which in the case of small or slowly varying $\mathrm{Im}[\varepsilon]$ around the resonance, simplifies to

$$\Re[\varepsilon(\omega)] = -2\varepsilon_m \tag{1.96}$$

This relationship is called the *Fröhlich condition* and the associated mode (in an oscillating field) is referred to as the *dipole surface plasmon* of the metal nanoparticle; for a sphere located in air, made of a Drude metal with a dielectric function given by Eq. (1.23), the Fröhlich condition is fulfilled at the frequency $\omega_0 = \omega_p/\sqrt{3}$. It is worth noting that: (i) the Fröhlich condition expresses also the strong dependence of the resonance frequency of surface plasmon on the dielectric environment, in particular, the resonance red-shifts as ε_m is increased; (ii) the resonant enhancement of α is affected by the incomplete vanishing of its denominator, due to the fact that $\Im[\varepsilon(\omega)] \neq 0$.

The spatial configuration of the electric field $\mathbf{E} = -\nabla\Phi$ can be evaluated from potentials (1.93) and (1.94):

$$\mathbf{E}_{in} = \frac{3\varepsilon_m}{\varepsilon + 2\varepsilon_m}\mathbf{E}_0 \tag{1.97}$$

$$\mathbf{E}_{out} = \mathbf{E}_0 + \frac{3\mathbf{n}(\mathbf{n}\cdot\mathbf{p}) - \mathbf{p}}{4\pi\varepsilon_0\varepsilon_m}\frac{1}{r^3} \tag{1.98}$$

As expected, the resonant behavior of α yields also a resonant enhancement of both internal and dipolar fields: it is this field enhancement at the plasmon resonance that many of the prominent applications of metal nanoparticles in optical devices and sensors rely on.

The framework of electrostatics is no more suitable to study the EM fields radiated by a small particle excited at its plasmon resonance. For a small sphere with radius $a \ll \lambda$, its representation as an ideal dipole is valid only in the quasi-static regime, that is, when allowing for time-varying fields but neglecting spatial retardation effects over the particle volume. Under plane-wave illumination, its electric field $\mathbf{E}(\mathbf{r}, t) = \mathbf{E}_0 e^{-i\omega t}$ induces an oscillating dipole moment $\mathbf{p}(t) = \varepsilon_0\varepsilon_m\alpha\mathbf{E}$. Radiation of this dipole leads to scattering of the plane wave by the sphere, which can be represented as radiation by a point dipole. Estimation of the scattered electric field from an ideal dipole, with dipole moment $\mathbf{p}(t) = \varepsilon_0\varepsilon_m\alpha\mathbf{E}_0 e^{-i\omega t}$ allows to calculate the cross sections:

$$C_{sca} = \frac{k^4}{6\pi}|\alpha|^2 = \frac{8\pi}{3}k^4a^6\left|\frac{\varepsilon-\varepsilon_m}{\varepsilon+2\varepsilon_m}\right|^2 \tag{1.99a}$$

$$C_{abs} = k\Im\left[\alpha\right] = 4\pi ka^3\Im\left[\frac{\varepsilon-\varepsilon_m}{\varepsilon+2\varepsilon_m}\right] \tag{1.99b}$$

For small particles with $a \ll \lambda$, absorption, which scales with a^3, dominates over the scattering, which scale with a^6. Since expressions (1.99) are valid also for dielectric scatterers, these equations shows that it is very difficult to pick out small objects from a background of larger scatterers.

For metal nanoparticles, both absorption and scattering are resonantly enhanced when the Fröhlich condition is fulfilled. For a sphere of volume V and dielectric function $\varepsilon = \varepsilon_1 + i\varepsilon_2$ in the quasi-static limit, the explicit expression for the extinction cross section $C_{ext} = C_{abs} + C_{sca}$ is:

$$C_{ext} = 9\frac{\omega}{c}\varepsilon_m^{3/2}V\frac{\varepsilon_2}{\left[\varepsilon_1+2\varepsilon_m\right]^2+\varepsilon_2^2} \tag{1.100}$$

Thus, also the extinction cross section shows a resonant enhancement when the Fröhlich condition is fulfilled.

If it is necessary to study ellipsoids or coated ellipsoids, a detailed treatment can be found in [Bohren and Huffman (1983)], where both the polarizability and the dipole moment are derived as functions of some geometrical factor L_i:

$$\alpha_i = 4\pi abc\frac{\varepsilon-\varepsilon_m}{3\varepsilon_m+3L_i(\varepsilon-\varepsilon_m)} \tag{1.101}$$

$$\mathbf{p}_i = \varepsilon_m\alpha_i\mathbf{E}_0 \tag{1.102}$$

Here, the dipole moment is related to a field parallel to the axis that determines L_i. In particular, when metallic nanoparticles become elongated along a given axis (nanorods), the extinction cross section, calculated by Gans theory, is written by exploiting a generalization of the Mie theory:

$$C_{ext} = \frac{\omega}{3c}\varepsilon_m^{3/2}V\sum_j\frac{(1/P_j^2)\varepsilon_2}{\left[\varepsilon_1+\left[(1-P_j)/P_j\right]\varepsilon_m\right]^2+\varepsilon_2^2} \tag{1.103}$$

where P_j are the depolarization factors along the three axes A, B, C of the nanorod with $A > B = C$, and are defined as:

$$P_A = \frac{1-e^2}{e^2}\left[\frac{1}{2e}\ln\left(\frac{1+e}{1-e}\right)-1\right],$$

$$P_B = P_C = \frac{1-P_A}{2},$$

$$e = \left[1-\left(\frac{B}{A}\right)^2\right]^{1/2} = \left(1-\frac{1}{R^2}\right)^{1/2}$$

where $e = (1-(1/R)^2)^{1/2}$ and R is the aspect ratio of the nanorods.

If the metallic nanoparticles are put in a surrounding medium with gain, a new fruitful theory is born. The problem can be treated by analyzing the case of a sub-wavelength metal nanosphere embedded in a homogeneous medium exhibiting optical gain. The quasi-static approach can be followed, and the presence of gain can be incorporated by replacing the real dielectric constant ε_m of the insulator surrounding the sphere with a complex dielectric function $\varepsilon_2(\omega)$. Using this straightforward analytical model, Lawandy has shown that the presence of gain, expressed by $\Im[\varepsilon_2] < 0$, can lead to a significant strengthening of the plasmon resonance [Lawandy (2004)]. This is due to the fact that in addition to the cancellation of the real part of the denominator in the polarizability α (1.95), the imaginary part of ε_2 can, in fact, lead to a complete cancellation of all the terms in the denominator of α and thus to an infinite amplitude of the resonant polarizability.

Acknowledgments

The research leading to these results has received funding from the European Union's Seven Framework Programme (FP7/2007-2013) under grant agreement n° 228455.

References

Ashcroft, N. and Mermin, N. (1976). *Solid State Physics* (Saunders College, Philadelphia).

Bohren, C. F. and Huffman, D. (1983). *Absorption and scattering of light by small particles*, Wiley science paperback series (Wiley), ISBN 9780471293408, http://books.google.it/books?id=S1RCZ8BjgN0C.

Born, M. and Wolf, E. (1999). *Principles of Optics*, 7th edn. (Cambridge University Press).

Drude, P. (1900). Zur elektronentheorie der metalle, *Annalen der Physik* **306**, 3, pp. 566–613, doi:10.1002/andp.19003060312, http://dx.doi.org/10.1002/andp.19003060312.

Jackson, J. (1999). *Classical Electrodynamics* (Wiley), ISBN 9780471309321, http://books.google.it/books?id=GuYNngEACAAJ.

Johnson, P. B. and Christy, R. W. (1972). Optical constants of the noble metals, *Phys. Rev. B* **6**, pp. 4370–4379, doi:10.1103/PhysRevB.6.4370, http://link.aps.org/doi/10.1103/PhysRevB.6.4370.

Lawandy, N. M. (2004). Localized surface plasmon singularities in amplifying media, *Applied Physics Letters* **85**, 21.

Maier, S. (2007). *Plasmonics: Fundamentals and Applications: Fundamentals and Applications* (Springer), ISBN 9780387378251, http://books.google.it/books?id=yT2ux7TmDc8C.

Mie, G. (1908). Beitrge zur optik trber medien, speziell kolloidaler metallsungen, *Annalen der Physik* **330**, 3, pp. 377–445, doi:10.1002/andp.19083300302, http://dx.doi.org/10.1002/andp.19083300302.

Pendry, J., Holden, A. J., Robbins, D. J. and Stewart, W. J. (1999). Magnetism from conductors and enhanced nonlinear phenomena, *Microwave Theory and Techniques, IEEE Transactions on* **47**, 11, pp. 2075–2084, doi:10.1109/22.798002.

Pendry, J. B. (2000). Negative refraction makes a perfect lens, *Phys. Rev. Lett.* **85**, pp. 3966–3969, doi:10.1103/PhysRevLett.85.3966, http://link.aps.org/doi/10.1103/PhysRevLett.85.3966.

Riley, M. P. H., K. F. and Bence., S. J. (2002). *Mathematical Methods for Physics and Engineering* (Cambridge University Press).

Shelby, R. A., Smith, D. R. and Schultz, S. (2001). Experimental verification of a negative index of refraction, *science* **292**, 5514, pp. 77–79.

Smith, D. R., Padilla, W. J., Vier, D. C., Nemat-Nasser, S. C. and Schultz, S. (2000). Composite medium with simultaneously negative permeability and permittivity, *Phys. Rev. Lett.* **84**, pp. 4184–4187, doi:10.1103/PhysRevLett.84.4184, http://link.aps.org/doi/10.1103/PhysRevLett.84.4184.

Veselago, V. (1968). The electrodynamics of substances with simultaneously negative values of, *Soviet Physics Uspekhi* **10**, pp. 509–513.

Chapter 2

Synthesis and Surface Engineering of Plasmonic Nanoparticles

Roberto Comparelli,[a] Tiziana Placido,[a,b] Nicoletta Depalo,[a] Elisabetta Fanizza,[a,b] Marinella Striccoli,[a] and M. Lucia Curri[a]

[a]*CNR-IPCF Istituto per i Processi Chimici e Fisici Sez. Bari, c/o Dip. Chimica, Via Orabona 4, 70126, Bari, Italy*
[b]*Università degli Studi di Bari – Dip. Chimica Via Orabona 4, 70126, Bari, Italy*
r.comparelli@ba.ipcf.cnr.it

When light interacts with plasmonic nanoparticles, their free electrons are driven by the alternating electric field and collectively oscillate at a resonant frequency leading to a phenomenon known as surface plasmon resonance. The frequency and amplitude of the resonance are sensitive to particle size and shape, which determine how the free electrons are polarized and distributed on the surface. As a consequence, control of the size and shape of plasmonic nanoparticles represents the most powerful means to tailor and finely tune their optical properties. In this chapter, a few among the most important techniques enabling size/shape-controlled growth of metal nanoparticles will be reviewed. Some examples of post-synthesis nanoparticle purification and surface modification toward plasmonic applications will also be provided.

Active Plasmonic Nanomaterials
Edited by Luciano De Sio

ISBN 978-981-4613-00-2 (Hardcover), 978-981-4613-01-9 (eBook)
www.panstanford.com

2.1 Introduction

Plasmonic nanoparticles (NPs) are characterized by the possibility to couple their electron density with electromagnetic radiation of wavelengths far larger than particle size. On the contrary, in a bulk metal there is a maximum limit on size wavelength that can be effectively coupled based on the material size [34]. NPs exhibit relevant scattering, absorbance, and coupling properties based on their geometry and relative position.

Surface plasmon resonance (SPR) is the most outstanding optical property of metallic nanostructures. It consists of a collective oscillation of conduction electrons in a metal when the electrons are perturbed from their equilibrium positions, for instance, by an electromagnetic wave (light). Upon interaction with light, the free electrons of a metal are driven by the alternating electric field to coherently oscillate at a resonant frequency relative to the lattice of positive ions [96]. Since the penetration depth of an electromagnetic wave in a metal surface is limited (<50 nm for Ag and Au), only plasmons caused by surface electrons are significant and are commonly referred to as *surface plasmons* [106]. When the wavelength of light is much larger than NP size, the electric field can make free electrons to move away from the metal particle in one direction, creating a dipole that can switch direction with the change in electric field (Fig. 2.1).

Light in resonance with the surface plasmon oscillation causes the oscillation of free electrons in the metal. As the wavefront of the light passes, the electron density in the particle is polarized

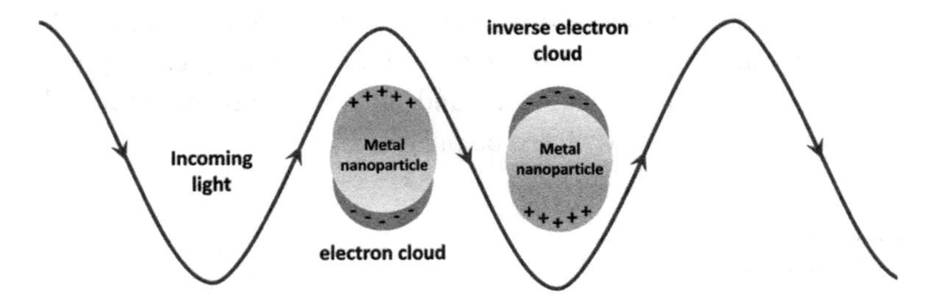

Figure 2.1 Depiction of interaction of incoming light with metal NPs and generation of SPR.

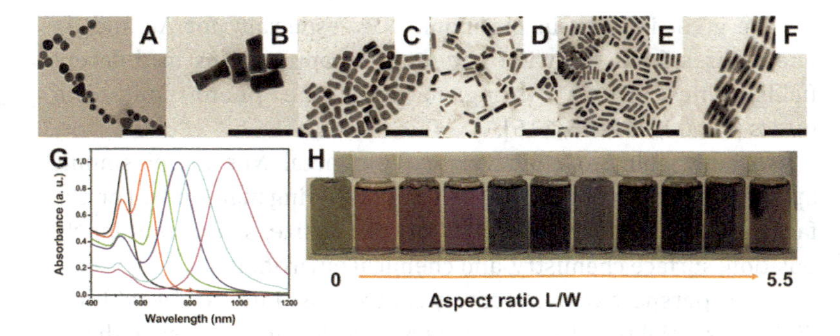

Figure 2.2 TEM image of Au NPs (A) and Au NR with increasing aspect ratio (B–F); scale bar 100 nm. (G) From black to purple, absorbance spectra of the samples reported in A–F, respectively. The longitudinal plasmon band shift at higher wavelength as NP aspect ratio increases. (H) Color of Au NR colloidal water solution at increasing aspect ratio (L/W) defined as the ratio between length and width.

and oscillates in resonance with the light frequency, resulting in a standing oscillation. The resonance condition can be investigated by absorption and scattering spectroscopy and is found to depend on shape, size, and dielectric constants of both metal and surrounding material. As the shape or size of the NP changes, surface geometry changes consequently, inducing a shift in the electric field density at the surface. This finally results in a change in the oscillation frequency of the electrons and, hence, different cross sections for the optical properties, including absorption and scattering. In fact, the optical properties of spherical metal NPs show just a weak dependence on their size; however, when anisotropy is present, such as in nanorods (NRs), the optical properties of such NPs dramatically change (Fig. 2.2).

SPR can be observed when the wavelength of incident light far exceeds the particle diameter. In the "intrinsic size" region (< 5 nm), noble metal NPs do not show any plasmon absorption, while the absorption of particles larger than 50–60 nm ("extrinsic size" region) is broad and covers most of the visible region. Such properties are mainly observed in Au, Ag, and Cu NPs, because of the presence of free conduction electrons, which are absent in the individual atoms as well as in the bulk [34].

The great interest in plasmonic NPs, especially for Au and Ag ones, arises from the wide potential for applications in different fields, which include sensing, nanomedicine, photovoltaics and optics [34, 60, 96, 106, 115].

The key point to effectively exploit metal NPs for plasmonic applications relies on the possibility of achieving high control on the factors affecting their plasmon absorption: size, shape, mutual NP position, surface chemistry, and chemical environment.

In the perspective to effectively control position and line width of SPR is essential the ability to grow NPs with a designed size, shape, and size distribution. Such an opportunity is offered by modern colloidal chemistry techniques, which allow material scientists to finely tune the synthetic conditions as a function of NP final application. In addition, a fine tuning of NP surface properties by a suitable post-synthetic surface engineering opens up the way to potentially exploit plasmonic NPs in any desired applications. This chapter will give a brief overview of the factors affecting the plasmon absorption and will report on the most recent advances in metal NPs size/shape-controlled syntheses (both in aqueous and organic environment) and post-synthetic surface modification procedures, in order to obtain original candidate structures for plasmonic applications.

2.2 Experimental Parameter Affecting Plasmon Absorption

Frequency and intensity of a plasmon resonance are primarily determined by (a) the intrinsic dielectric property of a given metal, (b) the dielectric constant of the medium in contact with the metal, and (c) the pattern of surface polarization. As such, any variation in the shape, size, and chemical environment of a metal particle causes a change in the plasmon resonance. Such a dependence offers the ability to tailor the SPR of metal NPs by means of a shape-controlled synthesis. The interactions of an electromagnetic wave with a NP can be understood by solving Maxwell's equations. However, Maxwell's equations can be exactly solved only for special cases such as a solid sphere, concentric spherical shells, a spheroid, and an infinite cylinder. For other particles with arbitrary geometrical shapes, some approximations are required. The discrete dipole approximation (DDA) method has been widely utilized to simulate the interaction of light with metal particles of arbitrary shape [106].

2.2.1 *Size Dependence of Plasmon Absorption: Mie's Theory*

Mie was the first to explain SPR in 1908 by solving Maxwell's equations for the absorption and scattering of electromagnetic radiation by small spherical particles. Here only a brief description of Mie's Theory will be provided, omitting the full mathematical discussion that the reader may find in relevant reviews on the topic [34, 96]. In Mie's theory, the total extinction coefficient of small metallic particles is given as the summation over all electric and magnetic multipole oscillations contributing to the absorption and scattering of the interacting electromagnetic field. For NPs much smaller than the wavelength of the absorbing light ($<$25 nm for Au NPs), only the dipole term is assumed to contribute to the absorption (dipole approximation) [171], and the related changes in the optical absorption spectra are referred to as *intrinsic size effects* [96]. The plasmon absorption is size-independent within such dipole approximation. Experimentally, a size effect on SPR is observed as the plasmon bandwidth increases with decreasing particle size. Indeed, it is well established that the bandwidth is inversely proportional to the radius of the particle for sizes smaller than 20 nm [171]. For larger NPs ($>$25 nm for Au NPs), the extinction coefficient explicitly depends on the NP size. For these NPs, the plasmon bandwidth increases with increasing size as the wavelength of the interacting light becomes comparable to the dimension of the NP. This behavior is referred to as an *extrinsic size effect* because the size dependence enters through the full expression of Mie's theory.

However, the absorption maximum of the surface plasmon oscillation also depends on the NP size. In the extrinsic size region, the peak position shifts to longer wavelengths or correspondingly to lower energies as NP size increases. The situation gets more complicated for smaller NPs for which intrinsic size effects should dominate. Experimentally, both a blue and a red-shift of the plasmon maximum have been observed with decreasing size [85, 96]. Furthermore, the magnitude of the wavelength shift in the absorption maximum for NPs in the intrinsic size region is small compared to the total line width of the plasmon resonance. Finally, the variety of theoretical approaches considering the dependence of the plasmon band maximum on the NP size reaches contradictory conclusions, even adding further confusion. Therefore, the peak position of the surface plasmon oscillation is generally not well suited for a discussion on a size effect.

2.2.2 *Shape-Dependence of Plasmon Absorption: Gans Theory and Discrete Dipole Approximation*

A much more drastic effect on the surface plasmon absorption can be observed when NP shape changes [42]. Indeed, for anisotropic metal NPs, electrons can oscillate to a different amplitude along the three axes of the nano-object. Such different oscillations result in a splitting of the plasmon band in two or more bands. For instance, the plasmon absorption of Au NRs splits into two bands corresponding to the oscillation of free electrons along and perpendicular to the long axis of the rods [115]. The transverse mode shows a resonance at about 520 nm, which is coincident with the plasmon band of spherical particles, while the resonance of the longitudinal mode is red-shifted and strongly depends on the NR aspect ratio R, which is defined as the length of the rod divided by its width.

According to Gans theory (first reported in 1912), the optical absorption spectrum of a collection of randomly oriented Au NRs with aspect ratio R can be modeled by using an extension of Mie's theory within the dipole approximation. Gans theory gives the exact solution to Maxwell's equation for spheroidal particles, while real NRs typically present rather a more cylindrical shape [34]. A popular alternative to solve Maxwell's equation for anisotropic particles is DDA. The DDA is an approximation of the continuum target by a finite array of polarizable points. The points acquire dipole moments in response to the local electric field. The dipoles of course interact each other via their electric fields, so that the DDA is also sometimes referred to as the coupled dipole approximation. For a finite array of

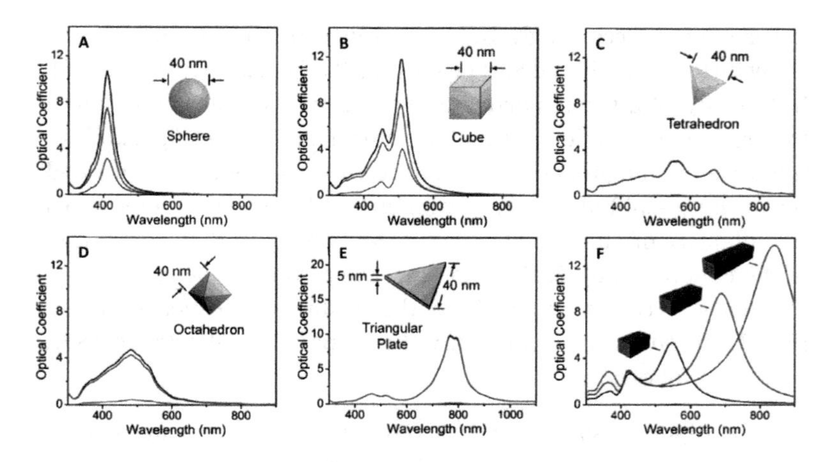

Figure 2.3 Extinction (black line), absorption (red line), and scattering (blue line) spectra calculated for Ag NPs of different shapes: (A) a sphere displaying a single dipole resonance peak and (B) a cube, (G) a tetrahedron, (D) an octahedron, and (E) a triangular plate. (F) Extinction spectra of rectangular bars with aspect ratios of 2 (black), 3 (red), and 4 (blue). Adapted with permission from Refs. [191] and [192]. Copyright © 2006 and 2007, American Chemical Society.

point dipoles, the scattering problem may be solved exactly, so the only approximation that is present in the DDA is the replacement of the continuum target by an array of N-point dipoles [106].

Figure 2.3 compares SPR spectra for Ag NPs of various shapes suspended in water calculated by DDA [191, 192]. The extinction, absorption, and scattering spectra for the 40 nm Ag sphere (Fig. 2.3A) were obtained using Mie's theory, whereas the DDA method was used for all the other shapes.

The nanosphere shows just one SPR signal, whereas the nanocube, due to several distinct symmetries for dipole resonance, shows more peaks (Fig. 2.3B) [192]. In addition, the position of the most intense peak for the nanocube is red-shifted compared with that of the sphere. Such a shift is caused by the accumulation of surface charges at the corners of the nanocube and is observed, in general, for any NP with sharp corners. In these systems, the increased charge separation reduces the restoring force for electron oscillation, which in turn results in a shift of the resonance peak to lower energy. Accordingly, both a tetrahedron and an octahedron show further red-shifted SPR peaks (Fig. 2.3C,D, respectively) [106]. In the case of a triangular plate, the spectra exhibit peaks shifted toward longer wavelengths and an intense SPR band (Fig. 2.3E). In this case, although sharp corners increase charge separation and result in red-shifted SPR, the symmetry of the particle determines the intensity of dipole resonances [78]. Control of the aspect ratio of the nanobar and NR allow to tune, the resonance peak positions of such NPs can be tuned from the visible to the near-infrared (Fig. 2.3F) [106, 190]. In summary, the shape of NPs affects their SPR spectra as follows:

- Dipole resonance peaks red-shift with increasing corner sharpness and particle anisotropy;
- The SPR peak intensity increases with particle symmetry;
- The number of SPR peaks is determined by the number of ways the nanostructure can be polarized.

2.2.3 *Other Parameters Affecting Plasmon Absorption*

2.2.3.1 Effect of surrounding medium: the Drude model

According to the Drude model, the position of the surface plasmon band of metal NPs is significantly influenced by the surrounding medium [112]. The Drude model was proposed in 1900 by Paul Drude to explain the transport properties of electrons in materials (especially metals). The Drude model considers the metal to be formed of a mass of positively charged ions from which a number of "free electrons" are detached. These may be thought to have become delocalized when the valence levels of the atom came in contact with the potential of the other atoms, that is, the surrounding medium. Therefore, depending on the surrounding medium, the surface plasmon band could oscillate between 500 nm and 550 nm for Au NPs.

Generally, NPs are dispersed in a medium, and interactions with such a medium can be classified into two main categories: (i) media that alter the refractive index around metal NPs and (ii) media formed of molecules that complex the metal surface. Media not possessing any active functional groups do not directly interact with metal surface; therefore, any modification of plasmon band can be safely ascribed only to change in the refractive index of the surrounding medium. In these systems, the surface plasmon band of metal NPs gradually shifts toward longer wavelengths with increasing refractive index of the solvent. Conversely, media chemically interacting with surface of metal NPs could alter the position of the plasmon band due to a different distribution of free electron.

2.2.3.2 Effect of interparticle coupling

The surface plasmon oscillation in metallic NPs changes drastically when particles are densely packed in the medium so that the individual particles are electronically coupled to each other. It has been seen theoretically and experimentally that when the individual spherical Au NPs come into close proximity to each other, electromagnetic coupling of clusters becomes effective for interparticle distances smaller than five times the NP radius ($d \leq 5R$, where d is the center-to-center distance and R is the radius of the particles) and may lead to complicated extinction spectra depending on the size and shape of the formed NP aggregate by a splitting of single NP resonance [142].

As a consequence, their plasmon resonance is red-shifted by up to 300 nm, causing strong color change [124]. This effect is negligible if $d > 5R$ but becomes increasingly important at smaller distances. Aggregation causes a coupling of the Au NPs' plasma modes, which results in a red-shift and broadening of the longitudinal plasma resonance in the optical spectrum [136].

2.2.3.3 Effect of temperature

Kreibig has evaluated the temperature dependence of small Ag and Au clusters (diameters of 1–10 nm) in a glass matrix and found that the spectrum decreases in intensity and broadens when increasing the temperature from 1.5 K to 300 K. Such temperature dependence is more pronounced with decreasing NP size. The temperature dependence of the electron–phonon scattering is partially compensated by a thermal lattice contraction of the NPs with decreasing temperature. The latter becomes more evident as NP size decreases. In addition, changes of the band structure in NPs below 5 nm are assumed to be effective [85]. However, since the electron concentration *n*, which determines the plasmon frequency, decreases at increasing temperature, thus affecting the overall absorption spectrum, the temperature has to be raised by several hundred degrees in order to record a significant line broadening. Therefore, at room temperature, metal NPs show a negligible dependence of SPR on temperature variation [34].

2.3 Nanoparticle Characterization

The study of materials at nanoscale is essential, as investigation of their structures and understanding of their properties enable to properly design experimental methods to successfully control material characteristics. Indeed, a plethora of novel techniques and methods have been implemented and are constantly improved to achieve an increasingly high and accurate control on size, morphology, crystalline phase, and chemical nature of the resulting nano-objects. Accordingly, the techniques to characterize materials had to make a big step forward to face the increasing accuracy needed in material investigation and to understand the new properties arising from materials at nanoscale. Existing optical, structural, and morphologic characterization techniques have been strongly enhanced to fit nanomaterial characterization, and new techniques have been proposed to face the increasing demand of information. For instance, an absolutely innovative tool was developed in the 1980s, commonly known as scanning probe microscopy, which, besides granting the Nobel Prize in 1986 to its inventors Gerd Binnig and Heinrich Rohrer, provided for the first time atomic-scale resolution on a surface [11, 12].

However, classical optical techniques can be effective in providing information on NP size, shape, mutual position, and surface chemistry. Monitoring the position and the line shape of plasmon band by UV-Vis-NIR absorbance spectroscopy is a fast and cost effective way to investigate NP size, aspect ratio (if any), and aggregation [34]. Dynamic light scattering can investigate size and size distribution of a diluted solution of NPs and possibly explore their surface charge, when combined with zeta potential measurements [69]. Also FT-IR and NMR spectroscopy have enabled an effective characterization of the surface chemistry of NPs [126].

Nevertheless, the most powerful techniques to relate experimental conditions with the final properties of NPs deal with the resolution of their crystal structure with atomic-scale accuracy. X-ray diffraction (XRD) is the most popular technique to investigate the crystal structure although data interpretation can be a delicate point in the case of nanopowder [21]. It has been demonstrated that nanocrystals (NCs) are characterized by peculiar lattice constants taking into account strain effects occurring at NC surface. Thus, softwares for XRD data interpretation have now been improved to take into account the peculiar structural characteristics of nanocrystalline materials. Much more powerful excitation sources such as synchrotron radiation, for instance, lead to the development of a new set of techniques based on X-ray absorption spectroscopy (XANES: X-ray absorption near edge structure; EXAFS: extended X-ray absorption fine structure), which demonstrated very useful to obtain information, for instance on crystal plane exposed, possible presence of any impurity, ultimately allowing to understand the key points to direct the growth of NPs with tailored shape [43]. Recently, a new microscopy technique using X-rays as a "light" source has been exploited to characterize NC structure and the geometrical order of NC superstructures. Image contrast is based on X-ray scattering at wide angles (WAXS) or small angles (SAXS) and contains information on NC structure and their mutual organization, respectively [5, 44].

Electron microscopy, both in transmission (TEM) and in scanning (SEM) mode, is fully used in any research groups dealing with NC synthesis. Both techniques exploit an electron beam to obtain information on NC size, shape, aggregation, and chemical composition at different size scale. TEM is characterized by a much higher resolution and can also investigate the crystalline structure of a single NP by exploiting the diffraction of a convergent electron beam (CBED). Such a technique can be coupled with TEM images in bright field contrast mode to investigate the crystalline plane exposed, the presence of defects, and in case of anisotropy the preferential direction of growth [137].

Both TEM and SEM can analyze the chemical composition of NPs by energy dispersive spectroscopy, even providing a chemical mapping in false color of the elements detected in the sample. For a deeper discussion of nanomaterial characterization by electron microscopy, please see Refs. [28, 34].

2.4 Synthesis of Hydrophilic Plasmonic Nanoparticles

Many different techniques have been developed to generate plasmonic NPs. There are two general strategies to obtain materials at the nanoscale: "top-down" and "bottom-up" approaches.

Top-down method means that bulk materials are broken into small pieces or, alternatively, carved at nanoscale range to generate NPs of desired dimensions. Common top-down techniques are photolithography, electron beam lithography, and laser ablation [60].

Bottom-up strategies provide metal NPs starting from individual species that undergo a chemical or biological reduction to provide metal atoms that assemble one by one in nanosized structures. In this chapter, we will focus on common bottom-up methods that include templated chemical, electrochemical, sonochemical, thermal, and photochemical reduction techniques.

Chemical reduction method typically involves nucleation and successive growth of NPs. When these steps are completed within the same process, one generally refers to an "in situ" synthesis; alternatively synthesis can be mediated by the presence of seeds, purposely prepared, which gives place to "seed-mediated" method. For the in situ synthesis method, we will focus on the preparation of spherical Au NPs. For the seed-mediated method, we will concentrate on the preparation of Au NPs with a geometrical control [210].

2.4.1 *Synthesis of Spherical Nanoparticles*

The most popular method for metal NP synthesis is the reduction of metal salts. In general, the preparation of Au NPs by chemical reduction is based on (i) reduction of Au precursor (typically

$HAuCl_4$) from Au(III) to Au(0) using strong reducing agents and (ii) stabilization by agents such as trisodium citrate dihydrate, sulfur ligands (in particular thiolates), phosphorus ligands, nitrogen-based ligands (including heterocycles), oxygen-based ligands, dendrimers, polymers, and surfactant (in particular cetyltrimethylammonium bromide, CTAB). The in situ synthesized metal NPs are also used in the preliminary step of the seed-mediated NP growth or for further functionalization [210].

2.4.1.1 Turkevich method and its modifications

The pioneering work of Turkevich in 1951 [177], and also that of Faraday in 1857 [38], on metal NP synthesis, along with studies on mechanisms behind the growth kinetics of NPs, stimulated research in the synthetic, resulting in a plethora of reports collected in review articles and books [34, 210]. Turkevich synthesis is probably the most utilized method for noble metal NP synthesis. Such an approach allows to prepare monodisperse Au NPs by the addition of trisodium citrate to a boiling solution of $HAuCl_4$ leading to Au NP sizing of about 20 nm [177].

In 1973, Frens published an improvement of this procedure, yielding to Au NPs within a broad size range from 15 nm to 150 nm by controlling the trisodium citrate to Au ratio. However, particles larger than 20 nm were always polydispersed [39].

The mechanism of Au NPs formation using this synthetic route has been examined in details [139]. Kimling et al. indicated that a high concentration of citrate more rapidly stabilizes Au NPs of smaller sizes, whereas a low concentration of citrate leads to large Au NPs and even to aggregation of Au NPs [83]. A comprehensive research on the mechanism of the Turkevich–Frens method in multiple-step process has been published by Kumar et al., indicating as the initial step is represented by the oxidation of citrate that yields dicarboxy acetone. Then, Au(III) is reduced to Au(I) and Au(0), and Au(I) sets down on the Au(0) atoms to form the Au NP [86]. Thus, the actual Au NP stabilizer is dicarboxy acetone resulting from the oxidation of citrate, rather than citrate itself. In addition, the presence of a citrate salt modifies the pH of the system and influences the size and size distribution of the Au NPs [73]. On this basis, nearly monodispersed Au NPs with sizes ranging from 20 nm to 40 nm have been synthesized upon variation of the pH solution [143]. Other improvements of the Turkevich method involved the control of the reaction temperature [147], the introduction of fluorescent light irradiation [81], and the use of high power ultrasound [166]. Citrate-stabilized Au NPs are generally larger than 10 nm, due to the very modest reducing ability of trisodium citrate dihydrate. An original report from Puntes's group accounts for the use of D_2O as the solvent instead of H_2O in the synthesis of Au NPs, which resulted of 5 nm, thus pointing out that D_2O increased the reducing strength of citrate [125, 210]. Other reducing agents can be used in place of sodium citrate: ascorbic acid [64], sodium borohydride [105], hydrogen [35], hydrazin dihydrochloride [117], methanol [176], and polyethylene glycol [23]. As an alternative, citrate acts just as a stabilized agent and $NaBH_4$ is used as reducing agent at room temperature [13]. The effect of the order of addition of reactants, that is, adding $HAuCl_4$ to the citrate solution, has also been investigated, producing monodispersed Au NPs sizing less than 10 nm [161].

Similarly, the Turkevich protocol for synthesis of Au NPs has been applied for the preparation of Ag NPs although it leads to the formation of highly polydispersed large-size particles (60–200 nm) [77]. A different mechanism has been proposed for the formation of Ag NPs via citrate reduction route, since citrate ion is a mild reducing agent for Ag. Thus, first citrate reduces a few Ag ions to form Ag seeds. During this process, one of the intermediates formed in the reaction (Ag^{2+}) strongly complexes with citrate. This complexation leads to a decrease in the extent of reduction of Ag ion to the zero-valent state. Therefore, due to the presence of residual charge on the seed, the growth of the seed is inhibited after an optimal size, thus preventing further aggregation. This point onward, growth happens due to Ostwald ripening [135].

2.4.1.2 Seed-mediated method

The seed-mediated method is another popular technique for metal NP synthesis that has been used for more than a century. Compared with the in situ synthesis, the seed-mediated method allows an easier control on sizes and shapes of formed metal NPs. In the first step of the procedure, small-size (typically less than 5 nm) Au NP seeds are prepared. In the second step, the seeds are added to a "growth" solution containing metal precursor and the stabilizing and reducing agents, then the newly reduced Au(0) grows at the seed surface to form large-size Au NPs. The reducing agents used in the second step are generally mild ones in order to reduce Au(III) to Au(0) only in the presence of Au seeds as catalysts. Thus, the freshly reduced Au(0) can only assemble on the surface of the Au seeds, and no new particle nucleation occurs in solution. In the course of the seed-mediated growth synthesis of Au NPs, the formation of seeds takes a significant place correlated to the size, shape, and surface properties that are controlled by the amount and nature of reducing agent and stabilizer, and their ratio to the Au precursor. The earliest Au seeds sizing 12 nm, have been obtained using citrate reductant [14]. A few years later, Murphy's group reported the synthesis of 3.5 nm citrate-capped Au seeds by dropping an ice-cold aqueous solution of $NaBH_4$ into a solution of a mixture of $HAuCl_4$ and citrate [65].

This procedure of Au seed formation has been modified by El-Sayed using CTAB (hexadecyltrimethylammonium) bromide as the stabilizer instead of citrate [122]. Subsequently, CTAB-capped seeds have been regarded as the most primary nucleation process in the synthesis of Au NPs. Therefore, the seed-mediated strategy has emerged as a very efficient method to synthesize monodispersed Au NPs even with large sizes (up to 300 nm) and with well-defined shapes [9, 122, 132]. This method has been later used by Han's group for the synthesis of icosahedral Au NPs having controlled size (from 10 nm to 90 nm) (Fig. 2.4) [87]. Reaction conditions such as temperature, pH, ratio between Au precursors and seed concentration and citrate concentration have been demonstrated to affect the size distribution and shape of the Au NPs [9, 183].

2.4.1.3 Photochemical synthesis of spherical metal nanoparticles

The photochemical synthesis method has attracted much attention being a versatile and convenient process. The beginning of

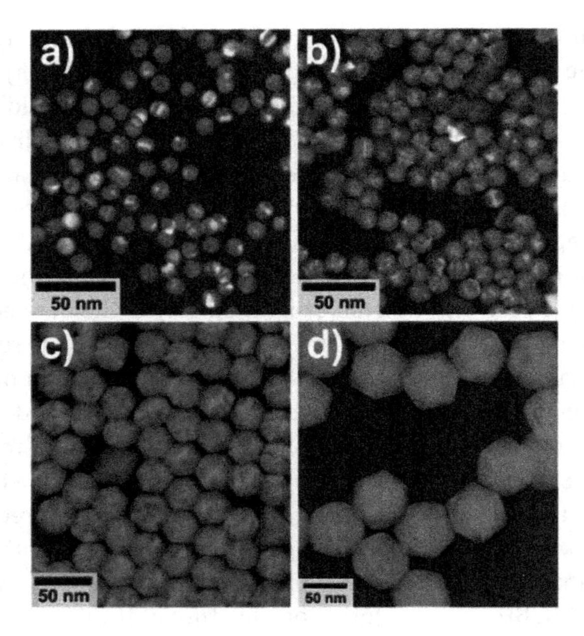

Figure 2.4 SEM images of Au NPs in solution (a) 11.0 ± 0.8 nm, (b) 13.3 ± 2.0 nm, (c) 32.2 ± 1.8 nm, and (d) 69.0 ± 3.7 nm. Reprinted with permission from Ref. [87]. Copyright © 2007, American Chemical Society.

photochemical synthesis of metal NPs dates back to a discovery in the 18th century, when Schulze discovered that certain Ag salts darkened by the irradiation of light. Nowadays, a variety of photoinduced synthetic methods for metal NPs and nanostructure have been developed to obtain well-defined monometallic and bimetallic NPs, and composite materials originated by the direct photoreduction of metal salt to the zero-valent metal or reduction of metal ions using the photochemically generated intermediates, such as excited molecules and radicals (photosensitization). The photochemical synthesis enables to fabricate metal NPs in various media, including microemulsion, polymer films, glasses, and cells. The photochemical formation of Ag NPs has been first reported in 1976 involving the direct excitation of $AgClO_4$ in aqueous and alcoholic solutions [53]. Conversely, the direct photoreduction of $AuCl_4^-$ in an aqueous solution provides Au NPs. Before the Au NP formation, the yellow color of Au^{3+} disappears, due to the formation of an intermediate, most probably Au^+ [56]. The common ionic surfactants in the photochemical process are CTAB (cationic) and sodium dodecyl sulphate (SDS, anionic) [149].

2.4.2 *Shape Control of Metal Nanoparticles*

As reported in Section 2.2.2, controlling NC shape provides an effective tool for optical tuning. Similarly, chemical reactivity is highly dependent on surface morphology. The bounding facets of the NC, the number of step edges and kink sites, as well as the surface-area-to-volume ratio can dictate unique surface chemistry that may prove useful in achieving highly selective catalysis [2]. Whereas highly symmetric spherical particles are characterized by a single scattering peak, anisotropic shapes such as rods [137], triangular prisms [75], and cubes [168] exhibit multiple scattering peaks in the visible wavelengths due to highly localized charge polarizations at corners and edges. For the noble metal systems, crystallographic control over the nucleation and growth of NPs has been most widely achieved using colloidal methods. These reactions are governed by thermodynamics (e.g., temperature, reduction potential) and kinetics (e.g., reactant concentration, diffusion, solubility, reaction rate). Surface energy considerations are crucial in understanding and predicting the morphology of noble metal NCs. Noble metals, which generally adopt a face-centered cubic (fcc) lattice, possess different surface energies for different crystal planes. This anisotropy results in stable morphologies where free energy is minimized by particles bound by the low-index crystal planes that exhibit closest atomic packing [172].

2.4.2.1 Gold nanorods

Among all the anisotropic metal NPs that have been synthesized, Au NRs have attracted the most attention as demonstrated by the enormous number or reviews on the topic [32, 34, 49, 62, 104, 113, 181]. The synthetic approaches to promote the formation of rod-like Au NPs commonly exploit hard template or micelles to drive metal particle growth along a preferential direction.

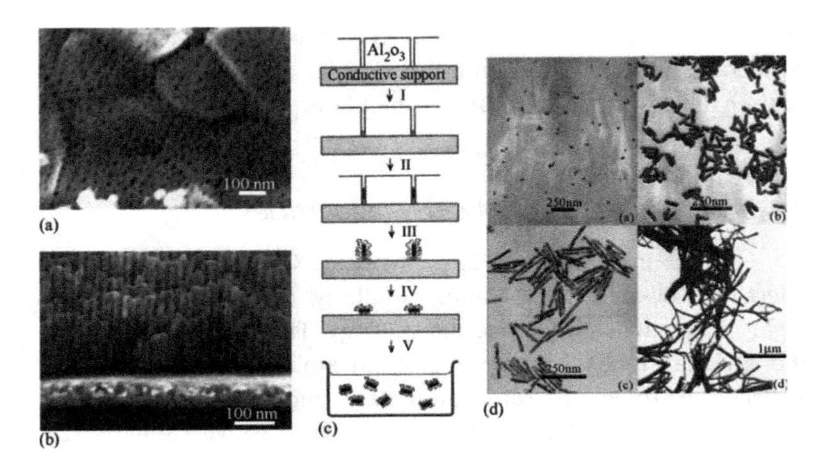

Figure 2.5 (a, b) FEG–SEM images of an alumina membrane. (c) Schematic representation of the successive stages during formation of gold nanorods via the template method. (d) TEM micrographs of gold nanorods obtained by the template method. Reprinted with permission from Ref. [133]. Copyright © 2005, Elsevier.

"Hard" template method. The hard template method has been first introduced by Martin and coworkers who have exploited rigid templates to direct the growth of anisotropic NPs [108]. The method is based on the electrochemical deposition of Au within the cylindrical pores of rigid matrices: for instance nanoporous polycarbonate or alumina template membranes. The redispersion of the template-synthesized Au NRs into water or organic solvents has been also shown [131]. Schematically, a small amount of Ag or Cu is sputtered onto the alumina template membrane to provide a conductive film for electro-deposition (stage I in Fig. 2.5). This is then used as a foundation for subsequent electrochemical deposition of Au NPs within the nanoporous of alumina (stage II). The next stage involves the selective dissolution of both the alumina membrane and the copper or Ag film, in the presence of a polymeric stabilizer such as poly(vinylpyrrolidone) (PVP) (III and IV in Fig. 2.5). In the last stage, the rods are dispersed either in water or in organic solvents by means of sonication or agitation. The diameter of the Au NPs thus synthesized coincides with the pore diameter of the alumina membrane [17]. The length of the NRs can be controlled through the amount of Au deposited within the pores of the membrane [179]. The fundamental limitation of the template method is in the low yield [133].

Micelle-based methods. In the last decade, micelles have been used as a "soft" template [66]. The general scheme involves the reduction of an Au precursor (typically tetrachloroauric acid or a gold sacrificial anode) in the presence of micelles of ionic surfactants [122]. Indeed, under specific conditions, surfactants are capable to form nonspherical micelles in water. CTAB is the most popular surfactant used for this purpose being able to form cylindrical micelles under proper experimental conditions. Such micelles participate in both the formation of NRs and their stabilization due to the formation of CTAB bilayer around the NR [136]. A variety of synthetic methods basically differing only in the strategies to trigger the Au precursor reduction make use of micellar template: electrochemical [205], seed-mediated [65, 66, 122], seed-less [63], photochemical [43, 111], combined chemical-photochemical [119], and microwave-assisted [213] approaches.

The **electrochemical route** to produce Au NRs in solution and in high yield has been first demonstrated by Wang and coworkers [22, 205]. The synthesis is conducted within a simple two-electrode-type electrochemical cell, as shown in the schematic diagram in Fig. 2.6. A gold metal plate is used as a sacrificial anode to produce $AuBr_4^-$, which is the actual Au precursor, while the cathode is a platinum plate. Both electrodes are immersed in an electrolytic solution containing CTAB and a small amount of a much more hydrophobic cationic surfactant tetradodecylammonium bromide ($C_{12}TAB$) which acts as a rod-inducing co-surfactant. CTAB serves not only as the supporting electrolyte but also as the stabilizer for the NPs, to prevent their further aggregation. Appropriate amounts of acetone and cyclohexane are also added into the electrolytic solution. Acetone is used for loosening the micellar structure facilitating the incorporation of the co-surfactant into the CTAB micelles, while cyclohexane enhances the formation of rod-like CTAB micelles [175].

$AuBr_4^-$ anions are complexed to the cationic surfactants and migrate to the cathode where reduction occurs. Sonication is needed

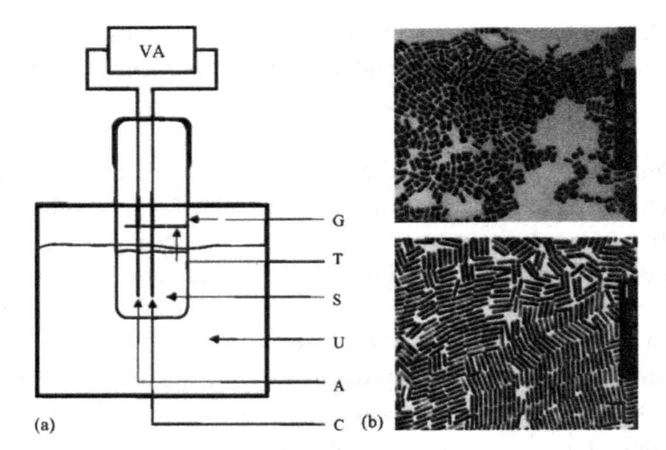

Figure 2.6 (a) Schematic diagram of the setup for preparation of Au NRs via the electrochemical method containing: VA, power supply; G, glassware electrochemical cell; T, teflon spacer; S, electrode holder; U, ultrasonic cleaner; A, anode; C, cathode. (b) TEM micrographs of Au nanorods with different aspect ratios 2.7 (top) and 6.1 (bottom). Scale bars represent 50 nm. Reprinted with permission from Ref. [133]. Copyright © 2005, Elsevier.

to shear the resultant rods as they form away from the surface or possibly to break the rod off the cathode surface. Another important factor controlling the aspect ratio of the Au NRs is the presence of an Ag plate inside the electrolytic solution, which is gradually immersed behind the Pt electrode. The redox reaction between Au ions generated from the anode and Ag metal leads to the formation of Ag ions. Wang and coworkers found that the concentration of Ag ions and their release rate determined the length of the NRs [133].

Seed-mediated routes in aqueous media have been among the most widely investigated strategies, since they allow the gram-scale preparation of metallic NRs at nearly room temperature [122, 132]. The general scheme recalls that already described in Section 2.4.1.2 for the growth of spherical NPs. Briefly, Au seeds (size <5 nm) can be produced by reduction of Au precursor by strong reducing agent (typically $NaBH_4$) in presence of CTAB. In a separate flask, Au(III) is reduced to Au(I) by a mild reductant (ascorbic acid for instance) in a CTAB micellar solution, then a certain amount of freshly prepared seed is added to the Au(I) solution to trigger the growth of Au

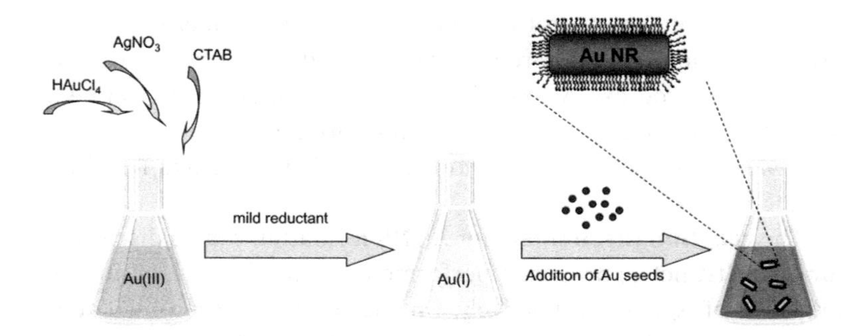

Figure 2.7 General scheme of the synthetic steps in seed-mediated growth of Au NRs assisted by Ag ions.

NRs (Fig. 2.7). It has been reported that addition of Ag ions to the growth solution and the presence and of acetone and cyclohexane to elongate CTAB micelles could assist the growth of NRs [14, 210]. Secondary nucleation during the growth stage is inhibited by a careful control of the growth conditions, and in particular by using ascorbic acid as a weak reducing agent, which cannot reduce Au(III) to Au(0). Interestingly, by tuning the ratio between Ag and Au, it is possible to influence not only the yield and aspect ratio control of the Au NRs but also the mechanism for Au NRs formation and correspondingly its crystal structure. As a consequence, the position of the longitudinal plasmon band can be tuned within the visible and near-infrared region.

A **seed-less photochemical approach** has been proposed in 2002 to prepare in one pot Au NRs in presence of selected surfactants [80] representing one of the most straightforward approaches to obtain Au NRs, thanks to the possibility to prepare NRs in just one step [137]. This process itself is highly promising for producing uniform NRs and very useful in shedding light on the growth mechanism of anisotropic metal NPs, due to its simplicity and the relatively slow growth rate of the NRs. In a typical synthesis, an aqueous solution of $HAuCl_4$ precursor and a mixture of surfactant, CTAB, and co-surfactant, TDAB, in the presence of aliquots of ketones as radical initiator (acetone, for instance) [111, 123] and cyclohexane (able to influence the micelle morphology) [22] and $AgNO_3$ is exposed to UV light ($\lambda \geq 254$ nm) and kept at room temperature. In addition, NP growth is allowed to proceed rather slowly, thereby permitting a detailed monitoring of the reaction progress. Also in the photochemical approach defined, small amounts of Ag^+ ions are crucial to promote the formation of NRs with controllable aspect ratios rather than spherical NPs [80, 137].

The **reaction mechanism** for Au NR formation can be discussed for two distinct approaches performed in the absence or in the presence of $AgNO_3$, respectively. On the basis of a seed-mediated procedure, Au NRs have been prepared, albeit with modest yield, with aspect ratios up to 25 [41]. More recently, this method has been modified by the introduction of Ag^+ ions to controllably drive anisotropic growth, which has led to the production of shorter Au NRs with uniform aspect ratio tunable between ~2 and ~5 in nearly quantitative yield [122, 137, 153]. Improving the performance of Au NR synthesis ultimately requires in-depth knowledge of the mechanisms by which Au NRs grow. In addition, assigning reagents specific roles during the synthesis is challenging, as mechanistic research is making it increasingly clear that reagents in the synthesis of Au NRs act synergistically to effect shape control [32, 34, 50, 62, 89, 99, 113].

It is well known that CTAB forms cylindrical micelles above its second critical micelle concentration [121, 205]. Anisotropic micelles, whose sizes depend on the surfactant concentration and ionic strength, are formed in the surfactant solution. As the seeding particles are added to growth solution, they are covered with surfactant molecules and incorporated into micelles. The reduction of Au precursor on seeding particles results in the formation of NRs with the geometry determined by the micelle anisotropy. Murphy et al. proposed the growth mechanism based on the predominant adsorption of surfactant molecules on lateral sides of the rod; as a result, diffusion growth can proceed only facing the edges ("zipping" mechanism) [41]. Investigations have been performed into how the size and surface chemistry of the seed [45], the chain length of the directing surfactant [41, 121], and the nature of the surfactant counter ion impacted the aspect ratio of the Au NRs [50, 104, 113].

For the seed-mediated approach, a one-to-one correlation between the initial seeds and final NCs has been established for a number of noble metals with the fcc structure, which is summarized in Fig. 2.8 [106, 197, 198].

Generally, from single-crystal seeds, octahedrons, cuboctahedrons, or cubes can be produced, depending on the relative growth rates along the <111> and <100> directions. If uniaxial growth is somehow induced (e.g., by selective adsorption of surfactants), the cuboctahedral and cubic seeds grow into octagonal rods and rectangular bars, respectively. From singly twinned seeds, right bipyramids enclosed by $\{100\}$ facets can be produced, which can also evolve into nanobeams. From multiply twinned seeds, icosahedrons, decahedrons, and pentagonal NRs can be obtained. Finally, when the seeds contain stacking faults, they grow into thin plates, with the top and bottom faces being $\{111\}$ facets. If the seed structure and the growth condition are thoughtfully adjusted, the shape of the obtained NCs can be readily controlled. For example, two kinds of Au crystal seeds are widely used in Au NC synthesis. One is citrate-capped, penta-twinned particles bound by $\{111\}$ facets, and the other is CTAB-capped single-crystal particles with a pseudo-spherical shape. If the single-crystal particles are used as seeds, single-crystalline [100]-oriented Au NRs are obtained in CTAB solution in the presence of Ag^+ ions [99]. When penta-twinned particles are used as seeds, penta-twinned, [110]-oriented Au NRs can be produced in CTAB solution without Ag^+ ions [114], whereas bipyramidal Au NCs are obtained in CTAB solution in the presence of Ag^+ ions [99].

However the role of Ag in the Ag-assisted seeded growth synthesis has been a particularly challenging area of mechanistic investigation. The $[Au^{3+}]$:$[Ag^+]$ molar ratio realized in the reaction environment has been correlated to the shape of the resulting Au nanostructures, whereby the growth of $\{100\}$ facets has been found to be somehow favored by the presence of Ag^+ ions. In addition, an excess of Ag ions could result in NRs with lower aspect ratio or could even prevent the growth of NRs [80, 99, 137]. The UV-driven photochemical synthesis of Au NPs in micelle template realized in a mixture of surfactants [namely,

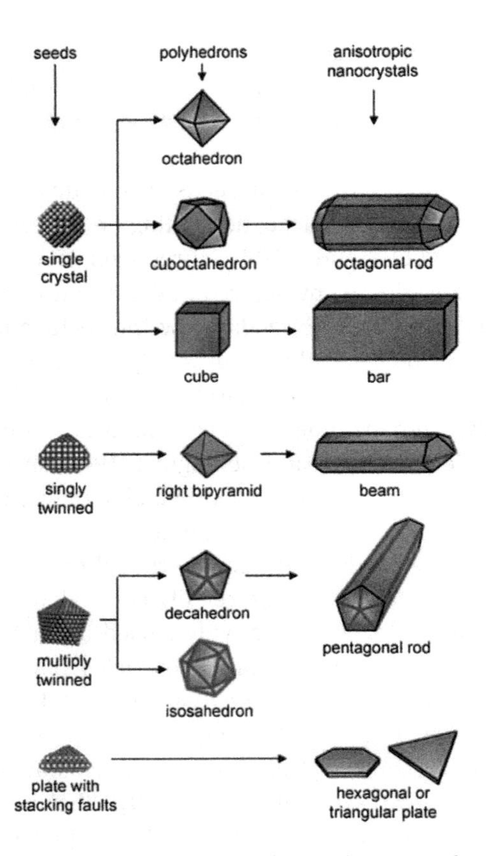

Figure 2.8 Schematic illustration of the evolution pathways from fcc metal seeds with different structures to shaped nanocrystals. The green, orange, and purple colors represent the {100}, {111}, and {110} facets, respectively. Twin planes are delineated in the drawing with red lines. Reproduced with permission from Ref. [198]. Copyright © 2011, Royal Society of Chemistry.

CTAB and tetrakis(decyl)ammoniumbromide] has been considered a convenient route to investigate the role of Ag ions in directing the growth of Au NR with tailored aspect ratio due to its long reaction time (17–88 h).

EXAFS experiments, HRTEM, CBED, and elemental analyses on the residual amount of Au and Ag ions in the reaction mixture as a function of reaction time have demonstrated the fate of Ag^+ ions and their chemical state and environment in the final product

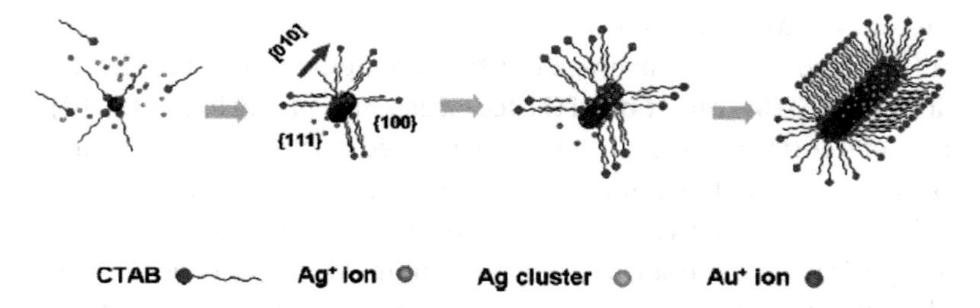

Figure 2.9 Proposed reaction mechanism for the Ag^+ ion-mediated formation of Au NRs in the photochemical synthesis approach. The growth is allowed to proceed preferentially along the {010} direction. Reprinted with permission from Ref. [137]. Copyright © 2009, American Chemical Society.

[137]. The photochemical route is typically characterized by an induction time for the nucleation of Au NPs involving a change in reaction mixture color from pale yellow to colorless, suggesting the reduction of Au(III) to Au(I) upon reaction with UV-generated ketyl radicals. The length of such an induction time depends on the reaction condition. Anyway, it is generally followed by the nucleation of Au NPs according to absorbance spectra and TEM images. Concomitantly with the appearance of Au SPR band in absorbance spectra, elemental analyses revealed that the amount of residual Ag and Au ions in the reaction mixture suddenly drops while Au NRs enrich in Ag. EXAFS analyses, HRTEM, and CBED demonstrated that Ag(0) islands are present on the long side of the rods (i.e., {100} and {110} facets). In addition, the elongate direction of NRs depend on the presence of Ag ions being [010] for NRs grown in presence of Ag and [101] without Ag ions.

As sketched in Fig. 2.9, the occurrence of an induction period prior to the NP nucleation, accompanied by defined spectral changes, supports the hypothesis that the first reaction step involves the reduction of Au(III) to Au(I). The invariant presence of Au species in solution during the entire induction period is consistent with CTAB-driven stabilization of Au(I) species, which are, therefore, allowed to accumulate until the critical supersaturation threshold is reached. From this point on, NP shape evolution can be easily followed by complementary TEM and UV-Vis absorption measurements. During growth, Ag(0) adsorbs onto {100} and {110} facets, with the average crystal development being preferred along the <010> direction of the fcc lattice. Such deposition of Ag(0) on preferential Au facets has been explained in literature according to an under potential deposition (UPD) mechanism [99, 137]. UPD is a phenomenon occurring when a metal submonolayer or monolayer is deposited onto the surface of a different metal. UPD can occur only when the work function of the depositing bulk metal is lower than the work function of the bulk metal substrate. The work function of Ag is lower than that of Au by more than 0.5 eV; therefore, the UPD of Ag can be expected to take place over Au. The UPD shifts of Ag^+ on Au surfaces can be expected to be in the order: $\{110\} > \{100\} > \{111\}$, thus accounting for the preferential adsorption of Ag onto {110} and {100} facets postulated earlier. When an excess of Ag is used, Ag(0) islands could be deposited even on the tip of the NR (i.e., {111}), thus decreasing the aspect ratio of the obtained NRs or even preventing the formation of anisotropic particle [137].

2.4.2.2 Silver nanorods/wires

Ag NRs and wires have also been synthesized in aqueous solutions, although their preparation is more critical and their reaction mechanism is still mostly unclear. For instance, Murphy and coworkers have reported a process to synthesize Ag NRs by reducing $AgNO_3$ with ascorbic acid in the presence of seed and CTAB [67]. Zhang et al. also have reported a seed-less synthesis of Ag nanowires using ascorbic acid as reducer in the presence of poly(methacrylic acid) (PMAA) [207]. Tian and coworkers have synthesized Ag NRs and nanowires by reduction of $AgNO_3$ with trisodium citrate in the presence of dodecylsulfonate [51, 59]. A method to make crystalline Ag nanowires in water, in the absence of a surfactant or polymer to direct NP growth, and without externally added seed crystallites has been proposed by Caswell et al. The reaction involves the reduction of Ag salt to Ag metal, at 100°C, by sodium citrate, in the presence of NaOH. Hydroxide ion concentration is key to producing nanowires, which are up to 12 microns long, instead of nanospheres [19].

2.4.3 *Other Shapes*

Among the shape-controlled strategies, seed-mediated growth is an efficient method for the synthesis of anisotropic Au NPs in a wide range of shapes, including nanocubes [153, 198], nanohexapods [79], nanoribbons [180], hollow nanocages [209], nanobranches [88], and nanopolyhedra [72].

For readers interested in much deeper discussion, a report on solution-based "chemical" route to multiple shaped-controlled Au NPs is suggested [153]. In a typical procedure, a $HAuCl_4$ solution is added to a CTAB solution followed by addition to the growth solution of an aqueous solution containing $AgNO_3$, ascorbic acid, and $HAuCl_4$. The different shape can be obtained by controlling the various combinations of [seed]/[Au (III)] ratio or the concentrations of CTAB and ascorbic acid (Fig. 2.10) [79, 104, 194]. Very recently,

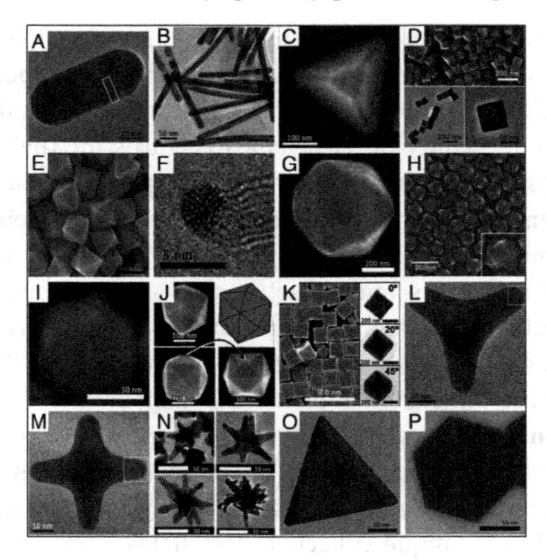

Figure 2.10 The major classes of noble metal NP shapes seen through TEM and/or SEM: (A) Au octagonal single-crystal rod, (B) Au pentagonally twinned rods, (C) Au tetrahedron NP, (D) Pd hexahedron (i.e., cube) NPs, (E) Au octahedron NPs, (F) decahedron, (G) Au icosahedron NP, (H) Au trisoctahedron NPs, (I) Au rhombic dodecahedron NP, (J) Pt tetrahexahedron NPs, (K) Au concave hexahedron NPs, (L) Au tripod NP, (M) Au tetrapod NP, (N) Au star NPs, (O) Au triangular plate/prism NP, and (P) Au hexagonal plate/prism NP. Reprinted with permission from Ref. [104]. Copyright © 2013, American Chemical Society.

Au NPs with hollow features have been reported [88, 209]. In the growth process, Au hollow nanocubes are prepared as seeds, and the reaction is initiated by the addition of the seeds to the growth solution containing the appropriate ratio of HCl, $HAuCl_4$, $AgNO_3$, ascorbic acid, and cetyltrimethylammonium chloride (CTAC) [210]. In addition, certain reactive solids with a specific shape can be used as sacrificial templates for the production of metal NCs, which then adopt the underlying shape of the template. In particular, Ag nanocubes can be used as sacrificial templates for the fabrication of Au nanocages and nanoframes through a galvanic displacement reaction [162].

2.5 Synthesis of Hydrophobic Plasmonic Nanoparticles

The synthesis of Au and Ag NPs directly in nonpolar organic media has been rapidly developing in recent years. The importance of the dispersibility of noble metal NPs in organic solvents mainly relies on the versatility of such nanostructures for their possible use as catalysts for organic reactions in nonpolar media, for their solvent-dependent optical properties, and also for their specific surface chemistry, which enables binding of metal NP surface to a variety of molecules and structures, in diverse environments [160]. Interestingly, hydrophobic Au- and Ag-based NPs are excellent candidates as building blocks for the bottom-up fabrication of two- and three-dimensional plasmonic nanoarchitectures, since oil-soluble metal NPs are susceptible to spontaneous assembly into hexagonal-close-packed monolayers upon solvent evaporation [129, 199, 204]. Such a new class of highly ordered assembled materials exhibits unique collective properties, including peculiar electromagnetic characteristics, and can find application in several research and technological fields, such as metamaterial fabrication, optical switches, plasmonic waveguides, plasmon diodes, solar cells, and transmitting optical nanoantennas [27, 52, 196]. In addition, it is worth to note that the fabrication of highly ordered assemblies of metal-based NPs requires high control of the particle size and an extremely narrow NP size distribution. Monodisperse nanostructures of different size and shape can be achieved by means of precisely controlled fabrication methods or post-synthetic size/shape sorting techniques [148].

2.5.1 *Hydrophobic Spherical Metal Nanoparticles*

Among the various methods developed to obtain monodisperse oil-dispersible Au nanospheres, the most common procedures involve two-phase synthesis and subsequent digestive ripening [165, 174]. Brust et al. have been the first to develop a two-step method useful for the synthesis of spherical Au NPs in organic media with a relatively narrow size distribution and a stabilizing self-assembled monolayer of thiols [16]. First, Au ion precursor ($HAuCl_4$) is transferred from water to toluene by using a phase-transfer catalyst (tetraalkylammonium bromide). Subsequently, addition of dodecanethiol to organic phase, followed by the reduction with sodium borohydride, yields dodecanethiolate-protected Au NPs. The ratio of Au to surfactant and the reaction temperature control particle size and dispersity. Later, the Brust protocol has been rapidly modified by using alkanethiol with different chain lengths [173], aromatic thiol [76], alkyl amine [15, 109], and dialkyl disulfide [82, 140], as capping agents able to coordinate Au and Ag NP surfaces. However, a significant drawback of this method is that the formed metal NPs contain surface impurities due to the presence of phase-transfer reagent, which prevents their further use [182]. Sastry et al. modified the Brust protocol for the synthesis of organic-based Au NPs by using a molecule (hexadecylaniline) able to perform both the phase transfer of Au ion precursor from water to chloroform and its spontaneous reduction to yield stable Au NPs [155].

Concomitantly, preparative methods have been developed to synthesize metal NPs directly in organic solvents. In the case of hydrophobic Au NPs, the challenging aspect lies in the limitation of suitable reagents, such as appropriate reducing agents and Au precursors. For example, the solubility of the commonly used Au salt ($HAuCl_4$) and reducing agent ($NaBH_4$) in the traditional organic synthetic medium, such as toluene, is very low. Nowadays, several strategies based on micelle solubilization [68], use of amine derivatives as mild reductant [57, 206], and on the synthesis of oil-soluble Au precursors [189, 212] have provided new possibilities. For example, Peng's group synthesized monodisperse Au NPs inducing the solubilization of the Au salt in organic solvents by means of appropriate surfactants such as fatty acids or aliphatic amines [68]. Moreover, Hiramatsu's work has reported the use of amines as reducing agents under heating for the synthesis of both Au and Ag NPs [57].

Similarly, the preparation of nearly monodisperse Ag NPs in a simple oleylamine-liquid paraffin system has been demonstrated [26]. However, oleic acid capped Ag NPs have been produced by using different synthetic approach, such as the thermolysis of suitable precursors in the presence of oleic acid [71, 95], chemical reduction [195], photochemical synthesis [90], green chemical synthesis [144], and ionic liquid [186] and solvent exchange methods [157]. Interestingly, it has been reported that the reduction of $AgNO_3$ with hot oleylamine in the presence of high-concentration chloride ions induces the formation of Ag nanocubes [167].

2.5.2 *Shape Control in Organic Media*

Although metal NPs with various shapes can be prepared in polar organic solvents, efforts are also required toward the shape-controlled synthesis of hydrophobic metal plasmonic NPs.

Several approaches using an organic medium to produce hydrophobic Au NRs have been recently proposed, although almost all the synthetic techniques take advantage of synthesis in aqueous environment. For example, Xia and coworkers reported on the formation of ultrathin Au NRs by mixture of an AuCl-oleylamine complex by using amorphous iron NPs in chloroform and aging for one week [93, 181]. In this example, iron NPs act as the reducing agent, while the oleylamine seems to direct the anisotropic growth of the rods. Concurrently, several other groups published the preparation of ultrathin Au nanowires in organic solvent [61, 107]. As an alternative, the direct transfer of preformed aqueous-based metal NPs to the organic phase has been proposed by using suitable post-synthetic surface engineering procedures as extensively reported in Section 2.7 [98, 163].

2.6 Nanoseparation and Purification Techniques: Size and/or Shape Sorting of Plasmonic Nanoparticles

Although much effort has been devoted to the controlled synthesis of plasmonic NPs, their structural polydispersity remains a critical issue. In several cases, the size distribution of the "as synthesized" metal NPs is characterized by a low degree of polydispersity; in many others, however, the particles need to be purified by post-synthesis procedures. In addition, for nonspherical particles, such as NRs, prisms, or cubes, solution-based methods yield target particles contaminated with differently shaped objects or inevitable byproducts [1]. Consequently, post-synthesis separation methods are required for the refining of NP populations according to their size, shape, and aggregation state [97]. Therefore, different separation techniques, such as electrophoresis, filtration and chromatographic methods, chemical or biochemical purification, have been adapted for nanoscience applications [84, 148]. Several examples of purification and separation procedures of water-dispersible metal NPs have been reported in literature. Although many literature reports describe the preparation of various nanostructures, only a few methods offer nearly quantitative yields of a defined shape. For example, it is possible to prepare Au NRs in high yield (>95%) and with relatively small-size deviations (~10% variation in aspect ratio) [122]. However, Ag NRs synthetized using the seed-mediated growth technique only represent a small sub-population of the synthesized particles, and there is currently no simple and reliable method to produce them as purely as Au rods [67]. For instance, Fig. 2.11a shows a typical TEM image of an Ag sample with 13% rods, 34% spheres (including hexagons), 44% triangles, and 9% other shapes.

2.6.1 *Gel Electrophoresis*

Gel electrophoresis (GE) has been proposed to separate metal NPs coated with thiolated polyethylene glycols terminating with carboxylic groups (SH-PEG-COOH) after their preparation. Four types of samples have been loaded onto one gel: an Ag sample, Au rods, Au spheres, and the mixture of the Au rods and Au

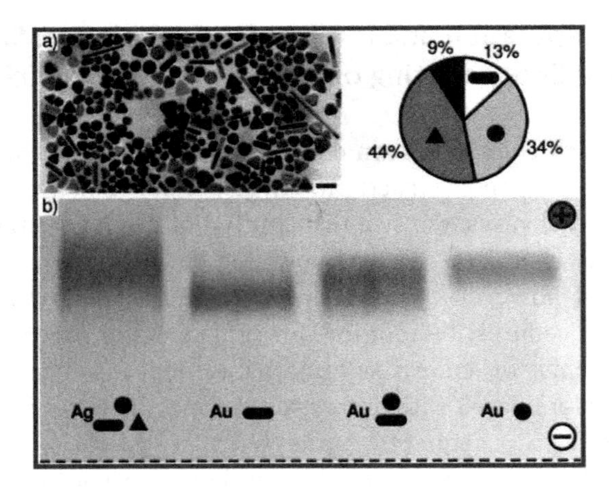

Figure 2.11 (a) Typical TEM picture of an Ag NPs sample (left, scale bar 100 nm) and the proportion of spheres, triangles, and rods (right) analyzed by assigning shapes to 600 NPs by eye. (b) True color photograph of a 0.2% agarose gel run for 30 min at 150 V (15 cm electrode spacing) in 0.5 × TBE buffer (pH = 9). The dashed line at the bottom indicates the position of the gel wells. The four lanes contain, from left to right, Ag NPs, Au NRs (40 × 20 nm), Au NRs and spheres mixed just before electrophoresis, and spherical Au NPs (D = 15 nm) as indicated in symbols. All NPs have been stabilized by a coating of 100% SH-PEG-COOH. The colors are due to the size- and shape-dependent optical properties of Au and Ag NPs and indicate separation according to NP morphology. Reprinted with permission from Ref. [55]. Copyright © 2007, American Chemical Society.

spheres. After GE, the gel shows different colors in the Ag lane and clear separation of Au spheres (red) mixed with Au rods (green) (Fig. 2.11b). The colors are due to the size- and shape-dependent optical properties of Au and Ag NPs and indicate their separation according to NP morphology [55].

2.6.2 *Centrifugation*

Centrifugation represents a common step of the procedure used to separate NPs. For example, spherical particles larger than ~60 nm can be removed by collecting the sedimentation at lower centrifugation speeds (~3000 rpm) [66]. Smaller size spheres (<~5 nm) can be separated as supernatant through repeated centrifugation at higher speed (6000–14500 rpm) [80]. Recently, sedimentation coefficient differences among NPs with different size and shape have been exploited for their sorting by centrifugation and sedimentation field-flow fractionation [29, 158].

An interesting approach involving depletion forces has been very recently demonstrated by Park et al. [128] and also by Hollamby et al. [58], who have reported shape and size separation of Au NPs exploiting micelle-induced depletion interactions between particles. The entropic, short-ranged depletion attractions between the NPs originated from the presence of micelles in solution. Remarkably, NPs of different sizes could be induced to aggregate by adjusting micelle concentration. This technique allows to separate spherical particles, as well as long and short rods from their mixture [84].

Interestingly, determination of NP size distribution together with density or molecular weight has been successfully achieved by two-dimensional analytical ultracentrifugation [18]. Furthermore, density gradient centrifugation has been proven as particularly effective for recovering refined NP populations, obtaining NPs with narrow diameter and shape distributions or a specific aggregation state [7, 37]. Such method has been successfully exploited also to separate NR with different aspect ratio [91]. Particles have also been centrifuged in a liquid column supporting a density gradient (such that the buoyant force varies within the tube). Density gradient can be created by carefully layering liquids at different concentration, one on top of the other; as a result, density increases from the top to the bottom of the tube. Then, particles in the different bands can be easily removed by a needle or a pipette syringe. The fractions can be purified by several centrifugation–redispersion cycles using an appropriate solvent [84]. Interestingly, Bai et al. recently adapted the density gradient ultracentrifugation technique to Au and Ag NPs in non-hydroxylic solvents using an organic density gradient [7].

2.6.3 *Size-Selective Precipitation*

Size-selective precipitation (SSP) is a simple technique that allows for separating NPs according to size-dependent physical and chemical properties, reactivity, and/or stability. Since these properties depend strongly on the surface chemistry of the NPs, SSP can be tailored to specific particle type/functionalization. Although most

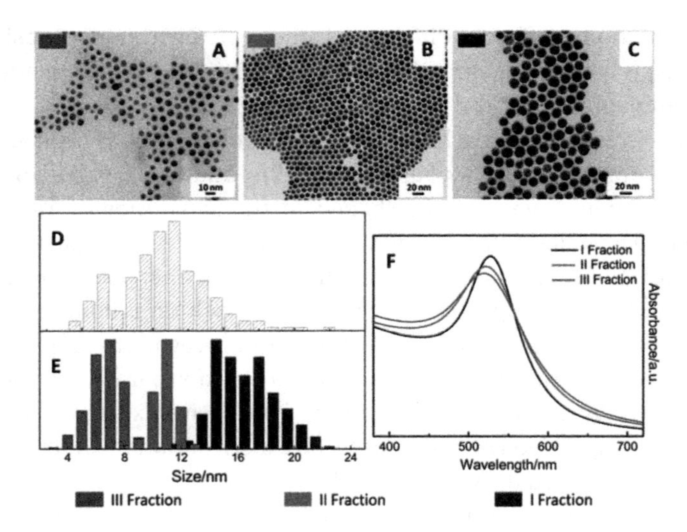

Figure 2.12 TEM micrographs (A–C) of Au size-selected fraction and size distribution histograms of as prepared Au NPs (D) and size sorted fractions (E). UV-Vis absorbance spectra (F). I Fraction (blue trace), II Fraction (red trace), III Fraction (green trace) obtained by size-selective precipitation starting from "as synthesized" Au NPs. Reprinted with permission from Ref. [37]. Copyright © 2013, Royal Society of Chemistry.

popular in aqueous conditions, selective precipitation has been also proven to work well in organic solvents [37, 170]. Very recently, oleylamine-capped Au NP size selection has been successfully carried out by antisolvent centrifugal fractional precipitation, based on the addition of a suitable volume of an antisolvent to a definite volume of as synthesized NP dispersion and by repeated cycle of antisolvent addition/centrifugation. Three monodisperse fractions of Au NPs have been collected, characterized by diameters ranging from 7 nm to 17 nm (Fig. 2.12) [37]. Alternatively, carbon dioxide (CO_2) has been used in place of a liquid antisolvent, enabling both NP purification and size-selective fractionation. For example, ligand-stabilized Ag NPs of a specific mean size were size selectively precipitated at desired locations, by using the pressure tunable solvent properties of CO_2-expanded liquids [110]. In a different approach, Williams et al. have used supercritical ethane in size-selective separation of alkanethiol-stabilized Au NPs. In contrast to selective precipitation techniques, this method relies on fractional, size-dependent redispersion of NPs in supercritical ethane. In particular, NPs with sizes ranging from 1 nm to 5 nm have been sorted at 318 K under adjustable pressures [193].

2.6.4 *Chromatographic Separations*

In chromatographic separation, a mobile phase containing a mixture to be separated passes through a stationary phase. The separation is then based on the differences in the partition coefficients between mobile and stationary phases for all components of the mixture. Size exclusion chromatography (SEC) is probably the most popular chromatographic technique used to fractionate NPs. SEC is based on the differences in the hydrodynamic volumes of the particles and not on specific interaction of these particles with the stationary phase. In fact, the addition of surfactant in the mobile phase has been shown to reduce the adsorption problem of packing materials in our recent study for the separation of different sizes of spherical Au NPs [187].

SEC is a rapidly developing technique, constantly adapting to new challenges and has been proposed for separating different shapes of nanometer-sized Au particles [188]. Recently, Al-Somali et al. introduced the so-called "recycling SEC," which significantly improved the resolution of the technique and made it more suitable for nanoscience applications. Such an approach allows for increasing the effective column length by returning the output back into column for several more runs. The resolution increases with each run and exhibits a square root dependence on the number of runs. With this technique, it has been possible to separate alkyl-thiol-stabilized Au NPs differing in size by only 6 Å [4].

2.6.5 *Filtration*

Filtration through a membrane is an alternative process useful for the purification and size-fractionation of NPs. In this class of methods, retention and elution of an analyte depend on the size of membrane pores. Akthakul et al. have fractionated Au NPs by using a thin polymeric membrane made of graft copolymer with hydrophobic poly(vinylidenefluoride) (PVDF) backbone and hydrophilic poly(oxyethylene methacrylate) (POEM) side chains. When a toluene solution containing polydispersed, octanethiol-modified Au NPs is filtered through the PEO/PVDF matrix, particles larger than 3.8 nm are retained on the membrane, while smaller ones pass through it freely. In an additional experiment, starting from octadecanethiol-capped Au NPs, the filtration cutoff decreased to 3.2 nm metal core diameter. Remarkably, the performance of the membrane could be changed by adjusting the polarity of the solvent swelling of PEO chains and changing the diameter of the nanochannels. Using binary solvent mixtures, it has been then possible to achieve a defined degree of swelling and NP diameter cutoff [3]. Membrane filtration is also an effective method for the size-fractionation of water-soluble NPs being effective in separating small Au NPs, binary mixtures, and polydispersed samples into several fractions according to different NP diameters [84, 169].

2.7 Surface Modification of Plasmonic Nanoparticles

Growing interest has been devoted to the integration of metal NPs, such as Au and Ag, in different matrices to exploit their unique size- and shape-dependent individual and collective properties into optic and sensing devices. In this perspective, engineering and functionalization of NPs are essential for their uses in several fields. One of the major drawback once trying to transfer metal NPs out of the solution is their tendency to coalesce if their surface is not suitably protected with ligands [178]. Indeed, the organic protecting layer at the metal NP surface is prone to photo-oxidation; therefore, devices based on these capping functionalities are likely to be susceptible to chemical degradation. In addition, the surface functionalization determines the interaction of the NPs with the environment since it is directly related to the colloidal stability of the NPs and controls assembly or targeting of the NPs [208].

2.7.1 *Post-Synthesis Functionalization*

Many groups have recently focused their research on chemical and biochemical post-synthesis functionalization of the metal NP surface for assembly [200], chemical and biological sensing [115], single molecule detection via surface-enhanced Raman spectroscopy (SERS) [127], in vitro and in vivo imaging and targeting [33], photothermal effect for therapeutic purposes [92], and gene delivery [150]. In addition, the definite geometry of particles enables to set up different approaches for both partial and complete surface functionalization [134]. As a general scheme, the native ligand should be replaced by a molecule with high affinity for metal surface and bearing a second functional group useful for further manipulation. For instance, cyano (–CN), mercapto (–SH), carboxylic acid (–COOH), and amino (–NH_2) are known to have a high affinity for Au and thus are useful as surface-protective functional groups [10, 30]. On the other hand, the nature of the surface functionality depends on the specific application planned for NPs. For instance, surface functionalization of Au NRs for biorelated applications could be realized by polyelectrolyte-coating, CTAB-capped Au NPs driven by electrostatic adsorption [116], chemical conjugation of amine-terminated biomolecule by carbodiimide coupling [46], Cu-catalyzed "click" addition of alkyne-terminated molecules onto azide-labeled surface [47], chemisorption of thiolated bioconjugates [94], chemisorption of amine-terminated bioconjugates by in situ dithiocarbamate (DTC) formation as alternative to simply thiols [211].

The CTAB bilayer at Au NP surface poses the major obstacle to surface functionalization due to its high packing. The peculiar geometry of NR yields a lower local density of CTAB at the tips of the rods than along their sides, thus allowing molecules possessing good affinity for Au to easily displace CTAB and bind preferentially to the tips of the Au rods [136]. In addition, the crystalline planes exposed along lateral faces of Au NRs and at the tip differ in their crystalline geometry. Such a peculiarity can be exploited for a selective functionalization with different groups. In view of this, it has been postulated that, because of the size of the CTAB head group, CTAB molecules would preferentially bind to the lateral faces, along the length of the rods rather than to the end (111) faces [20].

Several procedures have been optimized for functionalizing the entire surface of NRs. The original capping ligand used in the synthetic protocol can be replaced with molecules (usually a bifunctional ligand) bearing a specific functionality useful for further NC processing. For instance, mercapto-acids can interact with the NC surface by means of the thiol group, while the outermost exposed carboxylic moiety can be exploited in electrostatically driven assembling. Remarkably, functionalization with suitable bifunctional molecules has been reported to drive end-to-end assembly of the NRs [118, 185]. Such end-to-end assembly can be triggered by specific analytes and result in strong modification of NP color detectable even by the naked eye. Such an approach has been successfully exploited for biomedical [20, 46, 181] and environmental applications [6]. For instance, the limit of detection of Hg^{2+} ions by colorimetry has been pushed down to ppt level (3 ppt), exploiting chain-like assembly cysteine or pyrazole-functionalized Au NRs [136, 138]. Indeed, due to the close packing of CTAB molecules, pyrazole or cysteine molecules preferentially interact with Au NR tip (Fig. 2.13A–D). Such molecules bear functional group able to complex heavy metal ions and show high affinity for Hg^{2+}. Such interaction drives the formation of chain-like assemblies of Au NPs, which result in a shift of the longitudinal plasmon band up to 200 nm, thus resulting in strong color variation (Fig. 2.13E–G).

The assembly of plasmonic NPs driven by proper surface functionalization can have potential interesting applications also in optical metamaterials, photonics, electronics, sensors, and in the fabrication of microlenses. In this frame, interesting examples have been reported in the literature. Recently, optical metamaterial design using colloidal Au NP building blocks has been demonstrated [36]. In the solid state, post-assembly chemical functionalization of the NP surfaces with short ligands as ammonium thiocyanate (SCN) provides a tailorable dielectric-to-metal transition, allowing to engineer the properties of NC superstructures from nonplasmonic insulators to plasmonic, bulk-like conductors.

2.7.2 *Post-Synthesis Functionalization for Phase Transfer*

Since the synthesis of anisotropic NPs in organic solvents is still a challenge, the phase transfer of metal NRs prepared in water assisted by proper surface functionalization has been regarded as a convenient route to convey the peculiar properties of anisotropic

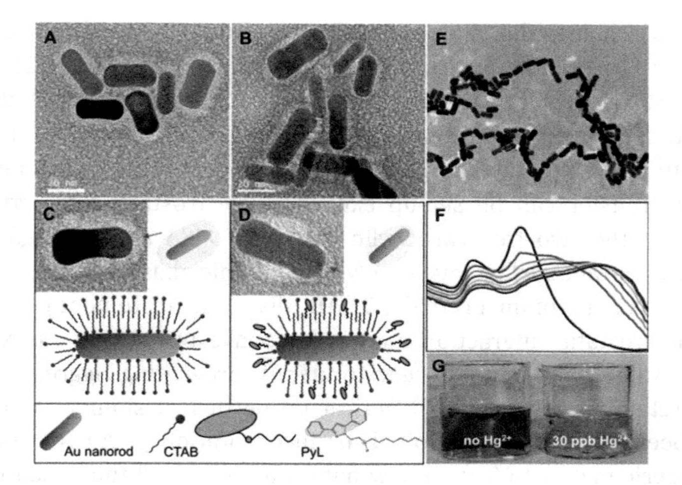

Figure 2.13 TEM micrographs recorded in negative staining mode of (A) "as prepared" and (B) Pyrazole ligand (PyL) functionalized Au NRs, respectively. In panels (C) and (D), high magnification TEM image recorded in negative staining mode of CTAB Au NRs and PyL Au NRs. The white shadow can be ascribed to the organic layer around the NRs. A strong modification of the organic layer can be observed at the tip of Au NRs upon functionalization with PyL. In the lower panels, a sketch of a possible PyL assembly at NR surface is reported. (E) TEM image of chain-like assembly occurring in presence of Hg^{2+} ions, (F) longitudinal plasmon band shift observed in absorbance spectra, and (G) color change in PyL Au NR solution after addition of Hg^{2+}. Adapted with permission from Ref. [136]. Copyright © 2013, American Chemical Society.

metal NPs in different apolar matrices. In this perspective, one of the most critical issues is retaining the original optical properties of the NPs. Indeed, a proper surface functionalization is necessary to protect the particle from aggregation and to prevent the position of the SPR to shift due to the different refractive index of the new chemical environment. Although alkyl thiols show strong affinity for Au surface and can stabilize NPs, they are generally advised against optical applications since they cause strong modification of the SPR due to alteration of electron density within metal NPs [70]. As an alternative, alkylamine can replace the native capping ligand to allow the resulting NCs to be exploited for incorporation in host polymer matrices [152]. Electrostatic interaction can be also used to promote the phase transfer of hydrophilic metal NPs into organic solvent retaining the original optical properties [201]. In the first step, a bifunctional ligand such as mercaptosuccinic acid (MSA) can be exploited to anchor Au NP surface through the thiol moiety in water phase. Under suitable pH conditions, a biphasic phase transfer can be set up exploiting electrostatic interaction between the two free carboxylic groups of MSA and an organic soluble cationic surfactants dissolved in organic phase (for instance, tetraoctylammonium bromide, TOAB, dissolved in toluene). Such an electrostatic interaction allows the phase transfer of Au NPs from water to organic solvent (toluene), preventing aggregation and retaining their original optical properties. A similar system has been successfully exploited for incorporation of Au NRs in a polymeric matrix to form a new nanocomposite and then used for the fabrication of components to be integrated in photonics and opto-electronic devices (Fig. 2.14A) [145].

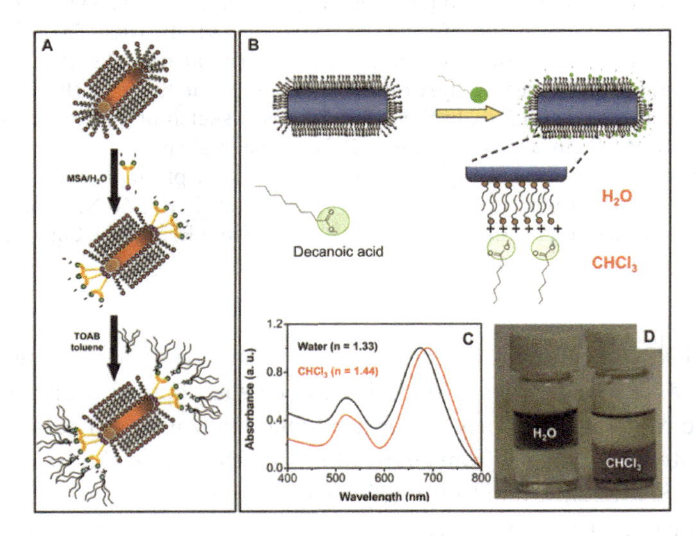

Figure 2.14 (A) Sketch of phase transfer mediated by MSA and TOAB. (B) Sketch of the decanoic acid ligand exchange procedure carried out on water-dispersible Au NRs. (C) Spectra of Au NRs before and after Au NRs. The slight shift of plasmon band could be ascribed to the different refractive index of the solvents. (D) Picture of vials containing Au NR dispersion in water and in chloroform, respectively, after phase transfer process. Adapted with permission from Ref. [31]. Copyright © 2013, Wiley-VCH.

Incorporation of metal NPs in polymers can provide simultaneous control of the organization (by guiding interparticle separation distance, orientation, and aggregation) and the integration of the NPs into functional materials. The formation of plasmonic nanocomposite materials is particularly attractive not only because the polymer matrix provides a convenient dielectric medium for encapsulation of the plasmonic NPs, but also because it enables the use of polymeric processing techniques (e.g., extrusion, molding, casting), of interest for large-scale manufacturing of these electromagnetic materials [40]. A convenient alternative to disperse Au NRs in organic solvents is functionalization with decanoic acid. This molecule is characterized by a carboxylic terminal group, able to electrostatically bind, once deprotonated, the positively charged CTAB bilayer. At the same time, the alkyl chain provides dispersibility in apolar solvents to obtain a dispersing medium with liquid crystals. Such an approach has been recently reported for dispersion of Au NRs in liquid crystals to set up a kind of nanothermometer to continuously monitor photoinduced temperature variations around Au NRs upon NIR light irradiation [31].

2.7.3 *Growth of a Silica Shell*

Several studies have focused on the silica-coating process of metal NPs with twofold purpose: (i) to investigate the fundamental properties of the nanostructures [141, 146] and (ii) to further explore self-assembling routes for the fabrication of artificial solids [178] or photonic crystals [74] as components of polymer solar cell [74] or to perform surface functionalization/conjugation for SERS, phototherapy, or colorimetric detection [24, 100].

Initially, growth of a silica shell has appeared, and has been further reported, as an ideal way to enhance the colloidal stability of NPs, as it provides tunable solubility in various solvents or even tailors their size- and shape-dependent optical properties. The importance of silica coating to metal NP includes prevention of coagulation, chemical inertness, and optical transparency, which allows spectroscopic monitoring of chemical reactions [178]. The silica shell can even provide a versatile platform for functionalization, which has opened up a wide range of new expectations in biomedical application for example as sensors marked and probes, to enhance both diagnostic and therapeutic technique. The easy control of silica growth, the possibility of achieving a controlled porosity, and high processability are other additional advantages make silica an ideal low cost material to be used to coat plasmonic NPs. In this regard, the silica shell plays the role of a spacer whose thickness can enable control of the energy transfer phenomena between the metal core and molecule anchored onto the silica surface. A thick silica coating onto plasmonic NPs, such as Ag and Au, can be used to separate chromophores decorating the core surface thus avoiding quenching processes that dominate when they are in direct contact with a metal NP. On the other hand, control of thickness may allow enhancement of various optical interactions, such as metal-enhanced fluorescence [25]. Moreover, the well-known tendency of silica to form ordered colloid crystals has been exploited to create two- and three-dimensional arrays [178]. The combination of nanochemical engineering and microfluidic self-assembly allows to obtain homogeneous optical metamaterials where plasmonic NPs are suitably separated by the silica shell [8].

Unlike metal oxide or semiconductor quantum dots, which can be easily coated with silica by following a conventional sol–gel Stöber method based on hydrolysis and condensation reaction of an alkoxides in basic aqueous solution containing alcohols [164], the growth of silica shell onto metal NPs has long been hindered due to the low chemical and electrostatic affinity between both components. The silica shell growth by addiction of tetraethoxysilane (TEOS) in a solution of seeds of core materials indeed requires core surface to be vitreophilic, which means to be receptive toward silica monomers or oligomers, involved in the coating process. Au and Ag have very little affinity for silica, which does not form a passivating oxide film in solution. Moreover, the NP stabilizers present in solution also contribute to render the Au surface vitreophobic. In addition, water-dispersible metal NPs (i.e., citrate- or CTAB-stabilized Au NPs) are not stable in the alcoholic solution (ethanol, methanol, 2-propanol) used for silica coating by the Stöber method. However, Han et al. have described a simple and fast preparation process based on the addition of extra sodium citrate to a boiling Au colloidal solution, thus allowing a direct and efficient coating of pre-treated citrate-stabilized Au NP with silica. This strategy has also the important advantage to be reproducible and produced a high quality of silica-coated Au NPs, with a controlled silica thickness [101]. In addition, Ag NPs in the size range of 40–100 nm, synthesized using low concentration of saccharide simultaneously as the reducing agent and electrostatic stabilizer, have been coated with silica from tetraalkoxysilanes by the sol-gel Stöber method, and the morphology of the obtained nanostructures has been found to depend on the alcoholic solvent and Ag particle concentration [120].

Among the plethora of reports that have reported strategies to coat metal NPs with silica shell, two large distinct classes will be defined here: protocols performed on hydrophilic plasmonic NPs and protocols performed on hydrophobic NPs. It is worth to note that, irrespective of the size, shape, and surface chemistry, the general approach guiding growth of a silica shell is based on a suitable NP surface processing, that is, by ligand-exchange reaction or polymer deposition, enabling the reaction with the silica shell precursors.

2.7.3.1 Growth of silica shell onto hydrophilic nanoparticles

Water-dispersible Au NPs can be prepared following different synthetic strategies, which mainly make use of citrate or CTAB, as stabilizing agents (see Section 2.4). Although direct coating of Au and Ag NPs has been described [101, 120], most of the reported coating routes start with priming the Au surface with coupling agents, usually a bifunctional molecule, surfactants, or polymers [103]. Such a priming molecule has the role to overcame the limited vitreophilic metal surface by increasing the affinity of the metallic surface toward silica and to provide colloidal NPs with sufficient stability to be transferred into ethanol or isopropanol, where subsequent growth of the silica shell can be successfully performed by conventional Stöber method.

Silica-coated Au NPs have been first reported by Liz-Marzán (Fig. 2.15) and a similar approach has been further similarly extended to Ag [178]. This silication method has been originally developed for citrate-stabilized Au NPs [102] and has been successfully applied to Au NRs [156, 184]. The method involved

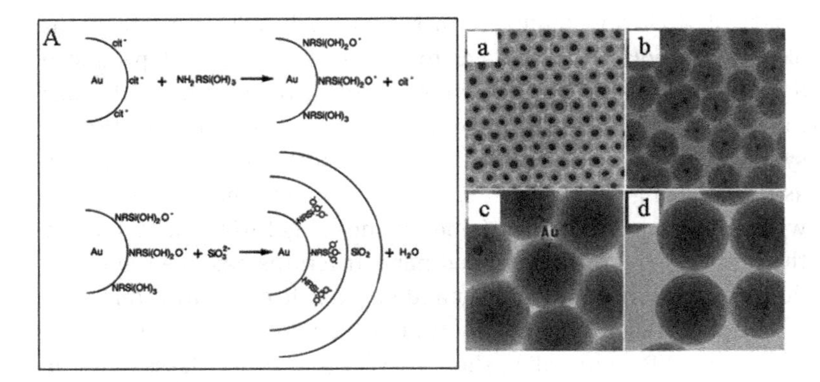

Figure 2.15 (A) Sketch of the surface reactions involved in the formation of a thin silica shell on citrate-stabilized Au NPs. Transmission electron micrographs of silica-coated Au NPs produced during the extensive growth of the silica shell around 15 nm Au NPs with TEOS in 4:1 ethanol/water mixtures. The shell thicknesses are (a, top left) 10 nm, (b, top right) 23 nm, (c, bottom left) 58 nm, and (d, bottom right) 83 nm. Reprinted with permission from Ref. [103]. Copyright ©1996, American Chemical Society.

the coating of Au NP with AMPS (3-aminopropyltrimethoxysilane), an amine-functionalized coupling agent, which coordinates to the Au surface via nitrogen. Other mercaptoalkoxisilanes such as 3-mercaptopropyl trimethoxysilane (MPTMS) have also been used as coupling agents to make the NPs surface vitreophilic. Subsequently, either the AMPS or the MPTMS alkyl siloxane groups undergo hydrolysis/condensation leading to the growth of silica shell.

Recently, a 11-mercaptoundecanoic acid bifunctional molecule has been used to cap the metal surface and provide a reproducible coating independent of the phase and morphology of the noble metal NPs [208]. The strong affinity of sulfur for metal surface is the driving force to effectively replace the capping agents (CTAB, citrate, oleylamine) from the surfaces of the metal NPs.

Beside ligand-exchange reaction, other strategies have provided alternative vitreophilic interfaces. These include using polyvinylpyrrolidone (PVP) to prime Au NPs avoiding the need for an initial sodium silicate shell growth or different silanes to prime Au NPs (Fig. 2.16) [48]. The stability of particles during

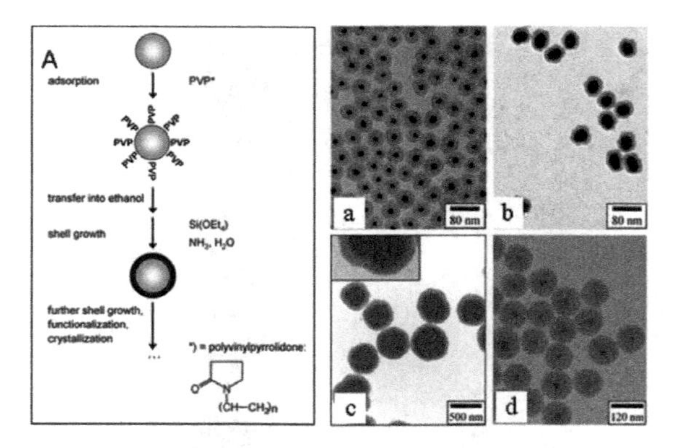

Figure 2.16 (**A**) Diagram of the general procedure for the coating of colloids with silica. In the first step, PVP is adsorbed onto the colloidal particles. Then these stabilized particles are transferred into a solution of ammonia in ethanol. A silica shell is grown by consecutive additions of TEOS. (a–b) TEM pictures of Au NPs coated with silica. (a) Au NPs (7 nm radius, polydispersity 15%) with an 18 nm shell (total polydispersity 9%) grown by two additions of TEOS. (b) Au NPs (20 nm radius, polydispersity 12%) with 12 nm shell (total polydispersity 9%). (c–d) TEM pictures of large Ag NPs (320 nm radius, polydispersity 15%) with 80 (12 nm silica shell (a; the inset shows an enlarged part of one of the silica coated particles) and of small Ag NPs (13 nm radius, polydispersity 23%) with a 47 nm silica shell (total polydispersity 5.9%) (b). Reprinted with permission from Ref. [48]. Copyright © 2003, American Chemical Society.

growth process and smoothness and homogeneity of the silica shells obtained has been found to be influenced by the length of the PVP used. Similar to Au NPs, PVP has been reported to easily wrap around Ag nanowires promoting the silica shell coating using the Stöber method (Fig. 2.17) [203]. When CTAB is the surfactant stabilizing the Au NP surface, its replacement with primers such as AMPS and MPTMS is difficult especially in the case of anisotropic rod-like NPs, where the displacement has been demonstrated to be more favorable at the tips, while it is much less straightforward, complicated at the flat side of the NRs. Pristine surfactant in high-aspect-ratio NRs has been reported by Murphy group to be quite easily replaced with MPTMS [114], but it becomes difficult for low-aspect-ratio NRs.

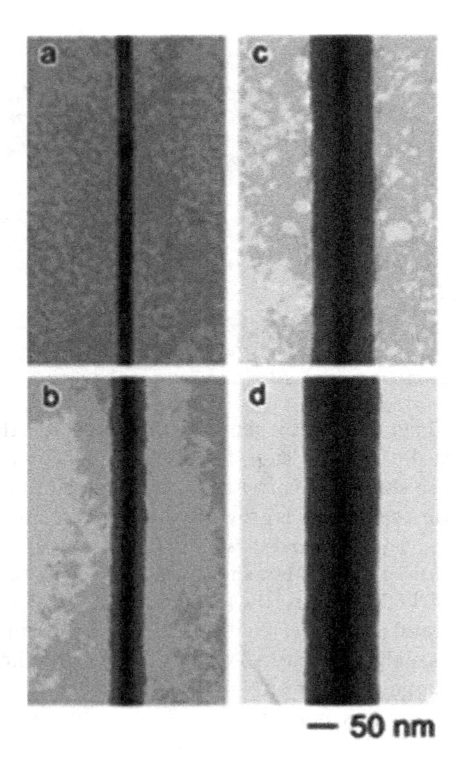

Figure 2.17 TEM images of Ag/silica coaxial nanocables obtained after the coating reaction had proceeded for (a) 5, (b) 10, (c) 30, and (d) 45 min. By controlling the reaction time, the sheath thickness can be easily tuned from 2 nm to 100 nm. Reprinted with permission from Ref. [203]. Copyright © 2002, American Chemical Society.

A novel strategy has been developed by Pastoriza-Santos et al. that exploits the positive charge at the double layer CTAB-capped Au NRs to assemble oppositely charged PSS (Fig. 2.18). A multilayer structure has been then grown by the alternating deposition of PAH and PSS, which ends up with a successful transfer of the NPs into isopropanol, where wrapping of PVP can occur avoiding rapid aggregation, and inducing subsequent growth of silica coating through the well-known Stöber method [130].

A two-phase silica-coating process has been developed to minimize the agglomeration of citrate-stabilized Au NPs by addiction of L-arginine in aqueous, which makes the NPs stable under the

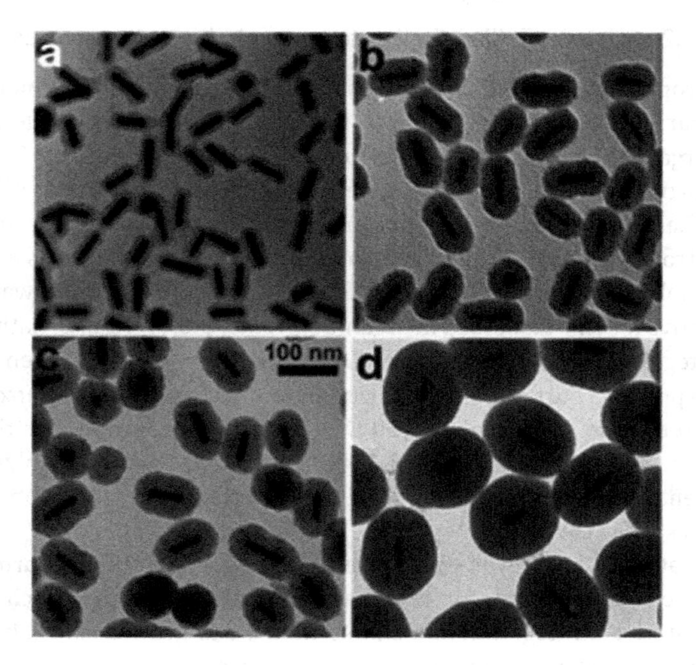

Figure 2.18 TEM images of silica-coated Au NRs, with silica shell thickness increasing from A to D. The scale is the same for all images. Reprinted with permission from Ref. [130]. Copyright © 2006, American Chemical Society.

slightly basic condition required in the next step. Cyclohexane is let layering as a second immiscible phase on top of the suspension and an MPTMS is added. The silane is slowly hydrolyzed under alkaline conditions and replaces the L-arginine, thereby producing a stable, vitreophilic Au colloid. Finally, TEOS is added to the organic phase. Depending on the TEOS concentration, silica shells of variable thickness can be obtained with a well-defined MPTMS interlayer [154].

Recently, silica-coated Au NPs have been synthesized without the additional coupling or passivating agents, and by fabricating the core–shell structures in ethanol–water or isopropanol-based solution. Such a strategy relies on the heating of Au NPs with a laser illumination near the plasmon resonance wavelength. The direct reaction of the NPs with TEOS can lead to Au-silicon bond, which seeds the growth of the silica shell [151].

2.7.3.2 Growth of silica shell onto hydrophobic nanoparticles

Hydrophobic metal NPs have been prepared by reduction of metal precursor by means of suitable reducing and stabilizing agents in organic solvent such as toluene and hexane (see Section 2.5). Unfortunately, the Stöber method cannot be applied for NPs synthesized in organic solvents, and a surface modification would be strongly required to improve their dispersibility in aqueous and alcohol solutions. AMPS and MPTMS have been found not always effective for NPs dispersed in organic solvents; therefore, alternative strategies and primers have been investigated. For example, Shen et al. reported a surface functionalization that makes use of methoxy-poly(ethylene glycol) silane (MPEG-sil) as a silica primer for the surface modification of metal NPs synthesized in nonhydrolytic solvents, which is effectively used as primer for the formation of silica shells [159].

Water in oil (reverse) microemulsion method for preparing silica-coated Au and Ag NPs has been also reported. Such a preparative strategy offers a simple method to grow a silica shell onto pre-synthesized organic soluble NPs even with a diameter of the core <20 nm [37, 54]. Silica shell has been indeed grown onto size-selected hydrophobic Au NPs with an average size ranging from 7 nm to 17 nm in diameter. According to the reported procedure, hydrophobic metal NPs can be easily dispersed in organic solvents such as cyclohexane. Addition of surfactant and aqueous ammonia solution results in a transparent red reverse microemulsion, where the silica shell growth takes place. The mechanism suggests that the hydrophobic metal NPs are phase transferred in the hydrophilic micellar pool of the reverse microemulsion, thanks to a direct coordination of surfactant molecules and hydrolyzed TEOS at the NPs surface (Fig. 2.19). The subsequent silica precursor condensation and formation of a siloxane network, occurring in the confined geometry of the micellar nanodroplet, results in a higher control over the silica nucleation and on the final bead size and shape [37].

Homonucleation of bare silica can be avoided by suitably setting the NP concentration to fit the micellar population. Therefore,

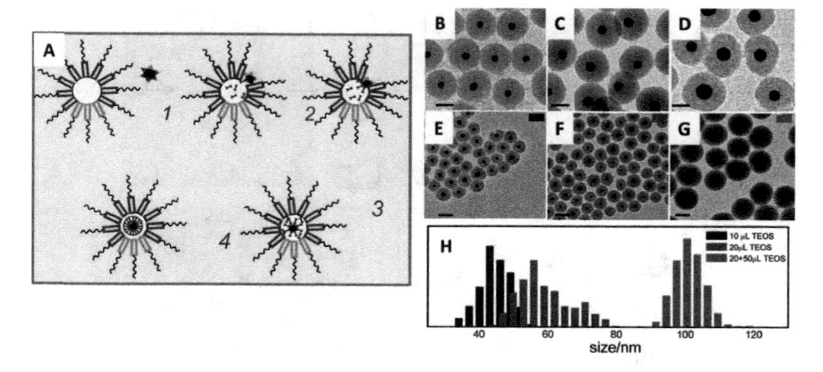

Figure 2.19 (A) Scheme of the mechanism of silica shell growth onto hydrophobic NPs in W/O microemulsion; TEM images of silica-coated size-selected Au NPs of 38 nm (B), 42 nm (C), and 48 nm (D) with Au NP core sizes of 7 nm (B), 11 nm (C), and 17 nm (D), respectively. Scale bar 20 nm; TEM micrographs (E–G) and statistical analysis (H) of silica-coated Au NPs prepared at increasing volume of TEOS 10 μL (E, H, black histogram), 20 μL (F–H, blue histogram), and 70 μL (G, H red histogram). Scale bar 50 nm. Adapted with permission from Ref. [37]. Copyright ©2013, Royal Society of Chemistry.

concentration of NPs is meant to be crucial to define morphology (i.e., multiplicity of NPs per silica) and yield of the final nanostructures. Moreover, the core–shell structure size has been found to depend on the Au NPs concentration. It has been also demonstrated that the density of the ligand agent such as oleylamine coordinating NPs surface is a crucial parameter to control the morphology of the final core–shell nanostructures, and mainly the multiplicity of the Au core inside each silica shell (Fig. 2.20) [37].

Recently, efforts have been devoted to the fabrication of homogeneous metal nanostructures based on dimers, trimmers, and tetramers, to exploit the original plasmonic properties arising from the particle–particle coupling, which depends on the interparticle distance. These nanostructures can find application in several research and technological fields such as metamaterial fabrication, optical switches, plasmonic waveguides, plasmon diodes, and transmitting optical nanoantennas [202].

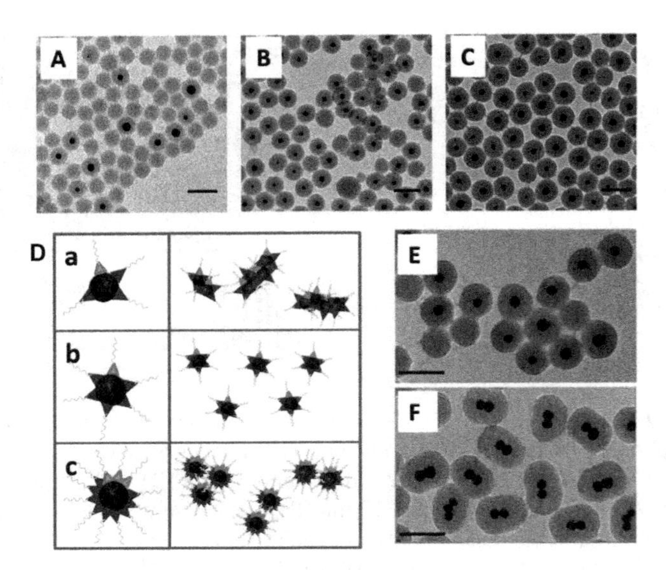

Figure 2.20 TEM images (A–C) and statistical analysis of the core–shell silica-coated Au NPs at increasing Au NP concentration: $4.5 \cdot 10^{13}$ NPs per mL (A), $9 \cdot 10^{13}$ NPs per mL (B), $1.3 \cdot 10^{14}$ NPs per mL (C). Scale bar 50 nm. (D) Sketches of oleylamine-capped Au NP assembled structures achieved upon NP surface treatments: Au NP extensively washed with methanol (a), Au NP incubation with 0.1 M (b) and 1 M (c) oleylamine. TEM images of monomer (E) and dimer (F) nanostructures separated by density gradient centrifugation. Adapted with permission from Ref. [37]. Copyright © 2013, Royal Society of Chemistry.

2.8 Conclusion

The overall description has clearly pointed out that modern colloidal chemistry routes, post-synthetic size sorting, and surface functionalization procedures have radically contributed to design and develop a plentiful kit of tools, particularly versatile and flexible, to tailor the unique optical properties of plasmonic NPs and convey them virtually in any chemical structure suitable for integration for application. The possibility of playing with size, shape of metal NPs and of controlling their mutual interaction and chemical environment allows to tailor surface plasmon resonance characteristics. This opportunity opens new perspective toward understanding the fundamental physics behind such a peculiar class of materials and, at the same time, for their technological applications in a wide range of fields, including (bio)sensing, nanomedicine, energy conversion, and optics.

Acknowledgments

This work has been partially funded by PRIN Programme 2010–2011 (n. 2010C4R8M8), by Research Project PON R&C 2007–2013 MAAT-Molecular Nanotechnology for Health and Environment (n. PON02_00563_3316357), by FIRB-Futuro in Ricerca (RBFR 122HFZ), project, by the EC-funded project METACHEM (Grant CP-FP 228762-2) and by the Apulia Regional Laboratories RELA-VALBIOR and Sens & Micro LAB (POFESR 2007–2013).

References

1. Adair, J.H., Suvaci, E. (2000). Morphological Control of Particles. *Curr. Op. Coll. Interf. Sci.* **5**, pp. 160–167.
2. Ahmadi, T.S., Wang, Z.L., Green, T.C., Henglein, A., El-Sayed, M.A. (1996). Shape-Controlled Synthesis of Colloidal Platinum Nanoparticles. *Science* **272**, pp. 1924–1925.
3. Akthakul, A., Hochbaum, A.I., Stellacci, F., Mayes, A.M. (2005). Size Fractionation of Metal Nanoparticles by Membrane Filtration. *Adv. Mater.* **17**, pp. 532–535.
4. Al-Somali, A.M., Krueger, K.M., Falkner, J.C., Colvin, V.L. (2004). Recycling Size Exclusion Chromatography for the Analysis and Separation of Nanocrystalline Gold. *Anal. Chem.* **76**, pp. 5903–5910.
5. Altamura, D., Lassandro, R., Vittoria, F.A., De Caro, L., Siliqi, D., Ladisa, M., Giannini, C. (2012). X-Ray Microimaging Laboratory (XMI-LAB). *J. Appl. Crystallogr.* **45**, pp. 869–873.
6. Aragay, G., Pons, J., Merkoçi, A. (2011). Recent Trends in Macro-, Micro-, and Nanomaterial-Based Tools and Strategies for Heavy-Metal Detection. *Chem. Rev.* **111**, pp. 3433–3458.
7. Bai, L., Ma, X., Liu, J., Sun, X., Zhao, D., Evans, D.G. (2010). Rapid Separation and Purification of Nanoparticles in Organic Density Gradients. *J. Am. Chem. Soc.* **132**, pp. 2333–2337.

8. Baron, A., Iazzolino, A., Ehrhardt, K., Salmon, J.-B., Aradian, A., Kravets, V., Grigorenko, A.N., Leng, J., Le Beulze, A., Tréguer-Delapierre, M., Correa-Duarte, M.A., Barois, P. (2013). Bulk Optical Metamaterials Assembled by Microfluidic Evaporation. *Opt. Mater. Express* **3**, pp. 1792–1797.
9. Bastùs, N.G., Comenge, J., Puntes, V. (2011). Kinetically-Controlled Seeded Growth Synthesis of Citrate-Stabilized Gold Nanoparticles up to 200 nm: Size Focusing vs. Ostwald Ripening. *Langmuir* **27**, pp. 11098–11105.
10. Bhattacharjee, R.R.M.T. (2007). Polymer-Mediated Chain-Like Self-Assembly of Functionalized Gold Nanoparticles. *J. Colloid Interface Sci.* **307**, pp. 288–295.
11. Binnig, G., Quate, C.F., Gerber, C. (1986). Atomic Force Microscope. *Phys. Rev. Lett.* **56**, pp. 930–933.
12. Binnig, G., Rohrer, H. (2000). Scanning Tunneling Microscopy. *IBM J. Res. Dev.* **44**, pp. 279–293.
13. Brown, K.R., Fox, A.P., Natan, M.J. (1996). Morphology-Dependent Electrochemistry of Cytochrome c at Au Colloid-Modified SnO_2 Electrodes. *J. Am. Chem. Soc.* **118**, pp. 1154–1157.
14. Brown, K.R., Natan, M.J. (1998). Hydroxylamine Seeding of Colloidal Au Nanoparticles in Solution and on Surfaces. *Langmuir* **14**, pp. 726–728.
15. Brown, L.O., Hutchison, J.E. (2001). Formation and Electron Diffraction Studies of Ordered 2-D and 3-D Superlattices of Amine-Stabilized Gold Nanocrystals. *J. Phys. Chem. B* **105**, pp. 8911–8916.
16. Brust, M., Walker, M., Bethell, D., Schiffrin, D.J., Whyman, R. (1994). Synthesis of Thiol-Derivatised Gold Nanoparticles in a Two-Phase Liquid-Liquid System. *J. Chem. Soc., Chem. Commun.* **7**, pp. 801–802.
17. C. Hulteen, J., Martin, C.R. (1997). A General Template-Based Method for the Preparation of Nanomaterials. *J. Mater. Chem.* **7**, pp. 1075–1087.
18. Carney, R.P., Kim, J.Y., Qian, H., Jin, R., Mehenni, H., Stellacci, F., Bakr, O.M. (2011). Determination of Nanoparticle Size Distribution Together with Density or Molecular Weight by 2D Analytical Ultracentrifugation. *Nat. Commun.* **2**, pp. 335.
19. Caswell, K.K., Bender, C.M., Murphy, C.J. (2003). Seed-less, Surfactant-less Wet Chemical Synthesis of Silver Nanowires. *Nano Lett.* **3**, pp. 667–669.
20. Caswell, K.K., Wilson, J.N., Bunz, U.H.F., Murphy, C.J. (2003). Preferential End-to-End Assembly of Gold Nanorods by Biotin-Streptavidin Connectors. *J. Am. Chem. Soc.* **125**, pp. 13914–13915.

21. Cervellino, A., Giannini, C., Guagliardi, A., Ladisa, M. (2005). Nanoparticle Size Distribution Estimation by a Full-Pattern Powder Diffraction Analysis. *Phys. Rev. B* **72**, pp. 035412.
22. Chang, S.S., Shih, C.W., Chen, C.D., Lai, W.C., Wang, C.R.C. (1999). The Shape Transition of Gold Nanorods. *Langmuir* **15**, pp. 701–709.
23. Chen, D.-H., Huang, Y.-W. (2002). Spontaneous Formation of Ag Nanoparticles in Dimethylacetamide Solution of Poly(ethylene glycol). *J. Colloid Interface Sci.* **255**, pp. 299–302.
24. Chen, G., Wang, Y., Yang, M., Xu, J., Goh, S.J., Pan, M., Chen, H. (2010). Measuring Ensemble-Averaged Surface-Enhanced Raman Scattering in the Hotspots of Colloidal Nanoparticle Dimers and Trimers. *J. Am. Chem. Soc.* **132**, pp. 3644–3645.
25. Chen, J., Jin, Y., Fahruddin, N., Zhao, J.X. (2013). Development of Gold Nanoparticle-Enhanced Fluorescent Nanocomposites. *Langmuir* **29**, pp. 1584–1591.
26. Chen, M., Feng, Y.-G., Wang, X., Li, T.-C., Zhang, J.-Y., Qian, D.-J. (2007). Silver Nanoparticles Capped by Oleylamine: Formation, Growth, and Self-Organization. *Langmuir* **23**, pp. 5296–5304.
27. Cheng, Y., Wang, M., Borghs, G., Chen, H. (2011). Gold Nanoparticle Dimers for Plasmon Sensing. *Langmuir* **27**, pp. 7884–7891.
28. Comparelli, R., Fanizza, E., Striccoli, M., Curri, M.L. (2013). Characterization of inorganic nanostructured materials by electron microscopy, in *Inorganic Micro- and Nanomaterials: Synthesis and Characterization* (Di Benedetto, A., Aresta, M., eds), De Gruyter, Berlin, pp. 157–198.
29. Contado, C., Argazzi, R. (2009). Size Sorting of Citrate Reduced Gold Nanoparticles by Sedimentation Field-Flow Fractionation. *J. Chromatogr. A* **1216**, pp. 9088–9098.
30. Corbierre, M.K., Cameron, N.S., Sutton, M., Laaziri, K., Lennox, R.B. (2005). Gold Nanoparticle/Polymer Nanocomposites: Dispersion of Nanoparticles as a Function of Capping Agent Molecular Weight and Grafting Density. *Langmuir* **21**, pp. 6063–6072.
31. De Sio, L., Placido, T., Serak, S., Comparelli, R., Tamborra, M., Tabiryan, N., Curri, M.L., Bartolino, R., Umeton, C., Bunning, T. (2013). Nano-Localized Heating Source for Photonics and Plasmonics. *Adv. Optical Mater.* **1**, pp. 899–904.
32. Dreaden, E.C., Alkilany, A.M., Huang, X., Murphy, C.J., El-Sayed, M.A. (2012). The Golden Age: Gold Nanoparticles for Biomedicine. *Chem. Soc. Rev.* **41**, pp. 2740–2779.
33. Eghtedari, M., Liopo, A.V., Copland, J.A., Oraevsky, A.A., Motamedi, M. (2008). Engineering of Hetero-Functional Gold Nanorods for the in

vivo Molecular Targeting of Breast Cancer Cells. *Nano Lett.* **9**, pp. 287–291.

34. Eustis, S., El-Sayed, M.A. (2006). Why Gold Nanoparticles Are More Precious than Pretty Gold: Noble Metal Surface Plasmon Resonance and Its Enhancement of the Radiative and Nonradiative Properties of Nanocrystals of Different Shapes. *Chem. Soc. Rev.* **35**, pp. 209–217.
35. Evanoff, D.D., Chumanov, G. (2004). Size-Controlled Synthesis of Nanoparticles. 1. "Silver-Only" Aqueous Suspensions via Hydrogen Reduction. *J. Phys. Chem. B* **108**, pp. 13948–13956.
36. Fafarman, A.T., Hong, S.-H., Caglayan, H., Ye, X., Diroll, B.T., Paik, T., Engheta, N., Murray, C.B., Kagan, C.R. (2012). Chemically Tailored Dielectric-to-Metal Transition for the Design of Metamaterials from Nanoimprinted Colloidal Nanocrystals. *Nano Lett.* **13**, pp. 350–357.
37. Fanizza, E., Depalo, N., Clary, L., Agostiano, A., Striccoli, M., Curri, M.L. (2013). A Combined Size Sorting Strategy for Monodisperse Plasmonic Nanostructures. *Nanoscale* **5**, pp. 3272–3282.
38. Faraday, M. (1857). The Bakerian Lecture: Experimental Relations of Gold (and Other Metals) to Light. *Philos. Trans.* **147**, pp. 145–181.
39. Frens, G. (1973). Controlled Nulceation for Regulation of Particle-Size in Monodisperse Gold Suspension. *Nature: Phys. Sci.* **241**, pp. 20–22.
40. Gao, B., Rozin, M.J., Tao, A.R. (2013). Plasmonic Nanocomposites: Polymer-Guided Strategies for Assembling Metal Nanoparticles. *Nanoscale* **5**, pp. 5677–5691.
41. Gao, J., Bender, C.M., Murphy, C.J. (2003). Dependence of the Gold Nanorod Aspect Ratio on the Nature of the Directing Surfactant in Aqueous Solution. *Langmuir* **19**, pp. 9065–9070.
42. Ghosh, S.K., Pal, T. (2007). Interparticle Coupling Effect on the Surface Plasmon Resonance of Gold Nanoparticles: From Theory to Applications. *Chem. Rev.* **107**, pp. 4797–4862.
43. Giannici, F., Placido, T., Curri, M.L., Striccoli, M., Agostiano, A., Comparelli, R. (2009). The Fate of Silver Ions in the Photochemical Synthesis of Gold Nanorods: An Extended X-ray Absorption Fine Structure Analysis. *Dalton Trans.* pp. 10367–10374.
44. Giannini, C., Altamura, D., Aresta, B.M., Sibillano, T., Siliqi, D., De Caro, L. (2013). Lens-less scanning X-ray microscopy with SAXS and WAXS contrast, in *Inorganic Micro- and Nanomaterials: Synthesis and Characterization* (Di Benedetto, A., Aresta, M., eds.), De Gruyter, Berlin, pp. 137–156.

45. Gole, A., Murphy, C.J. (2004). Seed-Mediated Synthesis of Gold Nanorods: Role of the Size and Nature of the Seed. *Chem. Mater.* **16**, pp. 3633–3640.

46. Gole, A., Murphy, C.J. (2005). Biotine-Streptavidin-Induced Aggregation of Gold Nanorods: Tuning Rod-Rod Orientation. *Langmuir* **21**, pp. 10756–10762.

47. Gole, A., Murphy, C.J. (2008). Azide-Derivatized Gold Nanorods: Functional Materials for "Click-Chemistry." *Langmuir* **24**, pp. 266–272.

48. Graf, C., Vossen, D.L.J., Imhof, A., van Blaaderen, A. (2003). A General Method To Coat Colloidal Particles with Silica. *Langmuir* **19**, pp. 6693–6700.

49. Grzelczak, M., Liz-Marzán, L.M. (2013). Colloidal Nanoplasmonics: From Building Blocks to Sensing Devices. *Langmuir* **29**, pp. 4652–4663.

50. Grzelczak, M., Perez-Juste, J., Mulvaney, P., Liz-Marzan, L.M. (2008). Shape control in gold nanoparticle synthesis. *Chem. Soc. Rev.* **37**, pp. 1783–1791.

51. Gu, X., Nie, C., Lai, Y., Lin, C. (2006). Synthesis of Silver Nanorods and Nanowires by Tartrate-Reduced Route in Aqueous Solutions. *Mater. Chem. Phys.* **96**, pp. 217–222.

52. Guerrini, L., Graham, D. (2012). Molecularly-Mediated Assemblies of Plasmonic Nanoparticles for Surface-Enhanced Raman Spectroscopy Applications. *Chem. Soc. Rev.* **41**, pp. 7085–7107.

53. Hada, H., Yonezawa, Y., Yoshida, A., Kurakake, A. (1976). Photoreduction of Silver Ion in Aqueous and Alcoholic Solutions. *J. Phys. Chem.* **80**, pp. 2728–2731.

54. Han, Y., Jiang, J., Lee, S.S., Ying, J.Y. (2008). Reverse Microemulsion-Mediated Synthesis of Silica-Coated Gold and Silver Nanoparticles. *Langmuir* **24**, pp. 5842–5848.

55. Hanauer, M., Pierrat, S., Zins, I., Lotz, A., Sönnichsen, C. (2007). Separation of Nanoparticles by Gel Electrophoresis According to Size and Shape. *Nano Lett.* **7**, pp. 2881–2885.

56. Henglein, A. (1999). Radiolytic Preparation of Ultrafine Colloidal Gold Particles in Aqueous Solution: Optical Spectrum, Controlled Growth, and Some Chemical Reactions. *Langmuir* **15**, pp. 6738–6744.

57. Hiramatsu, H., Osterloh, F.E. (2004). A Simple Large-Scale Synthesis of Nearly Monodisperse Gold and Silver Nanoparticles with Adjustable Sizes and with Exchangeable Surfactants. *Chem. Mater.* **16**, pp. 2509–2511.

58. Hollamby, M.J., Eastoe, J., Chemelli, A., Glatter, O., Rogers, S., Heenan, R.K., Grillo, I. (2009). Separation and Purification of Nanoparticles in a Single Step. *Langmuir* **26**, pp. 6989–6994.
59. Hu, J.Q., Chen, Q., Xie, Z.X., Han, G.B., Wang, R.H., Ren, B., Zhang, Y., Yang, Z.L., Tian, Z.Q. (2004). A Simple and Effective Route for the Synthesis of Crystalline Silver Nanorods and Nanowires. *Adv. Funct. Mater.* **14**, pp. 183–189.
60. Huang, X., Neretina, S., El-Sayed, M.A. (2009). Gold Nanorods: From Synthesis and Properties to Biological and Biomedical Applications. *Adv. Mater.* **21**, pp. 4880–4910.
61. Huo, Z., Tsung, C.-K., Huang, W., Zhang, X., Yang, P. (2008). Sub-Two Nanometer Single Crystal Au Nanowires. *Nano Lett.* **8**, pp. 2041–2044.
62. Jain, P.K., Huang, X., El-Sayed, I.H., El-Sayed, M.A. (2008). Noble Metals on the Nanoscale: Optical and Photothermal Properties and Some Applications in Imaging, Sensing, Biology, and Medicine. *Acc. Chem. Res.* **41**, pp. 1578–1586.
63. Jana, N.R. (2005). Gram-Scale Synthesis of Soluble, Near-Monodisperse Gold Nanorods and Other Anisotropic Nanoparticles. *Small* **1**, pp. 875–882.
64. Jana, N.R., Gearheart, L., Murphy, C.J. (2001). Evidence for Seed-Mediated Nucleation in the Chemical Reduction of Gold Salts to Gold Nanoparticles. *Chem. Mater.* **13**, pp. 2313–2322.
65. Jana, N.R., Gearheart, L., Murphy, C.J. (2001). Seed-Mediated Growth Approach for Shape-Controlled Synthesis of Spheroidal and Rod-like Gold Nanoparticles Using a Surfactant Template. *Adv. Mater.* **13**, pp. 1389–1393.
66. Jana, N.R., Gearheart, L., Murphy, C.J. (2001). Wet Chemical Synthesis of High Aspect Ratio Cylindrical Gold Nanorods. *J. Phys. Chem. B* **105**, pp. 4065–4067.
67. Jana, N.R., Gearheart, L., Murphy, C.J. (2001). Wet Chemical Synthesis of Silver Nanorods and Nanowires of Controllable Aspect Ratio. *Chem. Commun.* pp. 617–618.
68. Jana, N.R., Peng, X. (2003). Single-Phase and Gram-Scale Routes toward Nearly Monodisperse Au and Other Noble Metal Nanocrystals. *J. Am. Chem. Soc.* **125**, pp. 14280–14281.
69. Jans, H., Liu, X., Austin, L., Maes, G., Huo, Q. (2009). Dynamic Light Scattering as a Powerful Tool for Gold Nanoparticle Bioconjugation and Biomolecular Binding Studies. *Anal. Chem.* **81**, pp. 9425–9432.

70. Jebb, M., Sudeep, P.K., Pramod, P., Thomas, K.G., Kamat, P.V. (2007). Ruthenium(II) Trisbipyridine Functionalized Gold Nanorods. Morphological Changes and Excited-State Interactions. *J. Phys. Chem. B* **111**, pp. 6839–6844.

71. Jeevanandam, P., Srikanth, C.K., Dixit, S. (2010). Synthesis of Monodisperse Silver Nanoparticles and Their Self-assembly through Simple Thermal Decomposition Approach. *Mater. Chem. Phys.* **122**, pp. 402–407.

72. Jeong, G.H., Kim, M., Lee, Y.W., Choi, W., Oh, W.T., Park, Q.H., Han, S.W. (2009). Polyhedral Au Nanocrystals Exclusively Bound by {110} Facets: The Rhombic Dodecahedron. *J. Am. Chem. Soc.* **131**, pp. 1672–1673.

73. Ji, X., Song, X., Li, J., Bai, Y., Yang, W., Peng, X. (2007). Size Control of Gold Nanocrystals in Citrate Reduction: The Third Role of Citrate. *J. Am. Chem. Soc.* **129**, pp. 13939–13948.

74. Jiménez-Solano, A., López-López, C., Sánchez-Sobrado, O., Luque, J.M., Calvo, M.E., Fernández-López, C., Sánchez-Iglesias, A., Liz-Marzán, L.M., Míguez, H. (2012). Integration of Gold Nanoparticles in Optical Resonators. *Langmuir* **28**, pp. 9161–9167.

75. Jin, R., Cao, Y., Mirkin, C.A., Kelly, K.L., Schatz, G.C., Zheng, J.G. (2001). Photoinduced Conversion of Silver Nanospheres to Nanoprisms. *Science* **294**, pp. 1901–1903.

76. Johnson, S.R., Evans, S.D., Mahon, S.W., Ulman, A. (1997). Alkanethiol Molecules Containing an Aromatic Moiety Self-Assembled onto Gold Clusters. *Langmuir* **13**, pp. 51–57.

77. Kamat, P.V., Flumiani, M., Hartland, G.V. (1998). Picosecond Dynamics of Silver Nanoclusters. Photoejection of Electrons and Fragmentation. *J. Phys. Chem. B* **102**, pp. 3123–3128.

78. Kelly, K.L., Coronado, E., Zhao, L.L., Schatz, G.C. (2003). The Optical Properties of Metal Nanoparticles: The Influence of Size, Shape, and Dielectric Environment. *J. Phys. Chem. B* **107**, pp. 668–677.

79. Kim, D.Y., Yu, T., Cho, E.C., Ma, Y., Park, O.O., Xia, Y. (2011). Synthesis of Gold Nano-hexapods with Controllable Arm Lengths and Their Tunable Optical Properties. *Angew. Chem. Int. Ed.* **50**, pp. 6328–6331.

80. Kim, F., Song, J.H., Yang, P. (2002). Photochemical Synthesis of Gold Nanorods. *J. Am. Chem. Soc.* **124**, pp. 14316–14317.

81. Kim, J.-H., Brian, W.L., Roarke, D.B., Brett, W.B. (2011). Controlled Synthesis of Gold Nanoparticles by Fluorescent Light Irradiation. *Nanotechnology* **22**, pp. 285602.

82. Kim, S.J., Kim, T.G., Ah, C.S., Kim, K., Jang, D.-J. (2003). Photolysis Dynamics of Benzyl Phenyl Sulfide Adsorbed on Silver Nanoparticles. *J. Phys. Chem. B* **108**, pp. 880–882.

83. Kimling, J., Maier, M., Okenve, B., Kotaidis, V., Ballot, H., Plech, A. (2006). Turkevich Method for Gold Nanoparticle Synthesis Revisited. *J. Phys. Chem. B* **110**, pp. 15700–15707.

84. Kowalczyk, B., Lagzi, I., Grzybowski, B.A. (2011). Nanoseparations: Strategies for Size and/or Shape-Selective Purification of Nanoparticles. *Curr. Op. Coll. Interf. Sci.* **16**, pp. 135–148.

85. Kreibig, U., Genzel, L. (1985). Optical Absorption of Small Metallic Particles. *Surf. Sci.* **156**, pp. 678–700.

86. Kumar, S., Gandhi, K.S., Kumar, R. (2006). Modeling of Formation of Gold Nanoparticles by Citrate Method. *Ind. Eng. Chem. Res.* **46**, pp. 3128–3136.

87. Kwon, K., Lee, K.Y., Lee, Y.W., Kim, M., Heo, J., Ahn, S.J., Han, S.W. (2006). Controlled Synthesis of Icosahedral Gold Nanoparticles and Their Surface-Enhanced Raman Scattering Property. *J. Phys. Chem. C* **111**, pp. 1161–1165.

88. Langille, M.R., Personick, M.L., Zhang, J., Mirkin, C.A. (2011). Bottom-Up Synthesis of Gold Octahedra with Tailorable Hollow Features. *J. Am. Chem. Soc.* **133**, pp. 10414–10417.

89. Langille, M.R., Personick, M.L., Zhang, J., Mirkin, C.A. (2012). Defining Rules for the Shape Evolution of Gold Nanoparticles. *J. Am. Chem. Soc.* **134**, pp. 14542–14554.

90. Le, A.-T., Tam, L.T., Tam, P.D., Huy, P.T., Huy, T.Q., Van Hieu, N., Kudrinskiy, A.A., Krutyakov, Y.A. (2010). Synthesis of Oleic Acid-Stabilized Silver Nanoparticles and Analysis of Their Antibacterial Activity. *Mater. Sci. Eng. C-Biomimetic Supramol. Syst.* **30**, pp. 910–916.

91. Li, S., Chang, Z., Liu, J., Bai, L., Luo, L., Sun, X. (2011). Separation of Gold Nanorods using Density Gradient Ultracentrifugation. *Nano Res.* **4**, pp. 723–728.

92. Li, Z., Huang, P., Zhang, X., Lin, J., Yang, S., Liu, B., Gao, F., Xi, P., Ren, Q., Cui, D. (2009). RGD-Conjugated Dendrimer-Modified Gold Nanorods for in Vivo Tumor Targeting and Photothermal Therapy. *Mol. Pharmaceutics* **7**, pp. 94–104.

93. Li, Z., Tao, J., Lu, X., Zhu, Y., Xia, Y. (2008). Facile Synthesis of Ultrathin Au Nanorods by Aging the AuCl(oleylamine) Complex with Amorphous Fe Nanoparticles in Chloroform. *Nano Lett.* **8**, pp. 3052–3055.

94. Liao, H., Hafner, J.H. (2005). Gold Nanorod Bioconjugates. *Chem. Mater.* **17**, pp. 4636–4641.
95. Lin, X.Z., Teng, X., Yang, H. (2003). Direct Synthesis of Narrowly Dispersed Silver Nanoparticles Using a Single-Source Precursor. *Langmuir* **19**, pp. 10081–10085.
96. Link, S., El-Sayed, M.A. (1999). Spectral Properties and Relaxation Dynamics of Surface Plasmon Electronic Oscillations in Gold and Silver Nanodots and Nanorods. *J. Phys. Chem. B* **103**, pp. 8410–8426.
97. Liu, F.-K. (2009). Analysis and Applications of Nanoparticles in the Separation Sciences: A Case of Gold Nanoparticles. *J. Chromatogr. A* **1216**, pp. 9034–9047.
98. Liu, L., Kelly, T.L. (2013). Phase Transfer of Triangular Silver Nanoprisms from Aqueous to Organic Solvent by an Amide Coupling Reaction. *Langmuir* **29**, pp. 7052–7060.
99. Liu, M.Z., Guyot-Sionnest, P. (2005). Mechanism of Silver(I)-Assisted Growth of Gold Nanorods and Bipyramids. *J. Phys. Chem. B* **109**, pp. 22192–22200.
100. Liu, S., Han, M.-Y. (2010). Silica-Coated Metal Nanoparticles. *Chem. Asian J.* **5**, pp. 36–45.
101. Liu, S.H., Han, M.Y. (2005). Synthesis, Functionalization, and Bioconjugation of Monodisperse, Silica-Coated Gold Nanoparticles: Robust Bioprobes. *Adv. Funct. Mater.* **15**, pp. 961–967.
102. Liz-Marzan, L.M., Giersig, M., Mulvaney, P. (1996). Homogeneous Silica Coating of Vitreophobic Colloids. *Chem. Commun.*, pp. 731–732.
103. Liz-Marzán, L.M., Giersig, M., Mulvaney, P. (1996). Synthesis of Nanosized Gold-Silica Core-Shell Particles. *Langmuir* **12**, pp. 4329–4335.
104. Lohse, S.E., Burrows, N.D., Scarabelli, L., Liz-Marzan, L.M., Murphy, C.J. (2013). Anisotropic Noble Metal Nanocrystal Growth: The Role of Halides. *Chem. Mater.* **25**, pp. 1250–1261.
105. Lohse, S.E., Eller, J.R., Sivapalan, S.T., Plews, M.R., Murphy, C.J. (2013). A Simple Millifluidic Benchtop Reactor System for the High-Throughput Synthesis and Functionalization of Gold Nanoparticles with Different Sizes and Shapes. *ACS Nano* **7**, pp. 4135–4150.
106. Lu, X., Rycenga, M., Skrabalak, S.E., Wiley, B., Xia, Y. (2009). Chemical Synthesis of Novel Plasmonic Nanoparticles. *Annu. Rev. Phys. Chem.* **60**, pp. 167–192.
107. Lu, X., Yavuz, M.S., Tuan, H.-Y., Korgel, B.A., Xia, Y. (2008). Ultrathin Gold Nanowires Can Be Obtained by Reducing Polymeric Strands of

Oleylamine-AuCl Complexes Formed via Aurophilic Interaction. *J. Am. Chem. Soc.* **130**, pp. 8900–8901.

108. Martin, C.R. (1994). Nanomaterials: A Membrane-Based Synthetic Approach. *Science* **266**, pp. 1961–1966.
109. Mayya, K.S., Caruso, F. (2003). Phase Transfer of Surface-Modified Gold Nanoparticles by Hydrophobization with Alkylamines. *Langmuir* **19**, pp. 6987–6993.
110. McLeod, M.C., Anand, M., Kitchens, C.L., Roberts, C.B. (2005). Precise and Rapid Size Selection and Targeted Deposition of Nanoparticle Populations Using CO_2 Gas Expanded Liquids. *Nano Lett.* **5**, pp. 461–465.
111. Miranda, O.R., Ahmadi, T.S. (2005). Effects of Intensity and Energy of CW UV Light on the Growth of Gold Nanorods. *J. Phys. Chem. B* **109**, pp. 15724–15734.
112. Mulvaney, P. (1996). Surface Plasmon Spectroscopy of Nanosized Metal Particles. *Langmuir* **12**, pp. 788–800.
113. Murphy, C.J., Gole, A.M., Stone, J.W., Sisco, P.N., Alkilany, A.M., Goldsmith, E.C., Baxter, S.C. (2008). Gold Nanoparticles in Biology: Beyond Toxicity to Cellular Imaging. *Acc. Chem. Res.* **41**, pp. 1721–1730.
114. Murphy, C.J., Sau, T.K., Gole, A.M., Orendorff, C.J., Gao, J., Gou, L., Hunyadi, S.E., Li, T. (2005). Anisotropic Metal Nanoparticles: Synthesis, Assembly, and Optical Applications. *J. Phys. Chem. B* **109**, pp. 13857–13870.
115. Murphy, C.J.G., Anand M., Hunyadi, S.E., Stone, J.W., Sisco, P.N., Alkilany, A., Kinard, B.E., Hankins, P. (2008). Chemical Sensing and Imaging with Metallic Nanorods. *Chem. Commun.*, pp. 544–557.
116. Ni, W., Chen, H., Su, J., Sun, Z., Wang, J., Wu, H. (2010). Effects of Dyes, Gold Nanocrystals, pH, and Metal Ions on Plasmonic and Molecular Resonance Coupling. *J. Am. Chem. Soc.* **132**, pp. 4806–4814.
117. Nickel, U., zu Castell, A., Pöppl, K., Schneider, S. (2000). A Silver Colloid Produced by Reduction with Hydrazine as Support for Highly Sensitive Surface-Enhanced Raman Spectroscopy. *Langmuir* **16**, pp. 9087–9091.
118. Nie, Z., Fava, D., Kumacheva, E., Zou, S., Walker, G.C., Rubinstein, M. (2007). Self-assembly of Metal-Polymer Analogues of Amphiphilic Triblock Copolymers. *Nat Mater* **6**, pp. 609–614.
119. Niidome, Y., Nishioka, K., Kawasaki, H., Yamada, S. (2005). Effects of Ammonium Salts and Anionic Amphiphiles on the Photochemical Formation of Gold Nanorods. *Colloid Surf. A-Physicochem. Eng. Asp.* **257–258**, pp. 161–164.

120. Niitsoo, O., Couzis, A. (2011). Facile Synthesis of Silver Core – Silica Shell Composite Nanoparticles. *J. Colloid Interface Sci.* **354**, pp. 887–890.

121. Nikoobakht, B., El-Sayed, M.A. (2001). Evidence for Bilayer Assembly of Cationic Surfactants on the Surface of Gold Nanorods. *Langmuir* **17**, pp. 6368–6374.

122. Nikoobakht, B., El-Sayed, M.A. (2003). Preparation and Growth Mechanism of Gold Nanorods (NRs) Using Seed-Mediated Growth Method. *Chem. Mater.* **15**, pp. 1957–1962.

123. Nishioka, K., Niidome, Y., Yamada, S. (2007). Photochemical Reactions of Ketones to Synthesize Gold Nanorods. *Langmuir* **23**, pp. 10353–10356.

124. Norman, T.J., Grant, C.D., Magana, D., Zhang, J.Z., Liu, J., Cao, D., Bridges, F., Van Buuren, A. (2002). Near Infrared Optical Absorption of Gold Nanoparticle Aggregates. *J. Phys. Chem. B* **106**, pp. 7005–7012.

125. Ojea-Jiménez, I., Romero, F.M., Bastús, N.G., Puntes, V. (2010). Small Gold Nanoparticles Synthesized with Sodium Citrate and Heavy Water: Insights into the Reaction Mechanism. *J. Phys. Chem. C* **114**, pp. 1800–1804.

126. Orendorff, C.J., Alam, T.M., Sasaki, D.Y., Bunker, B.C., Voigt, J.A. (2009). Phospholipid-Gold Nanorod Composites. *ACS Nano* **3**, pp. 971–983.

127. Orendorff, C.J., Gearheart, L., Jana, N.R., Murphy, C.J. (2006). Aspect Ratio Dependence on Surface Enhanced Raman Scattering using Silver and Gold Nanorod Substrates. *Phys. Chem. Chem. Phys.* **8**, pp. 165–170.

128. Park, K., Koerner, H., Vaia, R.A. (2010). Depletion-Induced Shape and Size Selection of Gold Nanoparticles. *Nano Lett.* **10**, pp. 1433–1439.

129. Park, Y.-K., Yoo, S.-H., Park, S. (2007). Assembly of Highly Ordered Nanoparticle Monolayers at a Water/Hexane Interface. *Langmuir* **23**, pp. 10505–10510.

130. Pastoriza-Santos, I., Perez-Juste, J., Liz-Marzan, L.M. (2006). Silica-Coating and Hydrophobation of CTAB-Stabilized Gold Nanorods. *Chem. Mater.* **18**, pp. 2465–2467.

131. Penner, R.M., Martin, C.R. (1987). Preparation and Electrochemical Characterization of Ultramicroelectrode Ensembles. *Anal. Chem.* **59**, pp. 2625–2630.

132. Pérez-Juste, J., Liz-Marzán, L.M., Carnie, S., Chan, D.Y.C., Mulvaney, P. (2004). Electric-Field-Directed Growth of Gold Nanorods in Aqueous Surfactant Solutions. *Adv. Funct. Mater.* **14**, pp. 571–579.

133. Pérez-Juste, J., Pastoriza-Santos, I., Liz-Marzán, L.M., Mulvaney, P. (2005). Gold Nanorods: Synthesis, Characterization and Applications. *Coord. Chem. Rev.* **249**, pp. 1870–1901.
134. Pierrat, S., Zins, I., Breivogel, A., Sonnichsen, C. (2007). Self-Assembly of Small Gold Colloids with Functionalized Gold Nanorods. *Nano Lett.* **7**, pp. 259–263.
135. Pillai, Z.S., Kamat, P.V. (2003). What Factors Control the Size and Shape of Silver Nanoparticles in the Citrate Ion Reduction Method? *J. Phys. Chem. B* **108**, pp. 945–951.
136. Placido, T., Aragay, G., Pons, J., Comparelli, R., Curri, M.L., Merkoçi, A. (2013). Ion-Directed Assembly of Gold Nanorods: A Strategy for Mercury Detection. *ACS Appl. Mater. Interfaces* **5**, pp. 1084–1092.
137. Placido, T., Comparelli, R., Giannici, F., Cozzoli, P.D., Capitani, G., Striccoli, M., Agostiano, A., Curri, M.L. (2009). Photochemical Synthesis of Water-Soluble Gold Nanorods: The Role of Silver in Assisting Anisotropic Growth. *Chem. Mater.* **21**, pp. 4192–4202.
138. Placido, T., Comparelli, R., Striccoli, M., Agostiano, A., Merkoci, A., Curri, M.L. (2013). Assembly of Gold Nanorods for Highly Sensitive Detection of Mercury Ions. *Sensors Journal, IEEE* **13**, pp. 2834–2841.
139. Polte, J., Ahner, T.T., Delissen, F., Sokolov, S., Emmerling, F., Thünemann, A.F., Kraehnert, R. (2010). Mechanism of Gold Nanoparticle Formation in the Classical Citrate Synthesis Method Derived from Coupled In Situ XANES and SAXS Evaluation. *J. Am. Chem. Soc.* **132**, pp. 1296–1301.
140. Porter, L.A., Ji, D., Westcott, S.L., Graupe, M., Czernuszewicz, R.S., Halas, N.J., Lee, T.R. (1998). Gold and Silver Nanoparticles Functionalized by the Adsorption of Dialkyl Disulfides. *Langmuir* **14**, pp. 7378–7386.
141. Quinsaat, J.E.Q., Nuesch, F.A., Hofmann, H., Opris, D.M. (2013). Dielectric Properties of Silver Nanoparticles Coated with Silica Shells of Different Thicknesses. *RSC Advances* **3**, pp. 6964–6971.
142. Quinten, M., Kreibig, U. (1986). Optical Properties of Aggregates of Small Metal Particles. *Surf. Sci.* **172**, pp. 557–577.
143. Rahman, M.R., Saleh, F.S., Okajima, T., Ohsaka, T. (2011). pH Dependence of the Size and Crystallographic Orientation of the Gold Nanoparticles Prepared by Seed-Mediated Growth. *Langmuir* **27**, pp. 5126–5135.
144. Raveendran, P., Fu, J., Wallen, S.L. (2003). Completely "Green" Synthesis and Stabilization of Metal Nanoparticles. *J. Am. Chem. Soc.* **125**, pp. 13940–13941.

145. Reboud, V., Leveque, G., Striccoli, M., Placido, T., Panniello, A., Curri, M.L., Alducin, J.A., Kehoe, T., Kehagias, N., Mecerreyes, D., Newcomb, S.B., Iacopino, D., Redmond, G., Torres, C.M.S. (2013). Metallic Nanoparticles Enhanced the Spontaneous Emission of Semiconductor Nanocrystals Embedded in Nanoimprinted Photonic Crystals. *Nanoscale* **5**, pp. 239–245.

146. Rodríguez-Fernández, J., Pastoriza-Santos, I., Pérez-Juste, J., García de Abajo, F.J., Liz-Marzán, L.M. (2007). The Effect of Silica Coating on the Optical Response of Sub-micrometer Gold Spheres. *J. Phys. Chem. C* **111**, pp. 13361–13366.

147. Rohiman, A., Anshori, I., Surawijaya, A., Idris, I. (2011). Study of Colloidal Gold Synthesis Using Turkevich Method. *AIP Conf. Proc.* **1415**, pp. 39–42.

148. Romo-Herrera, J.M., Alvarez-Puebla, R.A., Liz-Marzan, L.M. (2011). Controlled Assembly of Plasmonic Colloidal Nanoparticle Clusters. *Nanoscale* **3**, pp. 1304–1315.

149. Sakamoto, M., Fujistuka, M., Majima, T. (2009). Light as a Construction Tool of Metal Nanoparticles: Synthesis and Mechanism. *J. Photochem. Photobiol. C-Photochem. Rev.* **10**, pp. 33–56.

150. Salem, A.K., Searson, P.C., Leong, K.W. (2003). Multifunctional Nanorods for Gene Delivery. *Nat Mater* **2**, pp. 668–671.

151. Salminen, T., Honkanen, M., Niemi, T. (2013). Coating of Gold Nanoparticles Made by Pulsed Laser Ablation in Liquids with Silica Shells by Simultaneous Chemical Synthesis. *Phys. Chem. Chem. Phys.* **15**, pp. 3047–3051.

152. Sastry, M., Kumar, A., Mukherjee, P. (2001). Phase Transfer of Aqueous Colloidal Gold Particles into Organic Solutions Containing Fatty Amine Molecules. *Colloid Surf. A-Physicochem. Eng. Asp.* **181**, pp. 255–259.

153. Sau, T.K., Murphy, C.J. (2004). Room Temperature, High-Yield Synthesis of Multiple Shapes of Gold Nanoparticles in Aqueous Solution. *J. Am. Chem. Soc.* **126**, pp. 8648–8649.

154. Schulzendorf, M., Cavelius, C., Born, P., Murray, E., Kraus, T. (2010). Biphasic Synthesis of Au@SiO_2 Core-Shell Particles with Stepwise Ligand Exchange. *Langmuir* **27**, pp. 727–732.

155. Selvakannan, P.R., Mandal, S., Pasricha, R., Adyanthaya, S.D., Sastry, M. (2002). One-step Synthesis of Hydrophobized Gold Nanoparticles of Controllable Size by the Reduction of Aqueous Chloroaurate Ions by Hexadecylaniline at the Liquid-Liquid Interface. *Chem. Commun.* pp. 1334–1335.

156. Sendroiu, I.E., Warner, M.E., Corn, R.M. (2009). Fabrication of Silica-Coated Gold Nanorods Functionalized with DNA for Enhanced Surface Plasmon Resonance Imaging Biosensing Applications. *Langmuir* **25**, pp. 11282–11284.

157. Seo, D., Yoon, W., Park, S., Kim, J., Kim, J. (2008). The Preparation of Hydrophobic Silver Nanoparticles via Solvent Exchange Method. *Colloid Surf. A-Physicochem. Eng. Asp.* **313–314**, pp. 158–161.

158. Sharma, V., Park, K., Srinivasarao, M. (2009). Shape Separation of Gold Nanorods using Centrifugation. *Proc. Natl. Acad. Sci.* **106**, pp. 4981–4985.

159. Shen, R., Camargo, P.H.C., Xia, Y., Yang, H. (2008). Silane-Based Poly(ethylene glycol) as a Primer for Surface Modification of Non-hydrolytically Synthesized Nanoparticles using the Stöber Method. *Langmuir* **24**, pp. 11189–11195.

160. Si, S., Dinda, E., Mandal, T.K. (2007). In Situ Synthesis of Gold and Silver Nanoparticles by Using Redox-Active Amphiphiles and Their Phase Transfer to Organic Solvents. *Chem. Eur. J.* **13**, pp. 9850–9861.

161. Sivaraman, S.K., Kumar, S., Santhanam, V. (2011). Monodisperse Sub-10 nm Gold Nanoparticles by Reversing the Order of Addition in Turkevich Method – The Role of Chloroauric Acid. *J. Colloid Interface Sci.* **361**, pp. 543–547.

162. Skrabalak, S.E., Chen, J., Sun, Y., Lu, X., Au, L., Cobley, C.M., Xia, Y. (2008). Gold Nanocages: Synthesis, Properties, and Applications. *Acc. Chem. Res.* **41**, pp. 1587–1595.

163. Stewart, A., Zheng, S., McCourt, M.R., Bell, S.E.J. (2012). Controlling Assembly of Mixed Thiol Monolayers on Silver Nanoparticles to Tune Their Surface Properties. *ACS Nano* **6**, pp. 3718–3726.

164. Stöber, W., Fink, A., Bohn, E. (1968). Controlled Growth of Monodisperse Silica Spheres in the Micron Size Range. *J. Colloid Interface Sci.* **26**, pp. 62–69.

165. Stoeva, S.I., Zaikovski, V., Prasad, B.L.V., Stoimenov, P.K., Sorensen, C.M., Klabunde, K.J. (2005). Reversible Transformations of Gold Nanoparticle Morphology. *Langmuir* **21**, pp. 10280–10283.

166. Su, C.-H., Wu, P.-L., Yeh, C.-S. (2003). Sonochemical Synthesis of Well-Dispersed Gold Nanoparticles at the Ice Temperature. *J. Phys. Chem. B* **107**, pp. 14240–14243.

167. Sun, Y. (2013). Controlled Synthesis of Colloidal Silver Nanoparticles in Organic Solutions: Empirical Rules for Nucleation Engineering. *Chem. Soc. Rev.* **42**, pp. 2497–2511.

168. Sun, Y., Xia, Y. (2002). Shape-Controlled Synthesis of Gold and Silver Nanoparticles. *Science* **298**, pp. 2176–2179.

169. Sweeney, S.F., Woehrle, G.H., Hutchison, J.E. (2006). Rapid Purification and Size Separation of Gold Nanoparticles via Diafiltration. *J. Am. Chem. Soc.* **128**, pp. 3190–3197.

170. Taleb, A., Petit, C., Pileni, M.P. (1997). Synthesis of Highly Monodisperse Silver Nanoparticles from AOT Reverse Micelles: A Way to 2D and 3D Self-Organization. *Chem. Mater.* **9**, pp. 950–959.

171. Tam, F., Chen, A.L., Kundu, J., Wang, H., Halas, N.J. (2007). Mesoscopic Nanoshells: Geometry-Dependent Plasmon Resonances Beyond the Quasistatic Limit. *J. Chem. Phys.* **127**, pp. 204703.

172. Tao, A.R., Habas, S., Yang, P. (2008). Shape Control of Colloidal Metal Nanocrystals. *Small* **4**, pp. 310–325.

173. Templeton, A.C., Wuelfing, W.P., Murray, R.W. (1999). Monolayer-Protected Cluster Molecules. *Acc. Chem. Res.* **33**, pp. 27–36.

174. Teranishi, T., Hasegawa, S., Shimizu, T., Miyake, M. (2001). Heat-Induced Size Evolution of Gold Nanoparticles in the Solid State. *Adv. Mater.* **13**, pp. 1699–1701.

175. Tornblom, M., Henriksson, U. (1997). Effect of Solubilization of Aliphatic Hydrocarbons on Size and Shape of Rodlike C16TABr Micelles Studied by 2H NMR Relaxation. *J. Phys. Chem. B* **101**, pp. 6028–6035.

176. Toshima, N., Yonezawa, T. (1998). Bimetallic Nanoparticles-Novel Materials for Chemical and Physical Applications. *New J. Chem.* **22**, pp. 1179–1201.

177. Turkevich, J., Stevenson, P.C., Hillier, J. (1951). A Study of the Nucleation and Growth Processes in the Synthesis of Colloidal Gold. *Discuss. Faraday Soc.* **11**, pp. 55–75.

178. Ung, T., Liz-Marzán, L.M., Mulvaney, P. (1998). Controlled Method for Silica Coating of Silver Colloids. Influence of Coating on the Rate of Chemical Reactions. *Langmuir* **14**, pp. 3740–3748.

179. van der Zande, B.M.I., Böhmer, M.R., Fokkink, L.G.J., Schönenberger, C. (1999). Colloidal Dispersions of Gold Rods:? Synthesis and Optical Properties. *Langmuir* **16**, pp. 451–458.

180. Vidinha, P., Barreiros, S., Cabral, J.M.S., Nunes, T.G., Fidalgo, A., Ilharco, L.M. (2008). Enhanced Biocatalytic Activity of ORMOSIL-Encapsulated Cutinase: The Matrix Structural Perspective. *J. Phys. Chem. C* **112**, pp. 2008–2015.

181. Vigderman, L., Khanal, B.P., Zubarev, E.R. (2012). Functional Gold Nanorods: Synthesis, Self-Assembly, and Sensing Applications. *Adv. Mater.* **24**, pp. 4811–4841.
182. Vijaya Sarathy, K., U. Kulkarni, G., N. R. Rao, C. (1997). A Novel Method of Preparing Thiol-Derivatised Nanoparticles of Gold, Platinum and Silver Forming Superstructures. *Chem. Commun.* pp. 537–538.
183. Volkert, A.A., Subramaniam, V., Haes, A.J. (2011). Implications of Citrate Concentration During the Seeded Growth Synthesis of Gold Nanoparticles. *Chem. Commun.* **47**, pp. 478–480.
184. Wang, C., Ma, Z., Wang, T., Su, Z. (2006). Synthesis, Assembly, and Biofunctionalization of Silica-Coated Gold Nanorods for Colorimetric Biosensing. *Adv. Funct. Mater.* **16**, pp. 1673–1678.
185. Wang, Y., DePrince, A.E., K. Gray, S., Lin, X.-M., Pelton, M. (2010). Solvent-Mediated End-to-End Assembly of Gold Nanorods. *J. Phys. Chem. Lett.* **1**, pp. 2692–2698.
186. Wang, Y., Yang, H. (2006). Oleic Acid as the Capping Agent in the Synthesis of Noble Metal Nanoparticles in Imidazolium-Based Ionic Liquids. *Chem. Commun.* pp. 2545–2547.
187. Wei, G.-T., Liu, F.-K. (1999). Separation of nanometer gold particles by size exclusion chromatography. *J. Chromatogr. A* **836**, pp. 253–260.
188. Wei, G.-T., Liu, F.-K., Wang, C.R.C. (1999). Shape Separation of Nanometer Gold Particles by Size-Exclusion Chromatography. *Anal. Chem.* **71**, pp. 2085–2091.
189. Wilcoxon, J.P., Provencio, P.P. (2004). Heterogeneous Growth of Metal Clusters from Solutions of Seed Nanoparticles. *J. Am. Chem. Soc.* **126**, pp. 6402–6408.
190. Wiley, B., Sun, Y., Xia, Y. (2007). Synthesis of Silver Nanostructures with Controlled Shapes and Properties. *Acc. Chem. Res.* **40**, pp. 1067–1076.
191. Wiley, B.J., Chen, Y., McLellan, J.M., Xiong, Y., Li, Z.-Y., Ginger, D., Xia, Y. (2007). Synthesis and Optical Properties of Silver Nanobars and Nanorice. *Nano Lett.* **7**, pp. 1032–1036.
192. Wiley, B.J., Im, S.H., Li, Z.-Y., McLellan, J., Siekkinen, A., Xia, Y. (2006). Maneuvering the Surface Plasmon Resonance of Silver Nanostructures through Shape-Controlled Synthesis. *J. Phys. Chem. B* **110**, pp. 15666–15675.
193. Williams, D.P., Satherley, J. (2009). Size-Selective Separation of Polydisperse Gold Nanoparticles in Supercritical Ethane. *Langmuir* **25**, pp. 3743–3747.

194. Wu, H.-L., Kuo, C.-H., Huang, M.H. (2010). Seed-Mediated Synthesis of Gold Nanocrystals with Systematic Shape Evolution from Cubic to Trisoctahedral and Rhombic Dodecahedral Structures. *Langmuir* **26**, pp. 12307–12313.

195. Wu, Y., Li, Y., Ong, B.S. (2006). Printed Silver Ohmic Contacts for High-Mobility Organic Thin-Film Transistors. *J. Am. Chem. Soc.* **128**, pp. 4202–4203.

196. Wustholz, K.L., Henry, A.-I., McMahon, J.M., Freeman, R.G., Valley, N., Piotti, M.E., Natan, M.J., Schatz, G.C., Duyne, R.P.V. (2010). Structure-Activity Relationships in Gold Nanoparticle Dimers and Trimers for Surface-Enhanced Raman Spectroscopy. *J. Am. Chem. Soc.* **132**, pp. 10903–10910.

197. Xia, Y., Xiong, Y., Lim, B., Skrabalak, S.E. (2009). Shape-Controlled Synthesis of Metal Nanocrystals: Simple Chemistry Meets Complex Physics? *Angew. Chem. Int. Ed.* **48**, pp. 60–103.

198. Xiao, J., Qi, L. (2011). Surfactant-Assisted, Shape-Controlled Synthesis of Gold Nanocrystals. *Nanoscale* **3**, pp. 1383–1396.

199. Xu, L., Han, G., Hu, J., He, Y., Pan, J., Li, Y., Xiang, J. (2009). Hydrophobic Coating- and Surface Active Solvent-Mediated Self-assembly of Charged Gold and Silver Nanoparticles at Water-Air and Water-Oil Interfaces. *Phys. Chem. Chem. Phys.* **11**, pp. 6490–6497.

200. Yang, D.-P., Cui, D.-X. (2008). Advances and Prospects of Gold Nanorods. *Chem. Asian J.* **3**, pp. 2010–2022.

201. Yang, J., Wu, J.-C., Wu, Y.-C., Wang, J.-K., Chen, C.-C. (2005). Organic Solvent Dependence of Plasma Resonance of Gold Nanorods: A Simple Relationship. *Chem. Phys. Lett.* **416**, pp. 215–219.

202. Yang, P., Ando, M., Murase, N. (2010). Various Au Nanoparticle Organizations Fabricated through SiO_2 Monomer Induced Self-Assembly. *Langmuir* **27**, pp. 895–901.

203. Yin, Y., Lu, Y., Sun, Y., Xia, Y. (2002). Silver Nanowires Can Be Directly Coated with Amorphous Silica To Generate Well-Controlled Coaxial Nanocables of Silver/Silica. *Nano Lett.* **2**, pp. 427–430.

204. Yockell-Lelièvre, H., Gingras, D., Lamarre, S., Vallée, R., Ritcey, A. (2013). In Situ Monitoring of the 2D Aggregation Process of Thiol-Coated Gold Nanoparticles Using Interparticle Plasmon Coupling. *Plasmonics* **8**, pp. 1369–1377.

205. Yu, Chang, S.-S., Lee, C.-L., Wang, C.R.C. (1997). Gold Nanorods: Electrochemical Synthesis and Optical Properties. *J. Phys. Chem. B* **101**, pp. 6661–6664.

206. Yun, Y., Ya, Y., Wei, W., Jinru, L. (2008). Precise Size Control of Hydrophobic Gold Nanoparticles using Cooperative Effect of Refluxing Ripening and Seeding Growth. *Nanotechnology* **19**, pp. 175603.

207. Zhang, D., Qi, L., Yang, J., Ma, J., Cheng, H., Huang, L. (2004). Wet Chemical Synthesis of Silver Nanowire Thin Films at Ambient Temperature. *Chem. Mater.* **16**, pp. 872–876.

208. Zhang, Y., Kong, X., Xue, B., Zeng, Q., Liu, X., Tu, L., Liu, K., Zhang, H. (2013). A Versatile Synthesis Route for Metal@SiO_2 Core-Shell Nanoparticles using 11-Mercaptoundecanoic Acid as Primer. *J. Mater. Chem. C* **1**, pp. 6355–6363.

209. Zhang, Y., Xu, F., Sun, Y., Guo, C., Cui, K., Shi, Y., Wen, Z., Li, Z. (2010). Seed-Mediated Synthesis of Au Nanocages and Their Electrocatalytic Activity towards Glucose Oxidation. *Chem. Eur. J.* **16**, pp. 9248–9256.

210. Zhao, P., Li, N., Astruc, D. (2013). State of the Art in Gold Nanoparticle Synthesis. *Coord. Chem. Rev.* **257**, pp. 638–665.

211. Zhao, Y., PÃcrez-Segarra, W., Shi, Q., Wei, A. (2005). Dithiocarbamate Assembly on Gold. *J. Am. Chem. Soc.* **127**, pp. 7328–7329.

212. Zheng, N., Fan, J., Stucky, G.D. (2006). One-Step One-Phase Synthesis of Monodisperse Noble-Metallic Nanoparticles and Their Colloidal Crystals. *J. Am. Chem. Soc.* **128**, pp. 6550–6551.

213. Zhu, Y.-J., Hu, X.-L. (2003). Microwave-Polyol Preparation of Single-Crystalline Gold Nanorods and Nanowires. *Chem. Lett.* **32**, pp. 1140–1141.

Chapter 3

Amorphous Nanoparticle Assemblies by Bottom-Up Principles

Alastair Cunningham, Mahshid Chekini, and Thomas Bürgi
Department of Physical Chemistry, University of Geneva, 30 Quai Ernest-Ansermet, 1211, Geneva 4, Switzerland

Self-assembly of plasmonic nanoparticles is a fast, robust, and cheap route to obtain large-scale materials with promising optical properties. Although the materials obtained in this way are usually amorphous, the approach has the advantage that three-dimensional assemblies of small particles can be obtained with control of average distances at the nanoscale. In this chapter, we will focus on one particular strategy to assemble plasmonic particles that relies on the interplay between charged particles, surfaces, and polyelectrolytes. Several geometries will be discussed with an emphasis on the optical properties that are dictated by the coupling between plasmons in different geometries.

3.1 Introduction

The controlled fabrication of nanoscale structures is at the heart of established and emerging technologies [1–3]. Advances in data

Active Plasmonic Nanomaterials
Edited by Luciano De Sio

ISBN 978-981-4613-00-2 (Hardcover), 978-981-4613-01-9 (eBook)
www.panstanford.com

storage and information technology in general largely relied on the ability to down-scale structures. The tremendous success in miniaturization of structures was made possible by top-down techniques such as electron beam lithography (EBL) [4] and etching methods, which can vary from wet-chemical etches [5] to focused ion beam [6] and laser ablation [7]. There is no doubt that top-down techniques will continue to play a key role in future technologies that rely on the fabrication of small structures. However, there are also severe drawbacks and inherent limitations of the top-down approach. It is becoming increasingly difficult to fabricate ever smaller structures as physical limits are being approached. The minimum size of structures that can be fabricated no longer matches the expectations that are put forward for future technologies. In addition, top-down fabrication techniques tend to be slow, cumbersome, and expensive, and the samples that can typically be fabricated are small and two-dimensional. These are severe constraints for future technologies.

Bottom-up strategies have significant advantages with respect to top-down fabrication techniques, although the former are far from being mature. However, the lesson we can learn from nature is motivation enough to use self-assembly principles to build up complicated structures and functional devices [8]. In fact, there is much to learn from the fantastic possibilities of self-organization of molecules and polymers that nature uses to produce complex and functional structures at different length scales. Bottom-up fabrication techniques are not restricted to two-dimensional structures; they are typically cheap, and large-scale samples are easily produced. It is, therefore, not astonishing that self-assembling techniques are now increasingly applied in the search to fabricate structured nanoscale samples. In particular, self-assembly of nanoparticles into two- and three-dimensional structures is a field of current interest because the interaction or coupling between particles within defined structures leads to optical effects that can be used in fields such as metamaterials [9] and sensing [10].

Different strategies were considered for the organization of nanoparticles. Liquid crystals were used making use of the inherent order in a liquid crystal system. One strategy is to simply mix nanoparticles into a liquid crystal [11]. A major challenge of this approach is the solubility of the nanoparticle to achieve a sufficient density. Another approach is to fix liquid-crystalline molecules directly to the nanoparticles [12, 13]. Self-organization of gold nanoparticles has been shown in this way. The nanoparticles in such a system are just separated by the liquid-crystalline molecules anchored to the particle surface, and high nanoparticle density is thus achieved. A challenge here is to move toward larger nanoparticle sizes, which exhibit the desired optical properties, because it becomes increasingly difficult to organize particles using liquid-crystalline molecules as the particles become larger [14].

Inspired by nature, researchers have used the recognition of DNA to assemble defined nanoparticle architectures [15]. The formation of nanoparticle dimers and trimers was achieved by functionalizing gold nanocrystals with single-stranded DNA oligonucleotides of defined length and sequence and assembling them on a complementary single-stranded DNA template. Also chemically modified peptides were used to assemble gold nanoparticles along the peptide backbone [16]. In this way, a chiral helical arrangement was achieved and the material showed optical activity. The use of biomolecules to organize nanoparticles has become quite popular.

Block copolymers show typical self-organization at interesting length scales and are an attractive means to organize nanoparticles [17]. Using this philosophy, gold nanorods were organized by functionalization of their ends by polymer chains. Depending on the conditions such as composition of solvent or molecular weight of the polymer, it was possible to organize the rods either end-by-end or side-by-side [18].

The techniques mentioned above are by far not the only ones that have been used up to now to organize nanoparticles, and much more will become possible in the future. In the following, we want to focus on techniques that make use of electrostatic forces between charged nanoparticles, polymers, and surface to organize nanoparticles [19]. This approach is relatively easy to apply and provides simple control over some structural parameters that can be used to systematically tune optical properties of the resulting material [20]. This electrostatic self-assembly is based on the layer-by-layer technique introduced by Decher for polyelectrolytes [21], but it can, of course, be used for other charged constituents. The resulting structures are organized in some way but do not show long-range order; put in other words, the structures are amorphous. However, in many cases, long-range order is not required for obtaining certain properties, as will be illustrated in the following sections [22].

3.2 Electrostatic Forces

The type of self-assembly to be discussed is based on electrostatic interactions of charged particles and surfaces in aqueous solutions. The latter contains ions, and their distribution in the vicinity of a charged surface or nanoparticle is not homogenous. Figure 3.1 shows the distribution of ions around a charged surface, which is known as electrical double layer (EDL).

The interaction between two charged particles (or a particle and a surface) is characterized by attractive van der Waals forces

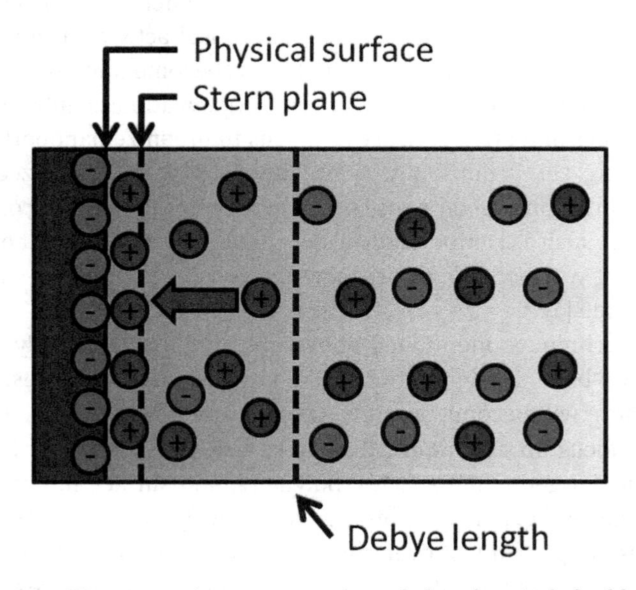

Figure 3.1 Diagrammatic representation of the electrical double layer showing the excess of ions of opposite charge to that of the surface in this zone.

and electrostatic forces (attractive for particles of opposite charge; repulsive for particles of the same charge). Particles of the same charge can form stable solutions, depending on the conditions that influence the competition between attractive and repulsive forces. At larger separations, repulsive interactions set up between the EDLs surrounding each particle dominate. The EDL is composed of solvent molecules and ions that are attracted to the charged surface of a nanoparticle to maintain charge neutrality. The characteristic length of the EDL, known as the Debye length, is shown in Fig. 3.1, along with the Stern layer, which refers to the layer of oppositely charged ions physisorbed at the surface. The Debye length is inversely proportional to the square root of the ion concentration in solution. The latter is, therefore, a very important parameter to influence the forces between charged particles.

3.3 Assembly of Nanoparticles on Flat Surfaces

Similar forces, as described above, between particles also exist between a charged surface and charged particles. For example, dipping a positively charged surface into a solution of negatively charged particles leads to spontaneous adsorption of the particles on the surface. At the same time, the negatively charged particles on the surface repel each other due to the electrostatic double layer. By playing with the ion concentration in solution, the average particle–particle separation can be adjusted. This is shown in Fig. 3.2 for citrate-stabilized gold nanoparticles on a positively charged surface.

By adjusting the NaCl concentration from very low (dialyzed sample) to 10 mM, the nanoparticle density can be changed from 420 particles/μm to 1450 particles/μm. Of course there are also limits to this. By further increasing the NaCl concentration, the repulsive forces between the nanoparticles become so small that the particles start to agglomerate.

The interaction between the particles can also be influenced by modifying the charge at their surface. Such a way is demonstrated in Fig. 3.3. Here the sample after deposition of citrate-stabilized gold nanoparticles was dipped in a solution of 1-dodecanethiol. The latter is not charged and has a strong affinity to the gold surface. Its

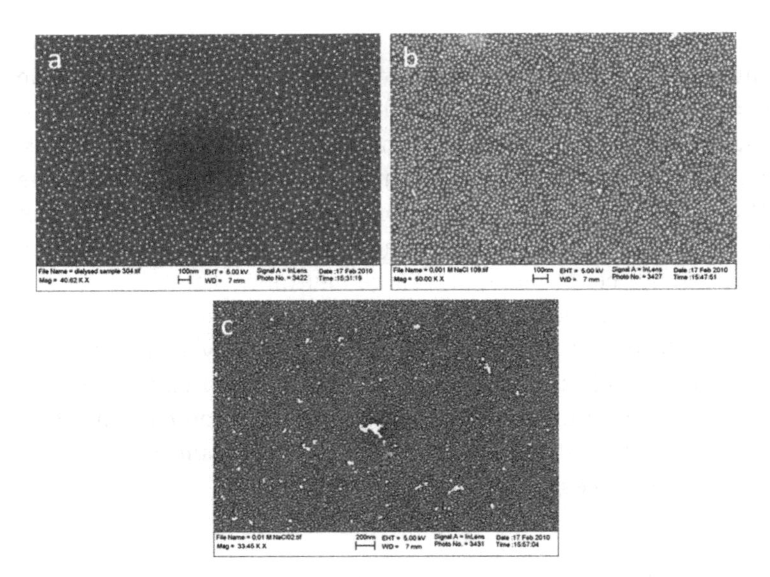

Figure 3.2 SEM images showing gold nanoparticle (~20 nm diameter) layers deposited from solutions with varying NaCl salt concentrations: (a) Dialyzed sample (very low ion concentration), (b) solution containing 1 mM NaCl, (c) solution containing 10 mM NaCl.

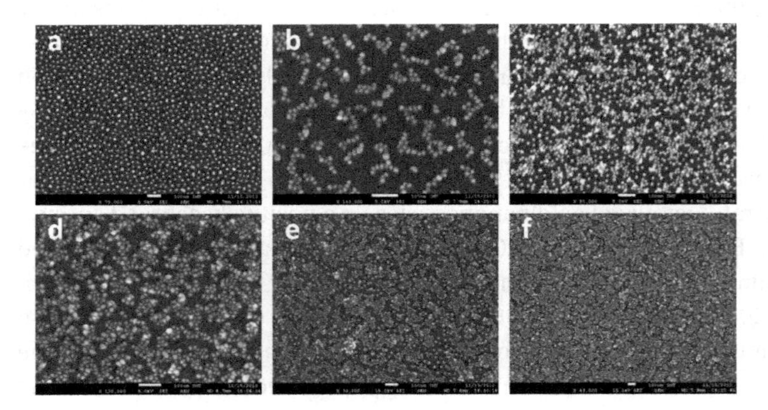

Figure 3.3 SEM micrographs depicting various stages of the deposition and reorganization process induced by adsorption of a thiol on the gold nanoparticles. Shown are the morphology observed after (a) the first deposition, (b) the first reorganization, (c) the second deposition, (d) the second reorganization, (e) the third deposition, and (f) the fourth deposition. An increase in particle density as well as the introduction of a degree of order can clearly be observed. All scale bars are 100 nm.

adsorption on the gold removes part of the weaker bound citrate and, therefore, also part of the charge. This leads to less repulsion between the particles, and the latter reorganize to form oligomers of particles that are in contact, probably separated only by a double layer of the adsorbed thiol, which contributes to the attractive forces between the particles. After the reorganization, further particles can be adsorbed by simply dipping the sample again into a nanoparticle solution. Then the sample can again be dipped into the thiol solution, and the whole process can be repeated to obtain a dense layer of nanoparticles.

The procedure outlined above not only increases the filling fraction but also increases the degree of order within the nanoparticle array. It is possible to quantify this order using the radial distribution function. The latter is a measure of the probability of finding a particle at a distance r away from a given reference particle. For a low density nanoparticle array before the reorganization and re-adsorption, the radial distribution function reveals the amorphous character of this situation (Fig. 3.4). The trough at short distances is due to the strong electrostatic repulsion between equally charged particles. A prominent peak indicates some very short-range order. In contrast, the radial distribution function after five reorganization and re-deposition cycles reveals much more structure with five peaks. The different radial distribution function

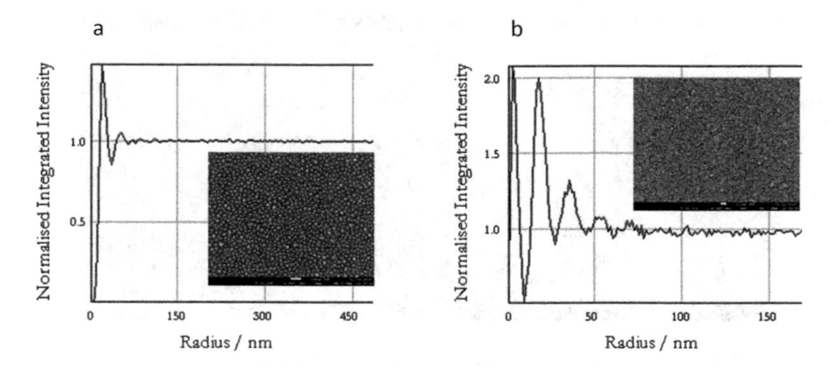

Figure 3.4 Radial distribution functions (a) before and (b) after reorganization and deposition processes. The corresponding SEM micrographs are shown in the insets.

is also reflected in different optical properties of the two samples, which is mainly due to the different coupling between neighboring gold nanoparticles.

The initial adsorption of the gold nanoparticles on the substrate, e.g. glass, relies on the functionalization of the latter to impart charge to the surface. One convenient way is to covalently link amine groups via silica surface chemistry. This results in the attraction of negatively charged nanoparticles in aqueous solution. This functionalization of the substrate surface can be combined with photolithography. By coating a functionalized substrate, glass or silicon, with a photoresist and exposing this to light through a suitably designed mask, it is possible to recreate the form of the mask in the resist. After this process, only certain areas of the functionalized substrate are exposed to the nanoparticle solution and, therefore, the nanoparticles will be adsorbed only on certain areas, as is demonstrated in Fig. 3.5. As can be seen, the

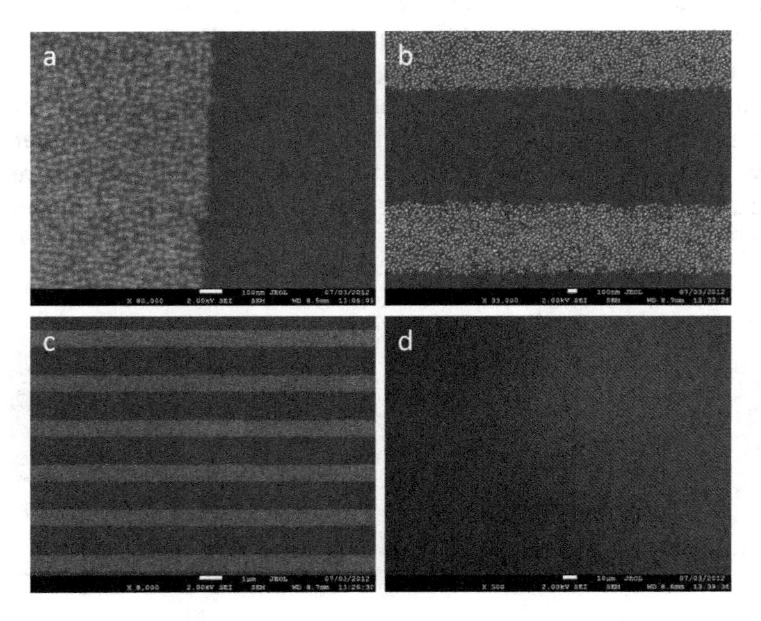

Figure 3.5 SEM images of a striped array of gold nanoparticles (~20 nm diameter) deposited on functionalized Si substrates pre-patterned using optical lithography techniques. Images at four different magnifications are shown: (a) 80000×, (b) 33000×, (c) 8000×, and (d) 500×.

samples exhibit very good sharpness and are basically free from defects. In the areas with no charge, not a single nanoparticle is adsorbed.

3.4 Building Up the Third Dimension

An interesting application of the charge-driven self-assembly of gold nanoparticles on substrates is the possibility to build up multilayers using polyelectrolytes as spacer and glue. For this the substrate, for example glass, is first functionalized to impart positive charge. After a first nanoparticle layer (negative charge), a positive polyelectrolyte layer is deposited by simply dipping the sample into a solution containing the respective polyelectrolyte. After washing the sample, a negative polyelectrolyte can be deposited in an analogous fashion, and this process can be repeated as many times as needed either manually or in an automated way using dedicated equipment. After a positive polyelectrolyte layer, a second nanoparticle layer can be deposited, as shown diagrammatically in Fig. 3.6. Although this approach yields amorphous samples, it allows one to control average parameters of the sample. Specifically, the average distance between nanoparticles within one layer can be controlled by the salt concentration during the deposition (see above). The distance between two nanoparticle layers can be controlled by the number

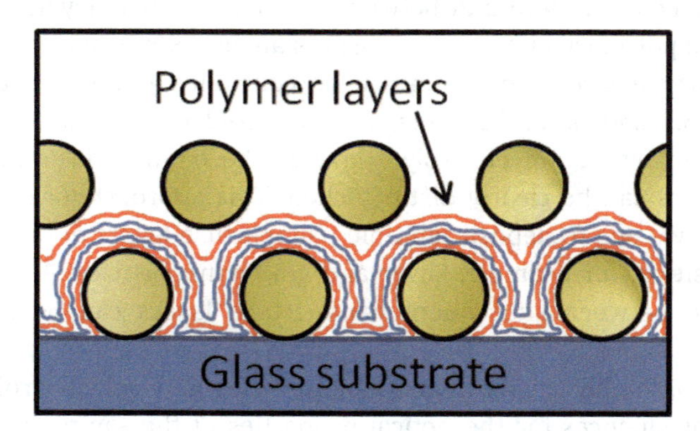

Figure 3.6 Diagrammatical representation of two gold nanoparticle layers separated by five polyelectrolyte layers.

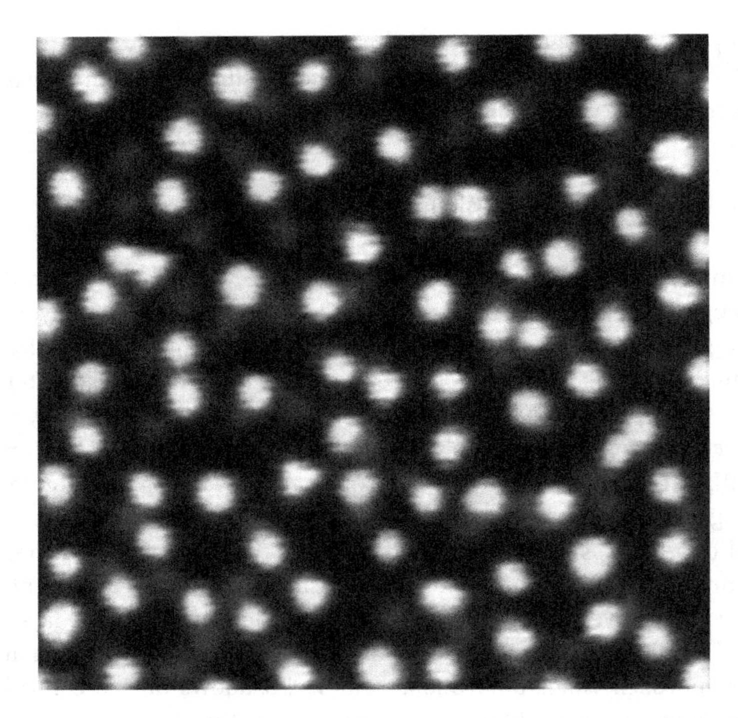

Figure 3.7 SEM image of two gold nanoparticle layers separated by 31 polyelectrolyte layers. The two layers can be clearly distinguished. The particles (about 20 nm in diameter) in the top layer are brighter.

of polyelectrolyte layers in between two nanoparticle layers. Note that one polyelectrolyte layer measures about 0.9 nm, and thus the nanoparticle layer separation can be controlled quite accurately. Figure 3.7 shows an SEM image of two gold nanoparticle layers separated by 31 polyelectrolyte layers. The nanoparticles in the two layers can be clearly distinguished. This approach, of course, also allows one to deposit multiple nanoparticle layers of different composition, for example, silver and gold nanoparticles. Also, the distance between subsequent nanoparticle layers can be easily varied.

The assembly of plasmonic nanoparticles and polyelectrolytes has consequences for the optical properties of the sample mostly because the neighboring nanoparticles can couple with each other. According to the plasmon ruler equation [23], the relative shift

($\Delta\lambda/\lambda_0$) of the plasmon band of two coupling nanoparticles, calculated as the ratio between the absolute plasmonic shift $\Delta\lambda$ and the single particle resonance peak maximum wavelength λ_0 (measured for isolated particles), exponentially depends on s/D:

$$\Delta\lambda/\lambda_0 = k^* \exp(-s/\tau D) \tag{3.1}$$

where s is the gap between particles of diameter D, k is the maximum plasmon shift for the particle dimer and τ is a decay constant that depends on the considered system. Such a behavior is indeed also observed for a bilayer of gold nanoparticles separated by different numbers of polyelectrolyte layers, as can be seen in Fig. 3.8. In these experiments, care was taken to avoid coupling between gold nanoparticles within one layer, by adjusting the average distance between the particles [20]. The coupling between nanoparticles within the two different layers is then varied by changing the number of polyelectrolyte layers between the nanoparticle layers. As can be seen in Fig. 3.8, the shift of the plasmon band is exponential with the distance between the coupling nanoparticles (number of polyelectrolyte layers). As expected from the plasmon ruler

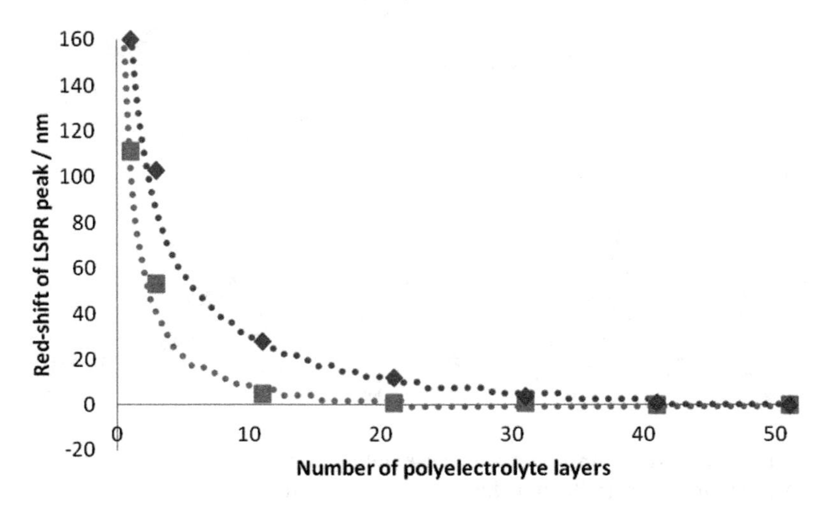

Figure 3.8 Red-shift of localized surface plasmon peak of two gold nanoparticle arrays as a function of the number of polyelectrolyte layers separating them. Experimental (solid points) and simulated data (dashed lines) are shown for arrays of ~20 nm (blue plots) and ~40 nm (red plots) in diameter.

equation, the decay length is larger for the bigger nanoparticles. The optical spectra of a sample consisting of two nanoparticle layers can be well reproduced by considering a nanoparticle dimer using rigorous simulations using Maxwell's equations. At very short inter-particle separations, two distinct plasmons can be observed, as is also predicted from qualitative plasmon hybridization models.

When going further into the third dimension by depositing multiple nanoparticle layers separated by different numbers of poly-electrolytes, the situation becomes more complex. The extinction spectra of assemblies of such nanoparticles are seen in Fig. 3.9. These samples contain multiple gold nanoparticle layers separated by, respectively, 1, 5, 11, and 25 polyelectrolyte layers. For the shortest inter-layer separation, two distinct plasmon peaks can be observed. Interestingly, the longer wavelength resonance becomes

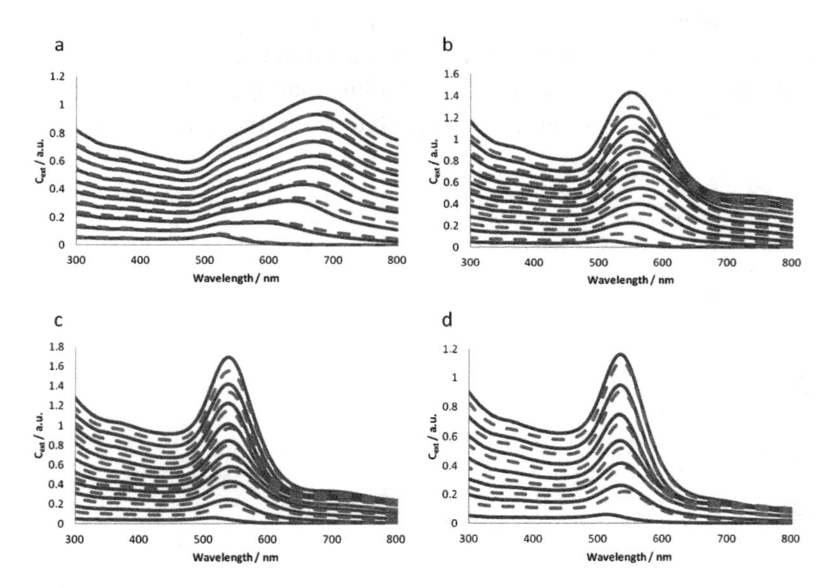

Figure 3.9 Extinction spectra of assemblies constructed from multiple gold nanoparticle (~20 nm diameter) layers and polyelectrolyte layers. Gold nanoparticle layers are separated by (a) 1 polyelectrolyte layer, (b) 5 polyelectrolyte layers, (c) 11 polyelectrolyte layers, and (d) 25 polyelectrolyte layers. In each case, the spectra taken directly after the deposition of the gold nanoparticle layers (red solid traces) and the requisite number of polyelectrolyte layers (blue-dashed traces) are shown.

increasingly prominent in relation to the shorter wavelength resonance (Fig. 3.9a). The samples containing 5 and 11 polyelectrolyte layers between the nanoparticles display an interesting trend. The plasmon resonance initially shifts to longer wavelengths upon the deposition of additional gold nanoparticle layers as would be expected and as was observed in all of the samples with only two nanoparticle layers. However, at a certain point, this resonance begins to shift back to higher frequencies. This behavior is more pronounced in the sample where the nanoparticle arrays were separated by five polyelectrolyte layers.

The essence of the optical behavior of a bilayer of gold nanoparticles can be characterized by considering a dimer model [20]. When exciting a nanoparticle dimer made from two identical particles, their coupling can be described by plasmon hybridization theory [24]. This interaction can be considered in its simplest form as the coupling of two nearby dipoles. Hybridization theory predicts two distinct classes of eigenmodes for a dimer. The first is associated with an in phase oscillation of the electric dipole in both spheres and is, therefore, termed a "bright" eigenmode since it can radiate into the far field. The second class requires a 180° out of phase oscillation of the electric dipoles of both spheres. These eigenmodes are termed "dark" since they cannot radiate into the far field. For a homo-dimer of gold nanoparticles, two bright modes are expected, which can indeed be observed when assembling two layers of gold nanoparticles with only one polyelectrolyte layer in between, as seen in Fig. 3.9a (second spectrum).

According to the plasmon hybridization theory, the situation changes qualitatively when the dimer is composed of two different nanoparticles, as illustrated in Fig. 3.10. An easy way to realize such a situation is the assembly of two nanoparticle layers, one containing gold nanoparticles and the other one containing silver nanoparticles [25]. Of course the extinction spectra of such samples, as shown in Fig. 3.11, largely depend on the number of polyelectrolyte layers deposited between the two nanoparticle layers, but two peaks can be clearly distinguished at approximately 380 nm and 395 nm, which correspond to the σ^* and π^* modes outlined in Fig. 3.10 [19]. In this case, it is difficult to conclusively identify which peak corresponds to which mode due to their relatively low separation

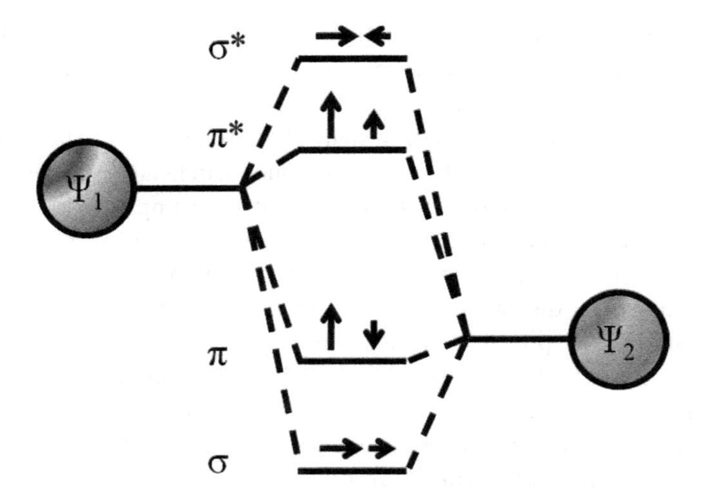

Figure 3.10 Plasmon hybridization diagrams showing the coupling between and the dipole moments on silver and gold nanoparticle in a dimer configuration.

and the potential influence of coupling between the plasmonic modes and inter-band absorption processes [26]. For samples separated by only one polyelectrolyte layer, two bands at longer wavelength can be observed at approximately 510 nm and 600 nm, corresponding to the π and σ bands. Thus, it appears that a dimer model seems to catch the essence of the optical response of the sample. This is furthermore corroborated by studying the angular dependence of the extinction spectra. By rotating the sample, the geometrical configuration of the two arrays (or the dimer) relative to the incident beam is altered with the two transverse modes, corresponding to light polarized perpendicularly to the main dimer axis, becoming increasingly important. As the sample is rotated and the transverse modes become more dominant in the spectral analysis, the longitudinal modes, those corresponding to light polarized along the main dimer axis, become correspondingly less prevalent.

It should also be noted that the build-up of layered nanoparticle architectures is not restricted to solid substrates such as glass. It can be applied to flexible substrates such as PDMS (polydimethylsiloxane), which opens further possibilities to change the arrangement

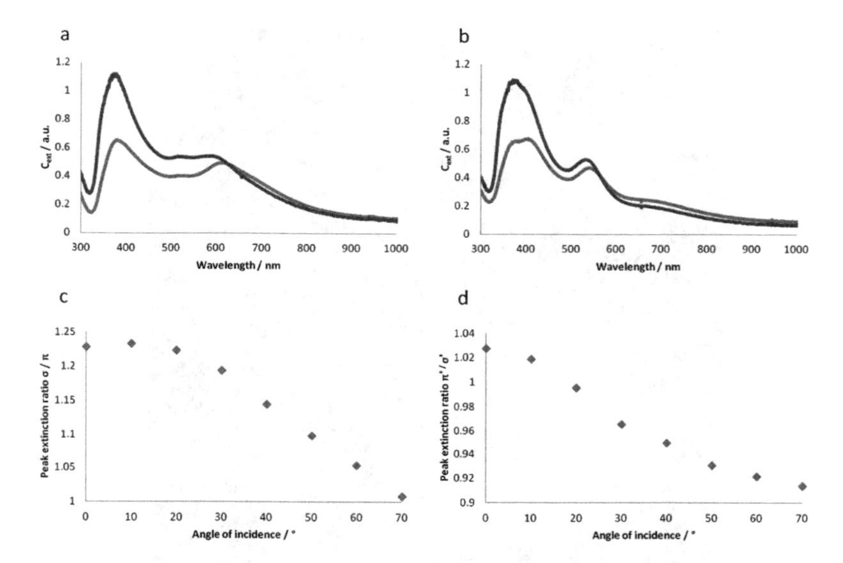

Figure 3.11 (a) Extinction spectra of gold and silver nanoparticle arrays separated by one polyelectrolyte layer at 0° (blue trace) and 70° (red trace). (b) Extinction spectra of gold and silver nanoparticle arrays separated by 21 polyelectrolyte layers at 0° (blue trace) and 70° (red trace). (c) Ratio of σ mode peak intensity to π mode peak intensity for gold and silver nanoparticle arrays separated by one polyelectrolyte layer. (d) Ratio of π^* mode peak intensity to σ^* mode peak intensity for gold and silver nanoparticle arrays separated by 21 polyelectrolyte layers.

of the nano-objects by just stretching the sample. Furthermore, the multilayer films can be removed from the substrate to obtain very thin self-supporting samples [19]. Such layered systems can be prepared by first building up a polymer multilayer based on hydrogen-bonding interactions. On top of this, the polyelectrolyte–nanoparticle systems can be grown as described above. By changing the pH, the hydrogen-bonded layer becomes unstable and the polyelectrolyte system can be removed from the substrate.

3.5 Core–Shell Structures

The principle of self-assembly using charged particles, surfaces, and polyelectrolytes does not only work on flat surfaces. It can, in

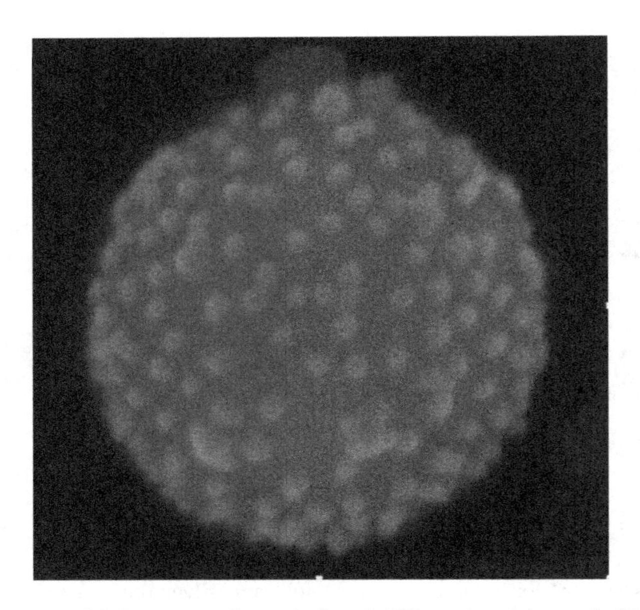

Figure 3.12 SEM image of an isolated SiO_2 microsphere (~270 nm diameter) decorated with gold nanoparticles (~20 nm diameter) using electrostatic forces.

an analogous way, also be applied to assemble nanoparticles on curved surfaces, and it can be used for the preparation of core–shell structures. Such materials have highly tunable optical properties, and it has been proposed that they will enable advancements toward materials with double negative properties at optical frequencies [27]. A core–shell structure can be realized by functionalizing a silica (SiO_2) microsphere using the same surface chemistry as is used for the flat glass slides described above to impart a positive charge onto the surface. Exposing this modified silica microsphere to negatively charged gold nanoparticles leads to their spontaneous adsorption on the surface of the microsphere. The resulting core–shell system consists of a dielectric sphere decorated by isolated gold nanoparticles, as can be seen in Fig. 3.12.

Due to the coupling properties of gold nanoparticles in this core–shell geometry, the extinction spectrum is quite different from the one of an isolated gold nanoparticle. In fact, a strongly red-shifted plasmon is observed at around 670 nm, as can be

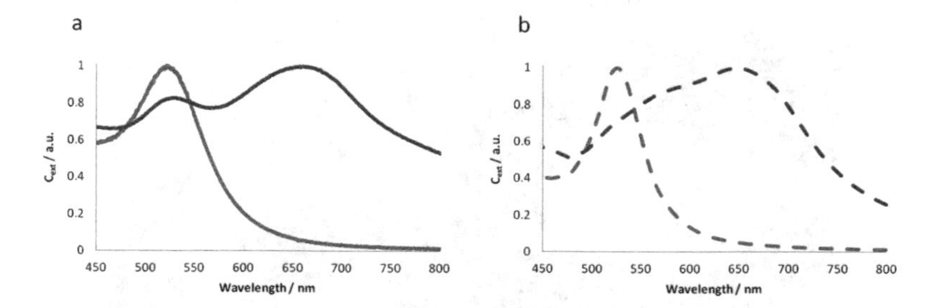

Figure 3.13 (a) Measured extinction spectrum of fabricated core–shell clusters in solution (red trace). For comparison, the extinction spectrum of a solution of gold nanospheres is shown (blue trace). (b) Simulated extinction spectra of core–shell clusters in solution (red-dashed trace) and a single gold nanosphere (blue-dashed trace).

seen in Fig. 3.13. Such core–shell structures exhibit a strong isotropic magnetic response [27]. This magnetism can be explained by assuming that the shell of metallic nanospheres acts as a medium with an extremely high permittivity at wavelengths slightly above the collective plasmonic resonance. The large permittivity in turn evokes Mie resonances. For the lowest order one, the electric displacement field rotates in a plane perpendicular to the polarization of the incident magnetic field, meaning that this mode can be associated with a magnetic dipole contribution. That the core–shell nanoclusters exhibit artificial isotropic magnetism was confirmed by theory [28]. Using simulations, the extinction spectrum can be decomposed, revealing the contribution of each multi-pole moment to the total scattered field of the structures. It was shown that the magnetic dipole contributes most to the observed extinction [28].

Such core–shell structures have other peculiar properties. The plasmonic nanoparticles can drastically reduce the scattering response of the core [29, 30]. The mechanism involves tuning the scattering response of the shell to be 180° out of phase with that of the core, effectively cancelling out any scattering from the core itself. Simulations revealed that the structure is relatively robust to changes in geometry—such changes resulting primarily in slight shifts in the frequency of operation. By treating the shell as an effective medium, it is possible to select at which wavelength the

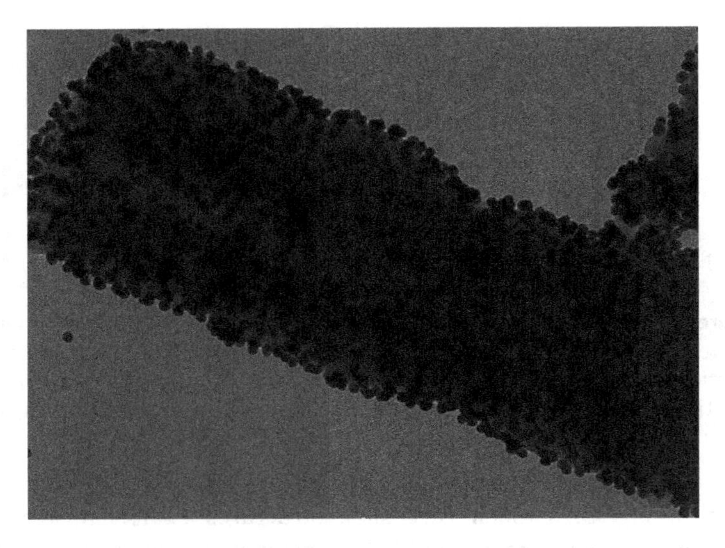

Figure 3.14 Gold nanoparticles (about 20 nm) assembled onto a mesoporous silica fiber.

cloaking device operates simply by tuning the relative sizes of the core and shell and/or the density of the nanoparticle shell.

This electrostatic approach can also be used to produce large-scale arrays of core–shell structures deposited on planar substrates. Such a step is required for the fabrication of functional optical devices and is a first step toward a hierarchical assembly [19]. New possibilities may also emerge from the assembly of plasmonic nanoparticles onto anisotropic and mesoporous substrates. An example is given in Fig. 3.14, which shows a mesoporous silica fiber decorated by gold nanoparticles.

3.6 Conclusion

The self-assembly of plasmonic nanoparticles is a simple route for the preparation of large-scale samples with promising and tunable optical properties. Although the samples are, in the general case, amorphous in nature, some short-range order can be obtained. For many applications, the amorphous nature may not be an obstacle. There are many ways to control average structural parameters in such samples. An easy and robust way is the use of electrostatic charges of surfaces, polyelectrolytes and nanoparticles to build up layered structures. Average distances between coupling nanoparticles can thus be easily obtained for different geometries. The bottom-up approach toward active plasmonic nanomaterials is still in its infancy, and we will certainly see much more exciting research and applications based on self-assembled nanoparticle systems notably in areas such as sensing, energy conversion and photocatalysis.

Acknowledgments

The authors thank Brittany Cindric for her contribution and the European Union's Seven Framework Programme (NANOGOLD project) for financial help.

References

1. Thompson, S. E., and Parthasarathy, S. (2006) Moore's law: The future of Si microelectronics. *Mater. Today* **9**, pp. 202–205.
2. Awazu, K., Fujimaki, M., Rockstuhl, C., Tominaga, J., Murakami, H., Ohki, Y., Yoshida, N., and Watanabe, T. (2008) A plasmonic photocatalyst consisting of sliver nanoparticles embedded in titanium dioxide. *J. Am. Chem. Soc.* **130**, pp. 1676–1680.
3. Boisselier, E., and Astruc, D. (2009) Gold nanoparticles in nanomedicine: Preparations, imaging, diagnostics, therapies and toxicity. *Chem. Soc. Rev.* **38**, pp. 1759–1782.
4. Acikgoz, C., Hempenius, M. A., Huskens, J., and Vancso, G. J. (2011) Polymers in conventional and alternative lithography for the fabrication of nanostructures. *Eur. Polym. J.* **47**, pp. 2033–2052.
5. Zhu, M. G., Chen, X. J., Wang, Z. L., Chen, Y., Ma, D. F., Peng, H., and Zhang, J. (2011) Structural and optical characteristics of silicon nanowires fabricated by wet chemical etching. *Chem. Phys. Lett.* **511**, pp. 106–109.
6. Reyntjens, S., and Puers, R. (2001) A review of focused ion beam applications in microsystem technology. *J. Micromech. Microeng.* **11**, pp. 287–300.

7. Zeng, H. B., Du, X. W., Singh, S. C., Kulinich, S. A., Yang, S. K., He, J. P., and Cai, W. P. (2012) Nanomaterials via laser ablation/irradiation in liquid: A review. *Adv. Funct. Mater.* **22**, pp. 1333–1353.
8. Lehn, J. M. (2002) Toward self-organization and complex matter. *Science* **295**, pp. 2400–2403.
9. Shalaev, V. M., Cai, W. S., Chettiar, U. K., Yuan, H. K., Sarychev, A. K., Drachev, V. P., and Kildishev, A. V. (2005) Negative index of refraction in optical metamaterials. *Opt. Lett.* **30**, pp. 3356–3358.
10. Willets, K. A., and Van Duyne, R. P. (2007) Localized surface plasmon resonance spectroscopy and sensing. *Ann. Rev. Phys. Chem.* **58**, pp. 267–297.
11. Hegmann, T., Qi, H., and Marx, V. M. (2007) Nanoparticles in liquid crystals: Synthesis, self-assembly, defect formation and potential applications. *J. Inorg. Organomet. Polym. Mater.* **17**, pp. 483–508.
12. Cseh, L., and Mehl, G. H. (2006) The design and investigation of room temperature thermotropic nematic gold nanoparticles. *J. Am. Chem. Soc.* **128**, pp. 13376–13377.
13. Frein, S., Boudon, J., Vonlanthen, M., Scharf, T., Barbera, J., Suss-Fink, G., Burgi, T., and Deschenaux, R. (2008) Liquid-crystalline thiol- and disulfide-based dendrimers for the functionalization of gold nanoparticles. *Helv. Chim. Acta* **91**, pp. 2321–2337.
14. Dintinger, J., Tang, B. J., Zeng, X. B., Liu, F., Kienzler, T., Mehl, G. H., Ungar, G., Rockstuhl, C., and Scharf, T. (2013) A self-organized anisotropic liquid-crystal plasmonic metamaterial. *Adv. Mater.* **25**, pp. 1999–2004.
15. Alivisatos, A. P., Johnsson, K. P., Peng, X. G., Wilson, T. E., Loweth, C. J., Bruchez, M. P., and Schultz, P. G. (1996) Organization of 'nanocrystal molecules' using DNA. *Nature* **382**, pp. 609–611.
16. Chen, C. L., Zhang, P. J., and Rosi, N. L. (2008) A new peptide-based method for the design and synthesis of nanoparticle superstructures: Construction of highly ordered gold nanoparticle double helices. *J. Am. Chem. Soc.* **130**, pp. 13555–13557.
17. Lin, Y., Boker, A., He, J. B., Sill, K., Xiang, H. Q., Abetz, C., Li, X. F., Wang, J., Emrick, T., Long, S., Wang, Q., Balazs, A., and Russell, T. P. (2005) Self-directed self-assembly of nanoparticle/copolymer mixtures. *Nature* **434**, pp. 55–59.
18. Nie, Z. H., Fava, D., Rubinstein, M., and Kumacheva, E. (2008) "Supramolecular" assembly of gold nanorods end-terminated with polymer "Pom-Poms": Effect of pom-pom structure on the association modes. *J. Am. Chem. Soc.* **130**, pp. 3683–3689.

19. Cunningham, A. (2012) *Bottom-up Organisation of Metallic Nanoparticles for Metamaterials Applications*, PhD thesis, University of Geneva.
20. Cunningham, A., Muhlig, S., Rockstuhl, C., and Burgi, T. (2011) Coupling of plasmon resonances in tunable layered arrays of gold nanoparticles. *J. Phys. Chem. C* **115**, pp. 8955–8960.
21. Decher, G. (1997) Fuzzy nanoassemblies: Toward layered polymeric multicomposites. *Science* **277**, pp. 1232–1237.
22. Rockstuhl, C., and Scharf, T. (eds) (2013) *Amorphous Nanophotonics*, Springer Verlag Berlin Heidelberg.
23. Jain, P. K., Huang, W. Y., and El-Sayed, M. A. (2007) On the universal scaling behavior of the distance decay of plasmon coupling in metal nanoparticle pairs: A plasmon ruler equation. *Nano Lett.* **7**, pp. 2080–2088.
24. Prodan, E., Radloff, C., Halas, N. J., and Nordlander, P. (2003) A hybridization model for the plasmon response of complex nanostructures. *Science* **302**, pp. 419–422.
25. Cunningham, A., Muhlig, S., Rockstuhl, C., and Burgi, T. (2012) Exciting bright and dark eigenmodes in strongly coupled asymmetric metallic nanoparticle arrays. *J. Phys. Chem. C* **116**, pp. 17746–17752.
26. Sheikholeslami, S., Jun, Y. W., Jain, P. K., and Alivisatos, A. P. (2010) Coupling of optical resonances in a compositionally asymmetric plasmonic nanoparticle dimer. *Nano Lett.* **10**, pp. 2655–2660.
27. Simovski, C. R., and Tretyakov, S. A. (2009) Model of isotropic resonant magnetism in the visible range based on core-shell clusters. *Phys. Rev. B* **79**, 045111, DOI: 10.1103/PhysRevB.79.045111.
28. Muhlig, S., Cunningham, A., Scheeler, S., Pacholski, C., Burgi, T., Rockstuhl, C., and Lederer, F. (2011) Self-assembled plasmonic core-shell clusters with an isotropic magnetic dipole response in the visible range. *ACS Nano* **5**, pp. 6586–6592.
29. Muhlig, S., Farhat, M., Rockstuhl, C., and Lederer, F. (2011) Cloaking dielectric spherical objects by a shell of metallic nanoparticles. *Phys. Rev. B* **83**, 195116, DOI: 10.1103/PhysRevB.83.195116.
30. Muhlig, S., Cunningham, A., Dintinger, J., Farhat, M., Bin Hasan, S., Scharf, T., Burgi, T., Lederer, F., and Rockstuhl, C. (2013) A self-assembled three-dimensional cloak in the visible. *Sci. Rep.* **3**, 2328, DOI: 10.1038/srep02328.

Chapter 4

Optimizing Surfactant on Nanoparticles

Hari M. Atkuri,[a,b] Ke Zhang,[b] and John L. West[b]

[a]*Cardinal IG Tech Center, 7201 W Lake St, Minneapolis, MN 55426, USA*
[b]*Liquid Crystal Institute Kent State University, Kent, Ohio 44242, USA*
hatkuri@cardinalcorp.com

In this chapter, we focus on how to use various techniques to coat nanoparticles with a surfactant and thus finally make a stable suspension. To use a particle suspension in a liquid crystal (LC) host, the particle suspension must have a minimum or, better, no presence of free surfactant (*free state*), while maximum amount of surfactant is bound to the particle surface (*bound state*). The optimization of the surfactant used on the nanoparticles is the subject of this chapter.

4.1 Introduction

When we use surfactant in a mill process to coat the particle surface, often the surfactant could assume two different states. These are the bound (to particle) state and the unbound (free) state. We may call the unbound state as single/dimer or free state. The unbound surfactant in a suspension should be minimized to clearly see the

Active Plasmonic Nanomaterials
Edited by Luciano De Sio

ISBN 978-981-4613-00-2 (Hardcover), 978-981-4613-01-9 (eBook)
www.panstanford.com

particle effect in an LC host, without uncertainties introduced by the presence of free surfactant in the mixture. Here we discuss the process of making surfactant-coated nanoparticles with an optimal amount of bound and minimal amount of free surfactant.

To facilitate this goal, we measure bound and unbound surfactant states present in a ferroelectric colloidal suspension. We find Fourier transform infrared spectroscopy (FTIR) to be a powerful technique for differentiating the two states of surfactant present in a colloidal suspension. FTIR is a technique used to obtain an infrared spectrum of absorption or emission of a solid, liquid, or gas and collect spectral data in a wide spectral range. In addition, we also report on how to measure the amount of surfactant that is in an unbound state in the suspension. Since it is difficult to measure the real amount of surfactant present in a particular bound/unbound state, we measure their relative ratio for the optimization process.

These measurements are critical for the production of enhanced LC-particle composites. The physical properties of these composites could be intriguing and achieved using various particles that are produced in completely different ways [1–4]. Such enhancement in LC properties may not be strictly predictable. We believe that one of the main reasons for such irreproducibility could be the difficulty in controlling the amount of free/unbound surfactant present in the colloidal suspension and thus in the LC-particle composite.

4.2 Role of Surfactant

To enhance the properties of LC-particle composites, we need colloidal ferroelectric particles to be coated with surfactant without any unbound free surfactant present. In other words, we need to produce coated nanoparticles without excess surfactant molecules, which will be introduced into the host LC medium.

In our studies we noted that adding surfactant to an LC host could negatively impact its physical properties, particularly lowering T_{NI}, $\Delta\varepsilon$, ΔS, etc. For example, adding 0.2 wt% of oleic acid to 5CB can decrease its clearing point by as much as 0.2°C. Figure 4.1 shows how the clearing temperature T_{NI} of unmodified 5CB can

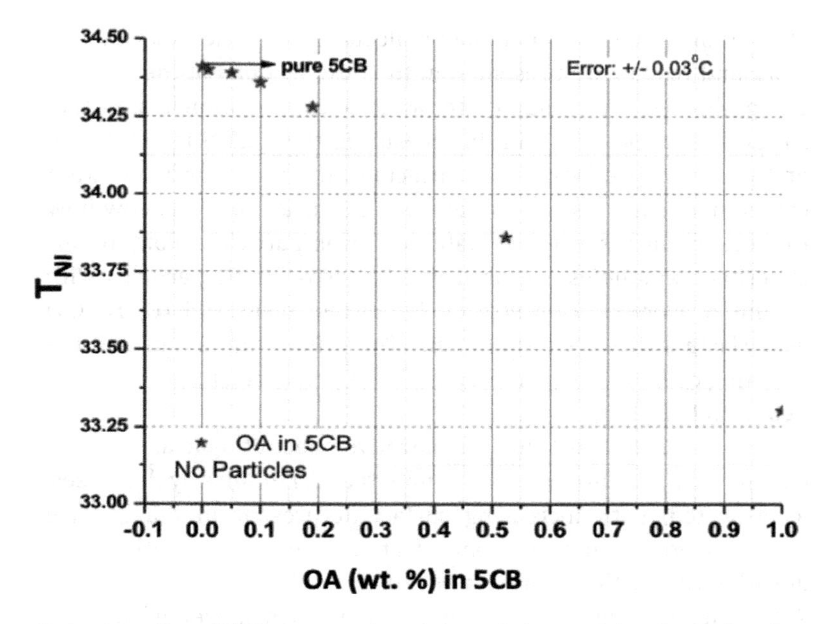

Figure 4.1 T_{NI} of 5CB varies as the amount of oleic acid (surfactant) in the host LC changes.

decrease by adding up to 1 wt% of oleic acid. As a result, the phase transition temperature, dielectric, and optic properties are significantly reduced. As we show later, we note that the surfactant in an unbound state offsets ferroelectric particle's effect on the host LC.

4.2.1 *States of Surfactant*

To control the amount of surfactant for stable suspensions and to cover only the surface area of nanoparticle, we noted that oleic acid could assume one of the three states (dimer, monomer, or bound) that we present later. These states correspond to oleic acid molecules that are free (single or dimer molecules) or bound states of surfactant (surfactant molecules bound to nanoparticle). We refer to single or dimer molecules states as *free states*.

Using this critical information of surfactant molecules that were present in nanoparticle dispersions, we used FTIR to measure the absorption of the free and bound molecule states. The experimental work and results are presented in the next sections of the chapter. Using the FTIR technique, we wanted to test which states of surfactant would be useful in measuring the quantities of bound and unbound (free) states of surfactant. In other words, we want coated nanoparticles without any excess surfactant and show how to measure each surfactant state in a given particle colloid. In our experimental studies, we used $BaTiO_3$ ferroelectric particles with an initial average grain size of 3 μm, and oleic acid ($C_{18}H_{34}O_2$) as surfactant dissolved in heptane. The primary materials and the resulting colloidal suspensions used and prepared are presented in details in Ref. [1, 3].

Figure 4.1 demonstrates the negative effect of oleic acid on the clearing temperature of 5CB: We note that ΔT_{NI} linearly decreases, as expected, with increasing surfactant present in the LC. We can correlate the percent concentration of oleic acid with the IR absorbance. As the IR absorption of free oleic acid (see Section 4.2.2) increases with the amount of oleic acid dispersed in 5CB, there is a linear correlation between oleic acid absorption and 5CB absorption.

4.2.2 *Surfactant IR Absorption Band*

The IR absorption bond C=O in the 1500–1750 cm^{-1} range is of particular interest for our current analysis. The absorption peaks of the C=O bond in a particle suspension can show three distinct states; they are single, dimer, and bound states. The C=O bond peaks at ~1700 cm^{-1}, ~1750 cm^{-1}, and ~1500–1600 cm^{-1} range represent the dimer form, the monomer form, and the complex conjugate forms, respectively. We show these surfactant states in Fig. 4.2. Consider that single/dimer states are interchangeable and can exist in the unbound state or the free state. Further, we compare the unbound state to the bound state obtained from FTIR data.

With milling of the particle-oleic acid colloid, as particle size simply decreased, the total available surface area increased. This allows more and more oleic acid molecules to bond with $BaTiO_3$. Then the A_{1700}/A_{1500} ratio (ratio of free surfactant absorption at 1700 cm^{-1} to bound state absorption at 1500 cm^{-1}) declines

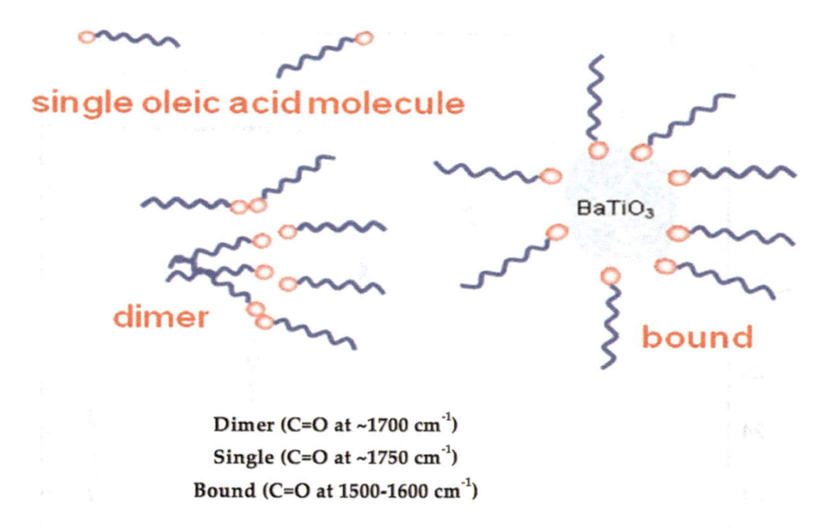

Dimer (C=O at ~1700 cm^{-1})
Single (C=O at ~1750 cm^{-1})
Bound (C=O at 1500-1600 cm^{-1})

Figure 4.2 Illustrating three possible states of oleic acid present in the $BaTiO_3$ particle dispersion.

with further mill. For example, Fig. 4.6 demonstrates the aspect of increasing bound surfactant state and decrease of free state (declining A_{1700}/A_{1500} ratio) with longer milling of up to 20 h and varying the amount of surfactant (oleic acid). Figure 4.6 shows the FTIR absorption for a given 10 g oleic acid used with 1 g of $BaTiO_3$ particles. At a certain point of milling, the ratio reaches a minimum and continues to stay constant. The minimum ratio occurs when the particle surface is completely covered and no free suspended oleic acid is available. Here, we see the higher the oleic acid concentration, the longer the milling time, as a result of smaller particle size required for the transition to occur. So it is clear that the more oleic acid dissolved in 5CB, the lower the clearing point. This is the necessary information we need to take advantage of our experimental set up. We present more details in the experimental section.

In the current surfactant monitoring, we focus on the duration of ball mill of $BaTiO_3$ particles, as discussed in Ref. [1], and the amount of oleic acid used (surfactant). Effects of varying oleic acid concentration in various mixtures with varying mill times are thoroughly studied.

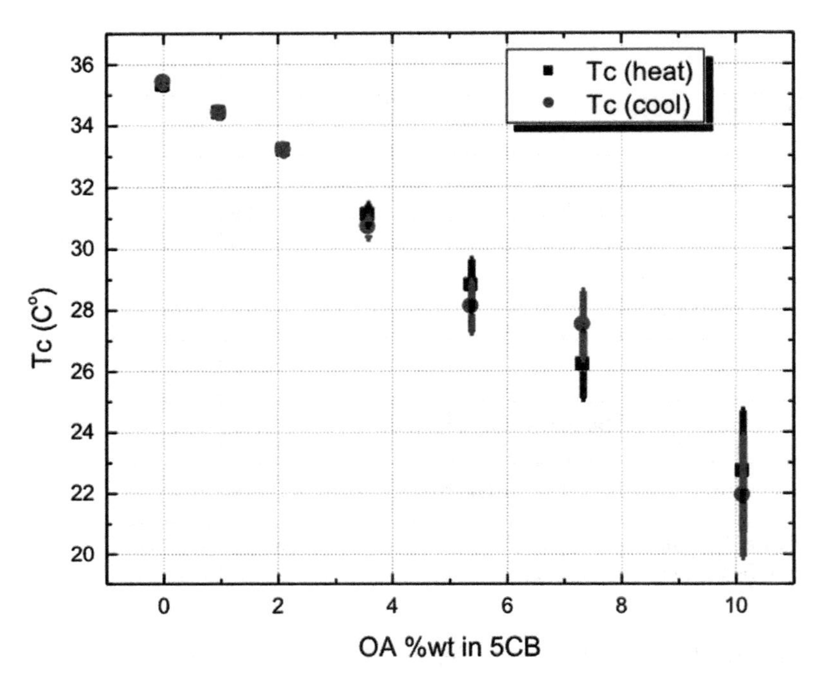

Figure 4.3 Dependence of T_{NI} to the oleic acid weight percent in the LC 5CB.

We think current FTIR technique may, however, be modified and applied to other particle dispersions (depending on the availability of a particular absorption bond) whenever a surfactant needs precise control or to understand free/bound surfactant states on a particle surface.

4.3 Experiment

In the mixture of ferroelectric $BaTiO_3$ particles and solvent (heptane), optimizing the exact amount of surfactant added is of paramount importance. We use the strong carbon–oxygen double bond (C=O) of oleic acid as a probe for surfactant monitoring analysis. The C=O bond is one of the useful absorption bonds, found in the range of 1500–1750 cm^{-1}. Its position can vary within this range depending on what sort of material component it is in.

Table 4.1 Materials used in the fabrication of ferroelectric nanoparticle dispersion

	$BaTiO_3$ (g)	Oleic acid (g)	Heptane (g)
1	1.0	1.0	10.0
2		1.5	
3		2.0	
4		3.0	
5		4.5	
6		10.0	

When oleic acid presents as a dimer, monomer, or complex conjugate with $BaTiO_3$, we obtain the C=O bond absorption peaks at ~1700 cm^{-1}, ~1750 cm^{-1}, and ~1500–1600 cm^{-1}, respectively [124–126] and measure their relative ratios.

As noted in Refs. [1–4], the particle size and, thus, the active surface area to bond with oleic acid are conveniently controlled with the help of planetary HEBM process, where we ball-mill oleic acid and $BaTiO_3$ in various ratios with heptane as solvent. In Table 4.1, we summarize the information of various oleic acid/$BaTiO_3$ ratios of particle suspensions used in our FTIR analysis.

We use FTIR as the critical experimental technique to precisely monitor the presence of free oleic acid that is freely suspended in dimer or single molecular (C=O at 1750/1700 cm^{-1}) state in an oleic-acid-coated ferroelectric particle colloid. Later we use particle colloid as ferroelectric nanoparticles in an LC host to prepare an LC-particle composite. We already presented the preparation of an LC-particle composite in Chapter 4. This is the very basic mechanism behind the enhancement of LC physical properties with ferroelectric nanoparticles dielectric-coupling with LC host possibly at a nanometer scale. So in our process, it is important to minimize or eliminate any free/unbound oleic acid in the LC-particle composite.

Figure 4.4 shows oleic acid in pure dimer form. The IR spectra of pure oleic acid (insert 1) and pure heptane (insert 2) are shown in Fig. 4.5. The three states of oleic acid in $BaTiO_3$ particle dispersion from corresponding IR absorption are shown in Fig. 4.6. No interference absorption is present in the 1500~1800 cm^{-1}

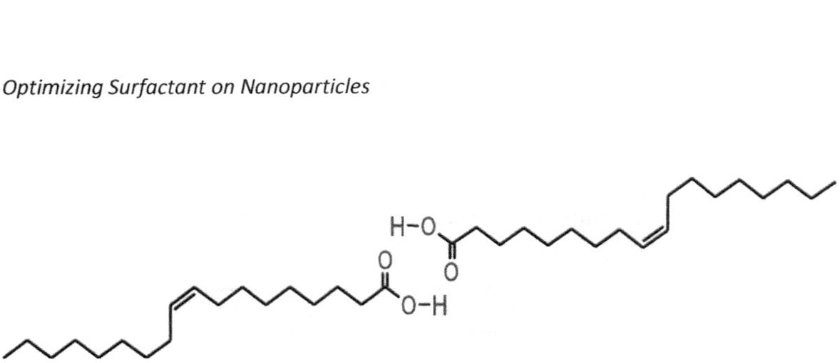

Figure 4.4 Pure oleic acid in dimer form.

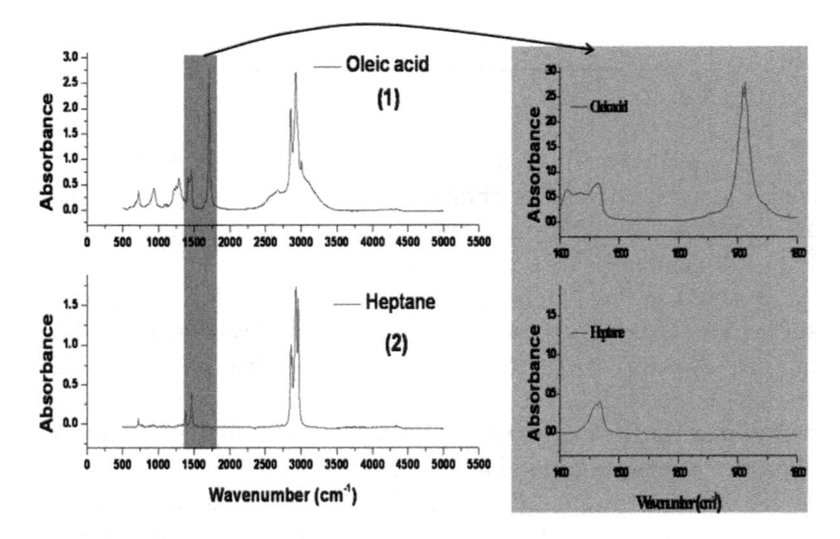

Figure 4.5 IR spectra of pure oleic acid (inset 1) and pure heptane (inset 2).

range of heptane spectra. The absorption free 1500~1800 cm^{-1} IR window particularly simplifies and accelerates the data analysis. In addition, this spectral window can also be useful to monitor the surfactant absorption onto the ferroelectric nanoparticle surface as the mill time increases.

4.4 Results and Discussion

In this section, we start with particle/surfactant bonding. During the ball-mill process, the particle size decreases and the total surface area increases. As a result, more oleic acid molecules bond

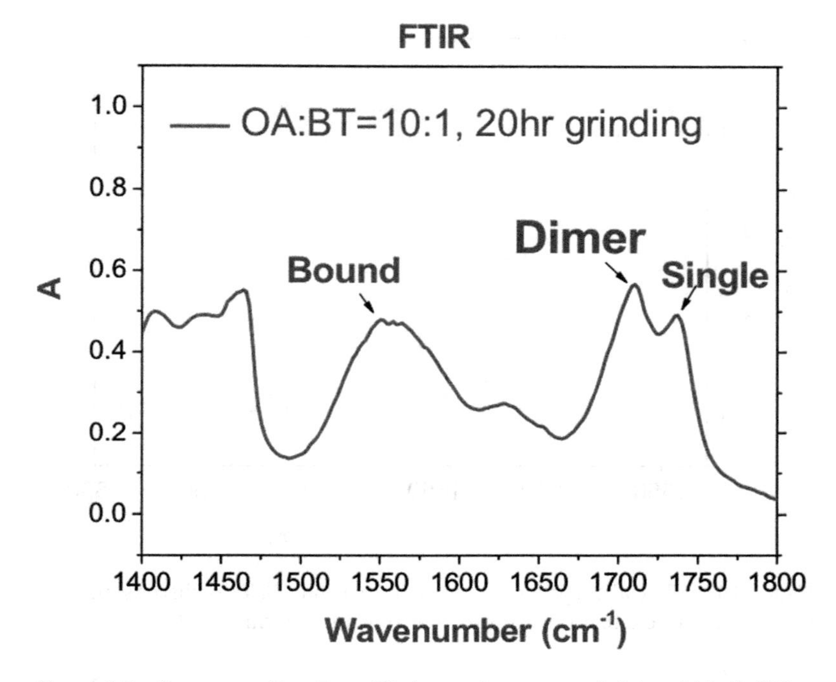

Figure 4.6 Corresponding three IR absorption states of oleic acid in $BaTiO_3$ particle dispersion.

to the particle and reflect in the 1550 cm^{-1} absorption (A_{1550}) peak increasing, while the 1700 cm^{-1} dimeric absorbance (A_{1708}) decreases, as shown in Fig. 4.6. At the same time, this absorption of heptane and oleic acid are shown in Fig. 4.1. On the other hand, Fig. 4.7 reveals the amount of oleic acid present in the "free state" from varying amounts of initial oleic acid used for the same mill time of 20 h. In a similar way, we also can have fixed oleic acid and $BaTiO_3$ ratio and observe a steadily increasing oleic acid absorption with mill time. In both events, the observations are similar, which are evident from Figs. 4.7 and 4.8 depicting varying oleic acid and $BaTiO_3$ ratios and mill times, respectively (keeping the other parameter unchanged). With various samples and mill time, we plotted various absorption peak ratios for various oleic acid and $BaTiO_3$ ratios. Figures 4.11–4.15 show various mill time vs. A_{1700}/A_{1500} or $A_{1700}/(A_{1700} + A_{1500})$. Note that A_{1700}/A_{1500}

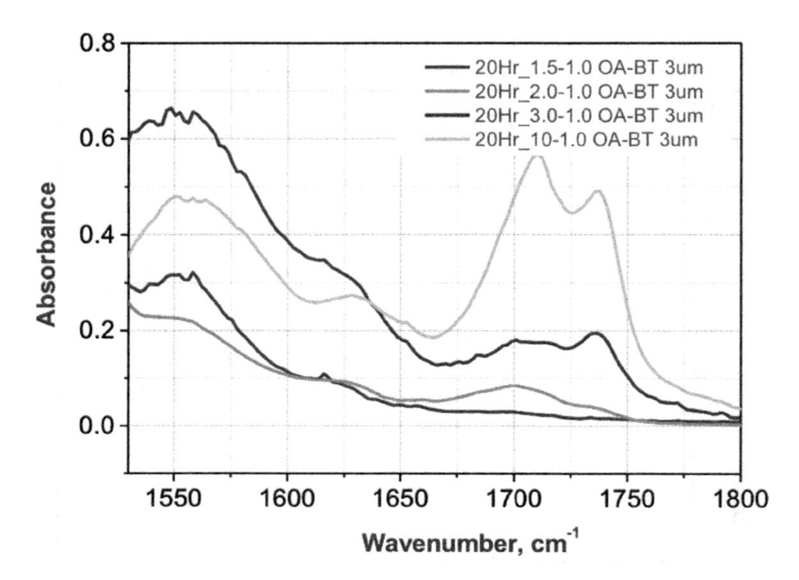

Figure 4.7 FTIR absorption spectra from various $BaTiO_3$ dispersions of different oleic acid to $BaTiO_3$ ratios at a constant mill time of 20 h.

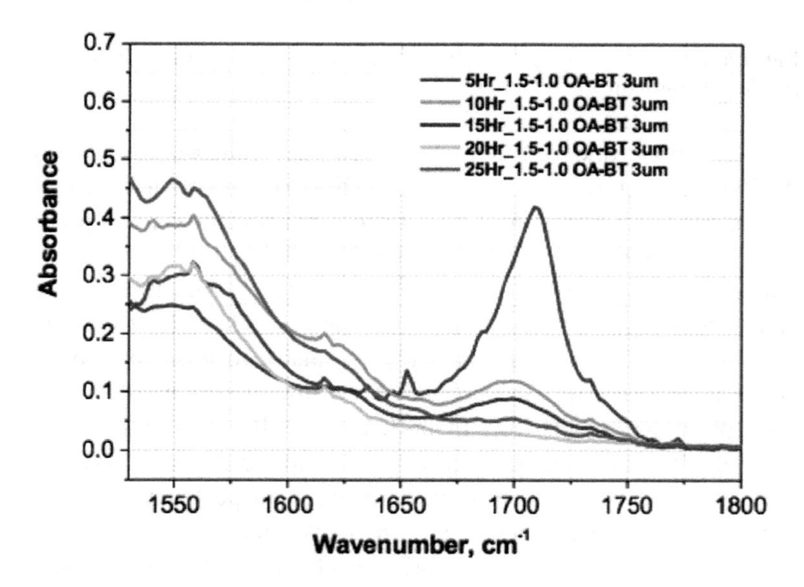

Figure 4.8 FTIR spectrum of $BaTiO_3$ dispersions milled for different durations; mill time varied from 5 h to 25 h and the oleic acid and $BaTiO_3$ ratio is kept constant.

represents the ratio of IR absorption of particle suspension in the dimer state to the bound state and $A_{1700}/(A_{1700} + A_{1500})$ represents the ratio of the bound state to the total surfactant absorption from the suspension. These plots are consistent with the observations we saw from Figs. 4.7 and 4.8. Also these plots provide us a saturation value for a given mill time and a given amount of oleic acid concentration at a fixed amount of $BaTiO_3$.

We add various particle suspensions to an LC host. Based on the relationship between the IR absorbance and observed T_{NI} values, we plot a graph that demonstrates the relationship of A_{OA}/A_{5CB} with the clearing point of 5CB in Fig. 4.9. The IR absorbance reflects the concentration of freely suspended oleic acid in 5CB. We can derive from the standard curve (line) that the oleic acid's effect on T_{NI} of 5CB. The two samples in Fig. 4.9 unambiguously demonstrate the positive effect of particles on the LC host. Therefore, the distance

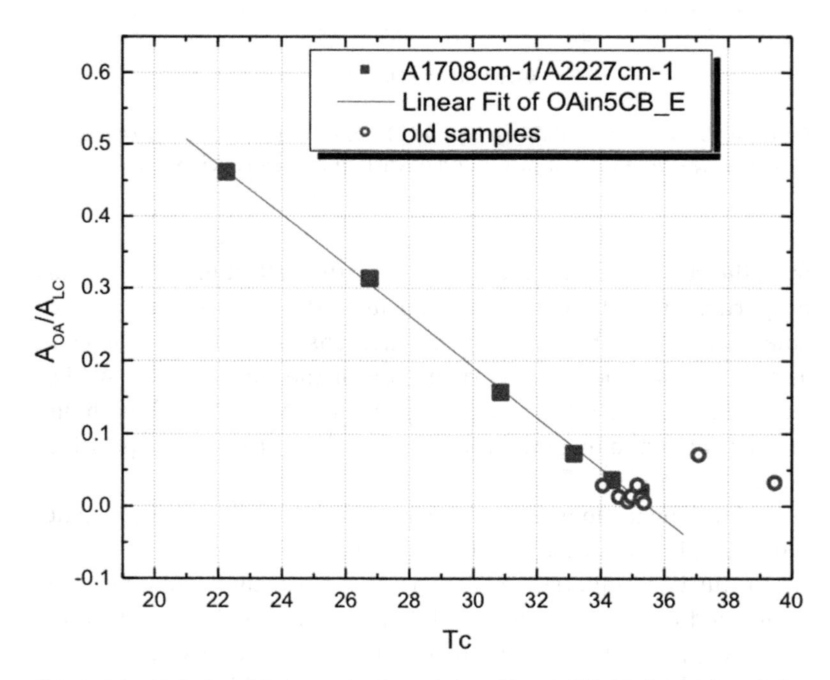

Figure 4.9 Relationship between free oleic acid and 5CB IR absorption ratio and the clearing point of 5CB/oleic acid mixture and numerous LC-particle composite mixtures.

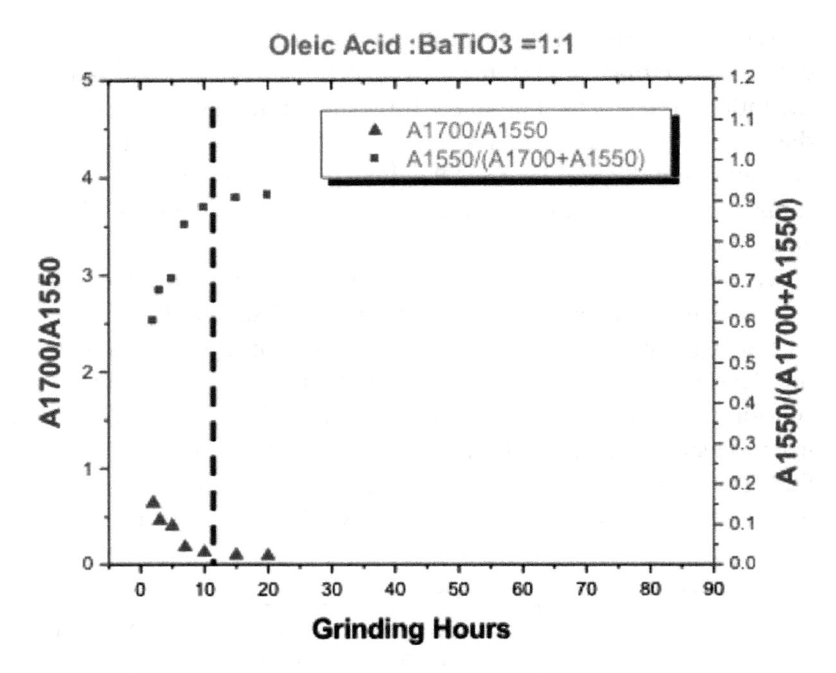

Figure 4.10 Oleic acid:$BaTiO_3$ = 1.0:1.0; mill duration vs. A_{1700}/A_{1550} or $A_{1550}/(A_{1700}+A_{1550})$, with saturation of variation in oleic acid state.

from the curve becomes the particle's distinct effect on the physical properties of the LC-particle composites. However, we may not be able to produce the samples that demonstrate similar behavior as the two samples on the right side of the linear curve in Fig. 4.8 (with change in clearing temperature, ΔT_{NI} >2°C) without further understanding the various factors that can cause the positive increase in T_{NI}.

A few suggestions and the challenges to achieve consistent reproducibility or preparation are discussed elsewhere.

To further understand the particle-surfactant bonding, we derive and analyze a series of plots showing the absorption ratio A_{1700}/A_{1550} vs. mill time and oleic acid concentrations. It shows a clear distribution of oleic acid, as shown in Fig. 4.10. For example, at oleic acid:$BaTiO_3$ = 1.0:1.0, the A_{1700}/A_{1550} ratio decreases with mill time reaching a minimum after ~15 h of milling.

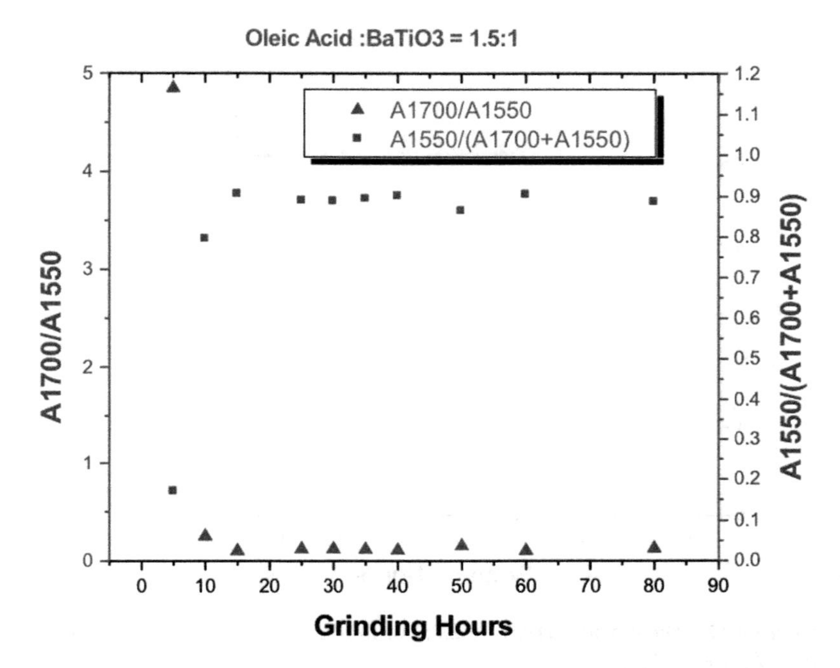

Figure 4.11 Oleic acid:$BaTiO_3$ = 1.5:1.0; mill duration vs. A_{1700}/A_{1550} or $A_{1550}/(A_{1700}+A_{1550})$.

Figure 4.10 (oleic acid:$BaTiO_3$ = 1.0:1.0) indicates that all the oleic acid molecules are bound to the $BaTiO_3$ after 15 h. Similarly, we monitor the ratio of $A_{1550}/(A_{1708}+A_{1550})$, which increases with mill time and saturates after 15 h. Figures 4.11–4.14 provide the plots for mill duration vs. A_{1700}/A_{1500} and $A_{1700}/(A_{1700}+A_{1500})$ for oleic acid and $BaTiO_3$ ratios 1.5:1.0, 2.0:1.0, 3.0:1.0, and 4.5:1.0, respectively. Further, we plot these observations in Fig. 4.15 and Fig. 4.16, which show oleic acid distribution from Figs. 4.10–4.14.

These results indicate the individual saturation time for each oleic acid concentration. Similarly, considering threshold saturation time for other ratios of mixtures, we obtained a predictable and linear relation between oleic acid concentration and threshold mill time (in hours). We plotted and presented the linear relation in Fig. 4.17. As expected, the saturation increases with oleic acid concentration since more time is needed to reduce particle size to accommodate all the oleic acid molecules. It is important that the

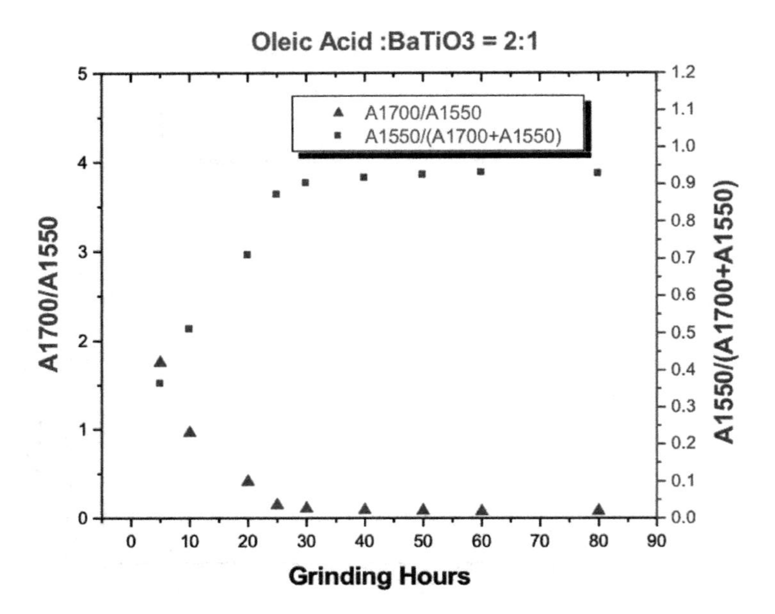

Figure 4.12 Oleic acid:$BaTiO_3$ = 2.0:1.0; mill duration vs. A_{1700}/A_{1550} or $A_{1550}/(A_{1700} + A_{1550})$.

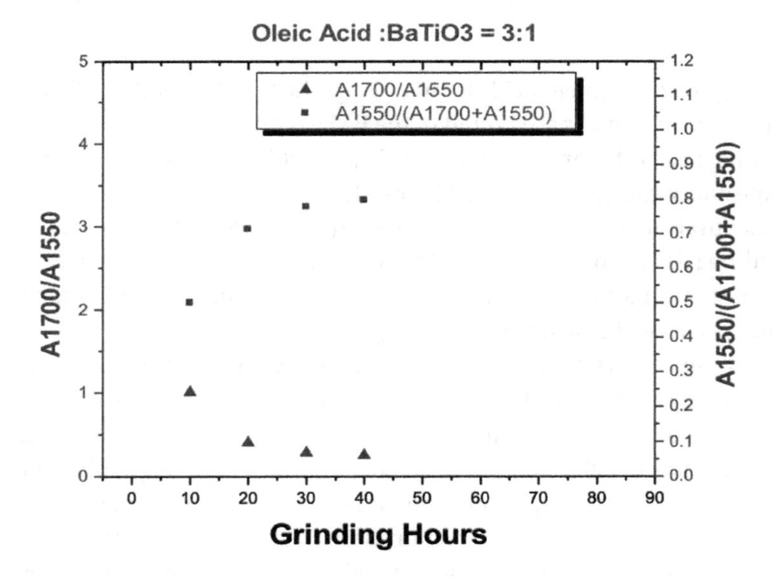

Figure 4.13 Oleic acid:$BaTiO_3$ = 3.0:1.0; mill duration vs. A_{1700}/A_{1550} or $A_{1550}/(A_{1700} + A_{1550})$.

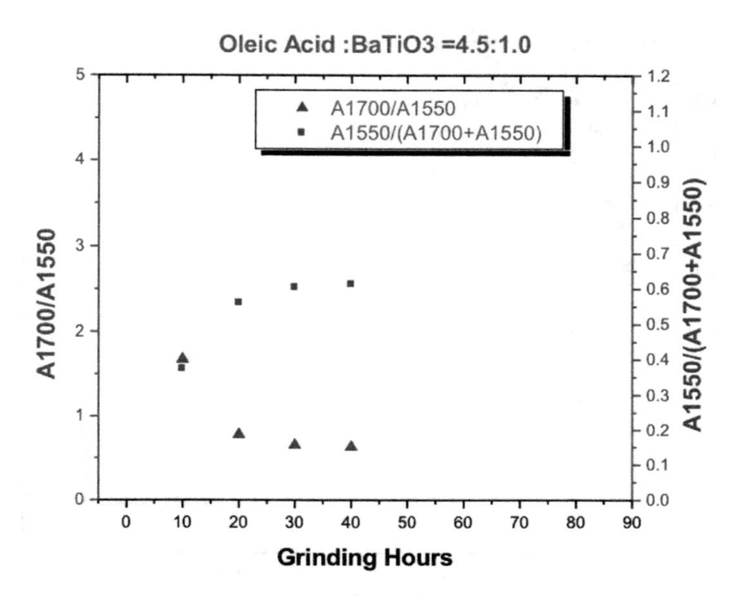

Figure 4.14 Oleic acid:$BaTiO_3$ = 4.5:1.0; mill duration vs. A_{1700}/A_{1550} or $A_{1550}/(A_{1700} + A_{1550})$.

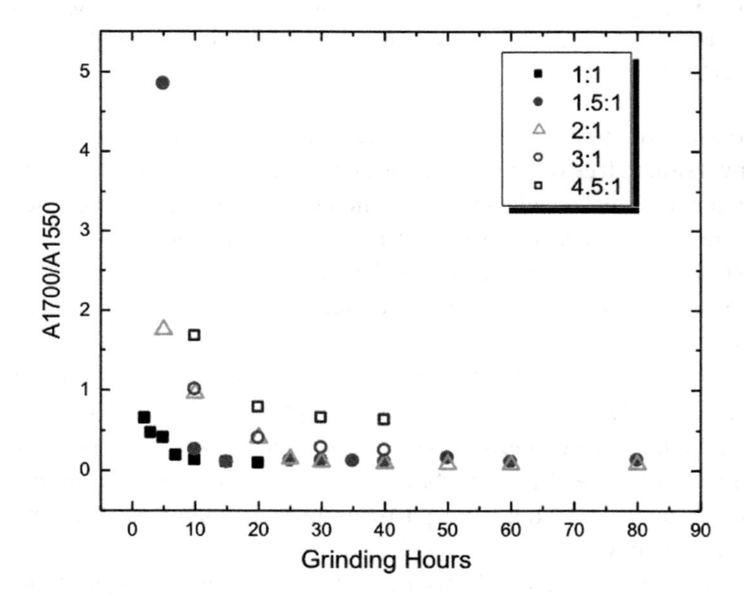

Figure 4.15 Dependence of oleic acid distribution with mill hours and oleic acid to $BaTiO_3$ ratio, absorbance ratio of free vs. bond oleic acid, A_{1700}/A_{1550}.

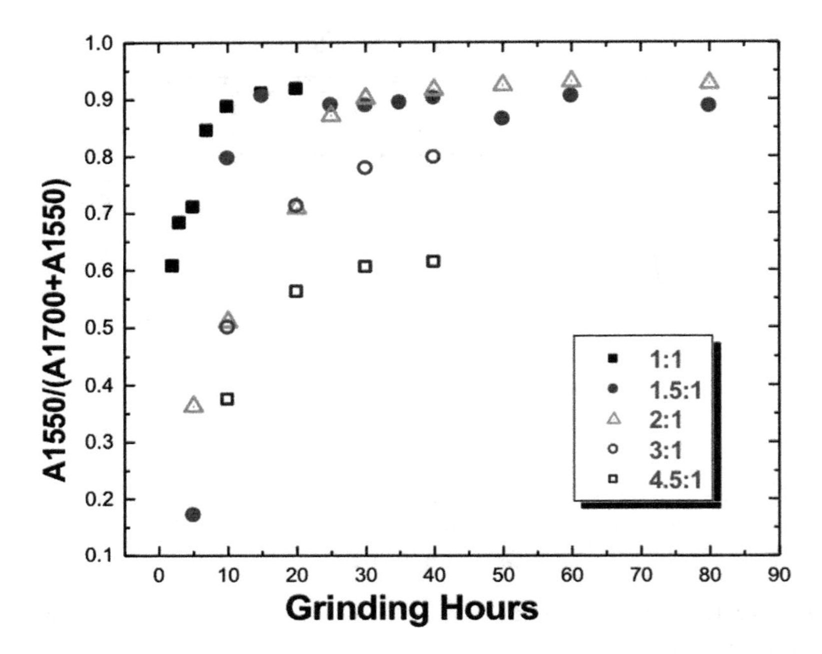

Figure 4.16 Dependence of oleic acid distribution with mill time for various oleic acid to $BaTiO_3$ ratios, absorbance ratio of bond vs. total oleic acid $A_{1550}/(A_{1700}+A_{1550})$.

particles are milled for a minimum amount of time, below which the conversion of free oleic acid into bound state would not be complete and any free oleic acid suppresses the LC order as we noted earlier. It helps to ensure no dimeric or free oleic acid is left behind in the LC host to deteriorate various physical properties of LC host.

Since it is difficult to remove the excess amount of unbound oleic acid without introducing impurities and deteriorating the LC host quality and physical properties, it is important to use the linear relation of Fig. 4.17 for producing and optimizing various particle colloids in advance before using it as particle dopant in an LC host. At the same time, it is important to note that longer mill produces smaller particles, where the particles may no longer retain their ferroelectricity.

Preserving the ferroelectricity of the particles throughout the process is important. So one of the important parameters we use to monitor the quality of particle suspension before dispensing it

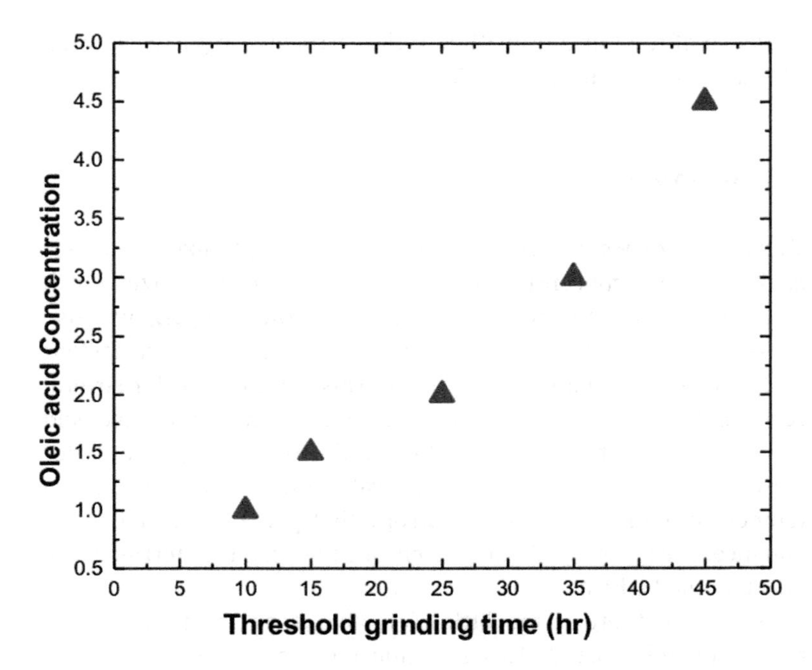

Figure 4.17 Final relation between the threshold mill time and oleic acid bonding to the $BaTiO_3$ nanoparticles surface.

into the LC host is the particle's ferroelectricity. It is widely agreed that below a critical size no ferroelectricity remains even in a single domain crystal. On the other hand, we found the ferroelectricity can be quickly destroyed by high-energy ball milling if no surfactant is present. Next, if not ferroelectric, these small particles would simply form aggregates and precipitate because of attractive van der Waal's forces. In the case of ferroelectric particles, in addition there will be additional electrostatic forces. Therefore, to prevent a gradual aggregation and particle sedimentation, the particles must be coated with a monomolecular film of a dispersing agent that acts as an elastic cushion.

For the aforementioned reasons, we are convinced it is best to optimize both the milling time and surfactant concentration to achieve small particle size, narrow size distribution and to maintain the ferroelectricity during milling. Further, we can observe the presence of the Curie transition of the ferroelectric particles using DSC as presented in Refs. [1, 3].

4.5 Summary

To summarize, we reported the importance of optimizing the oleic acid surfactant concentration for a particular particle size, which is directly connected to the mill time. In this chapter, we also presented FTIR as a method to monitor and optimize the surfactant on ferroelectric nanoparticle colloids. This is important for several reasons. Frist, if there is too much oleic acid, then free oleic acid is dissolved in the LC host, which reduces the T_{NI}, the order parameter, and other important physical properties such as clearing temperature and dielectric anisotropy that particles are meant to enhance. Free oleic acid can screen out the resulting ferroelectric effect on the LC host.

Second, if there is too little oleic acid, then the particles are not completely coated. This situation creates two problems: The surfactant is insufficient to prevent particle aggregation and, at the same time, achieve the needed ferroelectricity during grinding.

If there is just the right amount of oleic acid, then using our experimental results we can focus on milling the particles to obtain a specific particle size. It is, therefore, important to use just the right amount of oleic acid as well.

The key aspect of FTIR measurements was basically the difference in the carbonyl stretching frequency of free and bound oleic acid. The C=O stretching band from dimeric, monomeric, and conjugated oleic acid serves as a powerful probe to quickly and effectively identify and quantify the status of oleic acid in a ferroelectric suspension. If the particle colloid has any free oleic acid, the colloid would not be able to produce enhanced LC-particle composites. So we want to get the optimum point where the free oleic acid just disappears and completely converts to particle-bound oleic acid. We would not be able to completely predict and tell from a single measurement if too little or too much of oleic acid has been added. However, a series of FTIR measurements that are based on varying the amount of oleic acid and mill duration are critically necessary and provide a clear trend to reach an optimum point where the plateau occurs. It confirms where the oleic acid conversion is just achieved and any further mill process would be unnecessary. Unless the colloid mixture is free of unbound (free) oleic acid, we may observe no benefits from the ferroelectric nanoparticles. However, we can use the colloidal particles that are just covered with surfactant to make interesting LC-particle composites possible.

References

1. Atkuri, H., G. Cook, D.R. Evans, C.I. Cheon, A. Glushchenko, V. Reshetnyak, Y. Reznikov, J. West, and K. Zhang, Preparation of ferroelectric nanoparticles for their use in liquid crystalline colloids. *Journal of Optics A: Pure and Applied Optics*, 2009. **11**: pp. 024006.
2. Kurochkin, O., O. Buchnev, A. Iljin, S.K. Park, S.B. Kwon, O. Grabar, and R. Yu, A colloid of ferroelectric nanoparticles in a cholesteric liquid crystal. *Journal of Optics A: Pure and Applied Optics*, 2009. **11**(2): pp. 024003.
3. Atkuri, H.M., K. Zhang, and J.L. West, Fabrication of paraelectric nanocolloidal liquid crystals. *Molecular Crystals and Liquid Crystals*, 2009. **508**: pp. 183–190.
4. Lopatina, L.M., and J.V. Selinger, Theory of ferroelectric nanoparticles in nematic liquid crystals. *Physical Review Letters*, 2009. **102**(19): pp. 197802–197802.
5. Zhang, L., R. He, and H.C. Gu, Oleic acid coating on the monodisperse magnetite nanoparticles. *Applied Surface Science*, 2006. **253**(5): pp. 2611–2617.
6. Lee, D.H., and R.A. Condrate, FTIR spectral characterization of thin film coatings of oleic acid on glasses: I. Coatings on glasses from ethyl alcohol. *Journal of Materials Science*, 1999. **34**(1): pp. 139–146.
7. Lee, D.H., R.A. Condrate, and W.C. Lacourse, FTIR spectral characterization of thin film coatings of oleic acid on glasses - Part II - Coatings on glass from different media such as water, alcohol, benzene and air. *Journal of Materials Science*, 2000. **35**(19): pp. 4961–4970.

Chapter 5

A Brief Overview on Synthesis and Characterization of Nanomaterials

Ariel Meneses-Franco,[a] Eduardo Soto-Bustamante,[a] and Marcelo Kogan-Bocian[b]

[a] *Department of Organic and Physical Chemistry, University of Chile, Sergio Livingstone P. 1007, Independencia, Santiago, RM 8380492, Chile*
[b] *Department of Pharmacological and Toxicological Chemistry, Advanced Center for Chronic Diseases (ACCSDiS), University of Chile, Sergio Livingstone P. 1007, Independencia, Santiago, RM 8380492, Chile*
esoto@ciq.uchile.cl

Nanoscience began in the last century when characterization methods for observing the structure of nanomaterials were developed. One of the basic requirements in nanotechnology is the manipulation and characterization of materials at the nanoscale. While this is a fact, it is also true that nature itself, as a mechanism of defense, has continuously produced nanomaterials even before nanochemistry was known as a branch of chemistry. This chapter tries to provide some insights for the study of nanochemistry. Those interested in a better understanding of this thematics can read the reviews mentioned in the listed literature about the synthesis and characterization of nanomaterials.

Active Plasmonic Nanomaterials
Edited by Luciano De Sio

ISBN 978-981-4613-00-2 (Hardcover), 978-981-4613-01-9 (eBook)
www.panstanford.com

5.1 Introduction

In the last decades, research aimed at obtaining structures at nanometric scale has allowed the manipulation and exploitation of the unique properties of nanomaterials, and that has driven their use in such diverse areas as biomedicine, pharmaceutics, electronics, and in the manufacture of more and more complex technologies [1–6]. Studying and developing this branch of science have always been a challenge for many authors, who have been looking for better classification of obtention procedures as well as characterization methods. Among the suggested classification systems is the possibility to frame nanomaterials with respect to their properties. This is somewhat complex because a material may have two or more major features, or materials of various kinds may have comparable properties among them. On the other hand, if the methods of obtaining these materials are considered, we can visualize three major routes from which different techniques may be assessed: physical, chemical, and biotechnological [7, 8].

The physical routes are characterized by using techniques such as sublimation, laser ablation, electrodeposition, and even grinding. Typically, highly pure materials are obtained in some cases with highly homogeneous surfaces. However, both the procedures and the equipment involved may be very expensive from the point of view of technology and energy [9–12].

On the other hand, although chemical methods involving precipitation, recrystallization, solvothermolysis, or reactions such as complexations and oxidation-reduction have a great advantage in terms of access to materials and reagents, to achieve the necessary purity of materials obtained in some cases requires a strict control of the synthesis conditions involved, with consequent investment in time and resources to find them. Much of the research in the area of nanoscience focuses on obtaining new materials using well-known chemical synthetic routes. This has led to the development of an area known as nanochemistry, which in simple terms may be defined as the collection, handling, and study of the physicochemical properties of nanomaterials [13–15].

Finally, in recent years an increase in the development of production process for nanomaterials involving not just isolated biological molecules but also more complex biological agents such as bacteria, fungi, and plant cells has been reported. The synthesis in such cases is governed by complex metabolic pathways and biochemical reactions, which in some ways possess a limited control over the product obtained and the production efficiency. However, these methods resolve one of the biggest drawbacks of nanotechnology, especially at a biomedical level, that is, the biocompatibility of nanomaterials [16, 17].

5.2 Major Methods of Chemical Synthesis of Nanoparticles

5.2.1 *Precipitation*

Under precipitation, we can group all those methods based on controlling the kinetics of formation of insoluble compounds between two or more components. Precipitation methods are probably the cheapest and easiest techniques to obtain nanoparticles. Depending on the synthesis conditions, nanoparticles may exhibit uniform size and shape. Among the materials prepared on a large scale, we can include quantum dots, spions, or semiconductor oxides [18].

Typical examples for precipitation methods are those carried out with semiconductors. Here we found transition metal oxides, generally bi- or tetravalent, such as ZnO, SnO_2, ZrO_2, or TiO_2, which are widely used in photovoltaic systems. Generally, precipitates of oxides are produced by the hydrolysis of organometallic complexes in water or alkaline solutions. The particles can be obtained in very small diameter (1–5 nm), which is of great advantage for technological applications. However, newly formed precipitates often aggregate. To solve this problem, molecules that modify surface isoelectric points, such as weak organic acids or amphiphilic surfactants, which weaken the strength of particle–particle interactions, have been included in solutions. These molecules are generally oleic acid or polymers such as polyethylene glycol [19–22].

Spions or Uspions (small or ultra-small super-paramagnetic iron oxide nanoparticles) are obtained by means of alkaline precipitation of nano-particulated trivalent iron oxide in combination with other divalent transition metals. These are known as metal ferrites of general formula MFe_2O_4, where M can be a divalent cation such as Cu, Mn, Co, Ni, Fe, Zn, or even Ba. In this case, the co-precipitation of both salts is done in a strongly alkaline medium. In some cases, it is necessary to control the temperature of the process and also the presence of oxygen in the reaction, since these variables may control the structure of the material to be obtained. Usually nanomaterials obtained in this way are spinels, which allow the coupling of magnetic dipole moments between cations. These can be obtained in diameters below 10 nm. This size allows the appearance of super-paramagnetism, which is the reason why these are called Spions or Uspions [23–28]. As in the previous case, the addition of surfactants or polymers may prevent permanent aggregation and allows their biocompatibilization. Some commercial products already exist and are used in biomedicine as a contrast medium, formed by Spions obtained from precipitation methods and compatibilized with polyethylene glycol or other polymer such as chitosan. A few examples of commercial products are Feridex®, Resovist®, and Clariscan®, among others [29–31].

Another interesting case of synthesis by precipitation is that of production of the so-called quantum dots, which are semiconductors whose radiative decay occurs at visible wavelengths. Chemically, they are sulfides, selenides, or tellurides of transition metals. Early attempts to produce these in large scale involved adding sulfide solutions or even bubbling H_2S through solutions of salts such as chlorides or nitrates of Cd, Mo, Mn, or Zn. However, controlling precipitation was rather difficult and reproducibility was questionable. Then attempts were made to control kinetics by cation complexation with EDTA or other chelators. Currently, the synthetic routes include complexation with glutathione or other thiols and their thermal decomposition, which triggers a more controlled precipitation of clusters [32–35].

5.2.2 *Solvothermolysis and Recrystallization*

In recent years, hydro- and solvothermal techniques have become the most widely used tools for preparing nanoscale materials. These techniques allow obtaining highly monodispersed materials and with quite homogeneous structures, as well as generating processes for the preparation of hybrid and nanocomposites more or less reproducible. Curiously, the term comes from the geological concept of the action of water at high pressure and temperature, producing crystals and minerals naturally. Chemically, we can define a hydrothermal reaction process as a heterogeneous phase reaction where an aqueous solvent, under supercritical conditions, redissolved a suspension of particles, which will recrystallize in an organized manner with a well-defined structure. Similarly, we can apply the method to nonaqueous solvents, such as alcohols, alkalis, etc. All these solvents should be associated with solvothermal methods, which may also be understood as supercritical fluid technologies [36–37].

A number of materials have been prepared at nanoscale using these methods: metals, metal oxides, alloys, semiconductors, silicates, nanocrystals, ceramic nanotubes, among others. As is known, the properties of these materials in both electronics and catalysis depend largely on the materials' structural properties, shape, and size. From this point of view, solvothermal methods allow generating metal nanostructures with specific characteristics quite reproducible by controlling the reaction conditions. A typical example is the synthesis of silver nano-dendrites, magnetic nanoparticles of Co or Ni, either as nanorods or as nanoribbons. These methods also allow efficient generation of metal-coated nanocomposites [38–40].

Another interesting approach of solvothermal methods is the synthesis and structural modification of metal oxides such as TiO_2, ZnO, CeO_2, ZrO_2, CuO, Al_2O_3, Dy_2O_3, In_2O_3, Co_3O_4, NiO, etc. While these nanomaterials can generally be obtained by precipitation, as described above, some of its applications, especially in biomedicine, require low dispersion size and high redispersability. Solvothermal methods allow a controlled recrystallization of precipitated particles by a physical equilibrium between crystallization and re-dissolution, generating highly homogeneous materials. On the other hand, the possibility of adding surfactants allows modifying the shape of these species [41].

Since we are talking about nanocrystals, it is necessary to note that the different faces of the crystal structure possess different electron-density distributions. This makes surfactant molecules generally charged, or with probable electrostatic interactions, to interact differently with each face of the crystal. Such interactions govern crystallization, modifying the original structure on the surface of the crystal. In this way, metal oxides such as nanorods, nanoprisms, nanotubes, and nanowires are obtained [42, 43].

Another interesting application of these methods is the synthesis of single or mixed mineral nanocrystalline systems, such as carbonates, fluorite, hydroxyapatite, phosphate, and homologous species. Many of these minerals have more than one crystal structure, which will depend on the crystallization conditions imposed. For example, using solvothermal methods we can determine the cubic or hexagonal crystallization of fluorites with different shapes. These materials, mainly $NaYF_4$, $LiYF_4$, and YF_3, have wide application in optics and biomedicine. The species may be activated with luminescent cations to be used as biomarkers [44]. Minerals coming from carbonates or phosphates, such as nano-hedgehogs, nanotripods, or nanostars, can be obtained by these methods, with considerably increased active surface, a fundamental property for many applications [45–48].

Finally, an area within solvothermal methods that has become very important is the preparation of carbon structures and metal carbides. Within this scope, the most relevant prepared materials are carbon nanotubes. By controlling the synthesis conditions, it is possible to obtain them as either single-wall or multi-wall carbon nanotubes, also with the capability to modify and functionalize the surface [49–51].

5.2.3 *Sol–Gel Nanoparticles Synthesis*

In materials science, sol–gel synthetic methods can be used to obtain macromolecular systems from polymerizable precursors. Typically, the synthetic method is used to prepare metal nanoparticles or nanoceramic oxides. Nevertheless, taking advantage of the property that possesses both sol and gel in generating self-assembly systems has made it possible to develop methods to obtain different types of both inorganic and organic nanoparticles. Among them we can find a growing group of nanoparticles generated based on polymers.

In general terms, a colloidal suspension gradually progresses to form a biphasic, gel-like system of morphologies, which can vary from discrete particles to continuous polymer networks. After the formation of the precursor, a gelation step is generally initiated, and in the case of colloids, the particle volume fraction (or particle density) may be so low that it may require a significant amount of liquid to be initially removed. This can be accomplished in many ways. The simplest method is to let the solution to settle down first and then pour the remaining liquid. Centrifugation can also be used to accelerate the process of phase separation. Phase separation also occurs due to a change in the physicochemical characteristics of the solution, such as pH, ionic strength, etc., which in turn may generate a change in the characteristics of the aggregation of colloidal particles.

Removing the remaining liquid phase (solvent) required for a drying process will typically be accompanied by a significant amount of shrinkage or densification. The rate at which the solvent can be ultimately removed will determine the distribution of porosity in the gel. Subsequently, a heat treatment or baking, also known as sintering, process is often necessary, wherein the gel obtained is converted into a powder. Densification and grain growth will depend on the characteristics of the process (heating temperature and kinetics). One advantage of using this method in contrast to the more traditional techniques is that the densification process is generally conducted at a much lower temperature than the bulk material.

Sol precursor used to synthesize nanopowders can be deposited on a substrate to form a film by either dip coating or spin coating. Sol–gel nanomaterial synthesis is an inexpensive technique that allows the precise control of the chemical composition of the product. Even small amounts of dopants, such as organic dyes or rare earth elements, can be introduced into the sol by this process and uniformly distributed in the final product [52–54].

Among the materials commonly synthesized by this route are nanoceramics such as Al_2O_3, Y_2O_3, and glass ceramics such as phosphor-vanadates, hafnates or semiconductors such as SiO_2, ZnO, or TiO_2. Generally, the synthesis of these materials uses organometallic complexes, which in some cases are hydrolyzed to obtain the metal oxides. A major advantage of obtaining these oxides in this way is that separate co-precipitation of other crystalline species or components of the mixture is avoided, since at first the colloid has theoretically a fairly homogeneous composition before the gelation process [55, 56].

Other groups of materials widely obtained by this synthetic method are ferroelectric ceramics such as $BaTiO_3$, $LiNbO_3$, or PZT. The method used is the synthesis of precursor gel organic acids, which are subsequently submitted to heat treatment. The method completely eliminates water and also provides a product free of ionic impurities and carbonates. It is based on organometallic macromolecular complex formation and subsequent decomposition at high temperatures. In the process, water is removed to form complex oxides. Finally, above the temperature of amorphization of the crystal, the nanomaterial is brought to room temperature, obtaining high-purity nanopowders with tetragonal symmetry. With this synthetic method, it is possible to control certain experimental conditions such as the partial pressure of O_2 and the process temperature, which will have incidences on the properties and structure of the final material. Furthermore, it allows the preparation of nanosheets and coated surfaces with this kind of materials [57–61].

The sol–gel synthesis permits modification of other techniques such as co-precipitation, for example, in the production of superparamagnetic ferrites. In this case, polymers are used in a colloidal suspension acting as salts chelate, allowing to obtain a highly homogenous material with a better and efficient particle size control [62–64].

These processes have also made it possible to synthesize nanoparticles consisting of polymers or biopolymers such as collagen or chitosan, using the properties of the colloidal suspensions and the structural changes in their self-assembly. It has also been possible to coat inorganic nanoparticles with fairly structured polymers, thus increasing its stability and biocompatibility [65].

5.3 Biosynthetic Methods: The Use of Microorganisms and Plants

Nanoscience began in the last century when characterization methods for observing the structure of nanomaterials were developed. According to the National Nanotechnology Initiative [66], one of the basic requirements including the concept of nanotechnology is the manipulation and characterization of materials at the nanoscale. Although nanochemistry has been developed in recent decades, the nature itself has given proof of its existence before this period, when we observed the synthesis of nanomaterials in biological organisms. A large number of living organisms are able to generate nanoscale structures from biological processes, which usually are products of detoxification processes. These structures are generally made compatible with biomolecules.

Such processes and the materials obtained have attracted great attention in the last decade because they largely overcome major drawbacks in biomedical applications. The use of nanoparticles and nanostructures in biology, in general, needs to solve biocompatibility and toxicity problems. As already discussed, most of the biosynthetic pathways are involved in detoxification processes. From this point of view, it is logical to think that the final product should have rather lower cytotoxicity compared with the used initial precursors. Moreover, in general the so-obtained nanomaterials are made compatible with proteins, sugars, or membranes used in cells for the isolation of potentially toxic agents. On the other hand, obtained as a product of a metabolic pathway, the structure of the materials produced by the cell is not only governed by physicochemical laws. Also they often involve the action of enzymes and subcellular organizations that originate bio-nanocomposites with sometimes highly complex structures [67, 68].

To study this type of synthesis, we can take into account the use of microbiological entities as the production sources of nanomaterials. For instance, the main microorganisms used to produce nanomaterials are bacteria, and yeast or fungi. Here also plants as biological carrier will be discussed.

5.3.1 *Bacteria*

Bacteria are mainly used in the synthesis of metal nanoparticles such as Au, Ag, Pd, etc. One of the first used bacteria was *Pseudomonas stutzeri* AG259, present in silver mines. Its cultivation produces silver nanoparticles with low polydispersity. Some *Lactobacillus* strains present in butter could be used to synthesize gold nanoparticles as well as their composite alloys. After that it was found that the use of the cell supernatant of *Pseudomonas aeruginosa* was an effective route for the synthesis of gold nanoparticles, via an extracellular pathway. The use of *Plectonema boryanum* UTEX 485 in the presence of aqueous solutions of Gold $(S_2O_3)_2^{3-}$ or $AuCl_4^-$ at 200°C could produce in one day either nanocubes or nanohexagons of gold, respectively. Later, a mechanism was reported, which must be understood in two stages. First, an extracellular reduction of Au (III) into Au (I) occurs, which is deposited on the cell walls in the form of amorphous sulfide. This is then desulfurized and restructured in octagons or cubes. Also some extremophiles, such as *Thermomonospora* sp, may produce gold nanoparticles with high efficiency and practically monodispersed in rather extreme conditions [69–73]. Another example is the magnetotactic bacteria *Magnetospirillum magneticum*. Such organisms may generate magnetic nanoparticles such as maghemite (Fe_2O_3) or greigite (Fe_3S_4) [74, 75].

5.3.2 *Yeast and Fungus*

Yeast such as *Schizosaccharomyces pombe* and *Candida glabrata* have been successfully used in the synthesis of nanoparticles of transition metal sulfides. This is of great importance for the production of quantum dots, since the conventional methods, carried out in aqueous medium, to obtain them are slightly toxic. It is a good method for synthesizing CdS, CdSe, and even MnS. In general terms, the process responds to detoxification pathways triggered by the presence of heavy metals, which involves the production and use of large quantities of glutathione. The nanocrystals are generated extracellularly with an acceptable and adjustable polydispersity [76, 77].

Also it has been possible to obtain TiO_2 nanoparticles with good reproducibility, using $TiO(OH)_2$ or K_2TiF_6. They are synthesized by extracellular processes using surface-compatibilized polysaccharide. A breakthrough was the synthesis of ferroelectric $BaTiO_3$ using *Fusarium oxysporum* and barium acetate, taking into account some metabolic routes of fungus that can use acetates as energy source. In this case, it appears that the production process involves both extracellular and intracellular stages, producing a monodispersed ferroelectric material with diameters below 10 nm even with a high dielectric constant. The utility of *F. oxysporum* and *Verticillium* sp. has been demonstrated in the synthesis of other materials such as zirconia, SiO_2 in addition to metal nanoparticles such as gold and silver [78–81].

5.3.3 *Plants*

The role that some plants play in the detoxification of soils rich in heavy metals is well known. It has been possible to take advantage of this principle by developing synthetic routes especially for gold, silver, or bimetallic nanoparticles, using plants such as alfalfa or oats sativa. The first is able to absorb metallic silver from an agar culture and restructure it into nanoparticles of fairly regular sizes. Likewise, *Avena sativa* produces gold nanoparticles with sizes between 25 nm and 35 nm possessing carboxylated molecules onto the surface. This surface functionalization allows the nanoparticles to be well suspended in aqueous media, thus obtaining stable dispersions. *Pelargonium graueolens* and *Azadirachta indica* are other examples, which are capable of producing highly stable and crystalline gold, silver, and gold/silver nanoparticles. Unlike fungi that use, for instance, protein for stabilization, plants stabilize the nanoparticles with flavones and terpenoids. They possess significantly lower molecular weights. Then the nanomaterials obtained with plants are much better stabilized in various aqueous solutions, which differ from the initial culture medium [82–84].

5.4 Characterization of Nanomaterials

The emergence of nanoscience and nanotechnology is supported by the development of techniques for the characterization of nanoparticles. We can consider two types of techniques. The first one allows us to observe their shape and structure, such as transmission electron microscopy (TEM), and the second one gives us information related to the properties and individual characteristics of nanoparticles, for example dynamic light scattering (DLS).

The structural characterization of nanomaterials has been possible by integrating diverse and complementary techniques together. The most important set of techniques that allowed the development and handling of these materials are TEM, scanning and atomic force microscopy (SFM, AFM).

5.4.1 *Transmission Electron Microscopy*

An essential tool for the study of materials at nanoscale is TEM. The first transmission electron microscope was developed between 1931 and 1933 by Ernst Ruska and colleagues. The basic standpoint of this first electron microscope remains to this day. The first commercial transmission electron microscope was constructed by Siemens in 1939. This microscope, unlike optical microscopes, uses an electron beam to display an object, where the power amplifier of an optical microscope is limited by the wavelength of visible light (Fig. 5.1A). The electron beam in an electron microscope is generally produced by a filament, typically tungsten or nickel-chromium alloys, similar to that of an incandescent bulb (Fig. 5.1B), through a process known as incandescent thermionic emission or field emission. Electrons emitted are then accelerated by means of an electric potential (measured in kV) and are focused by electrostatic or electromagnetic capacitors. These electrons pass through the sample and impact in an electro-fluorescent display. When the beam passes through dense atoms of the sample, its projection acquires greater contrast. This effect is particularly useful for samples made up of atoms of high weight, such as Au, lanthanide metals of period 5 and above. Lower density samples consisting of carbon, silicon, and other light metals are required to increase the resolution or the use of contrasting materials such as uranium salts. Figure 5.1C shows a typical geometry for the sample holder, consisting Formvar and carbon film on a copper grid. Formvar is a polymer film widely used as support for specimens in electronic microscopy and consists of a polyvinyl alcohol reacted with formaldehyde.

High resolution transmission electron microscopy (HR-TEM) possesses the same technology, but the acceleration and, therefore, the penetration of the electron beam in the sample are increased by using higher voltages between 100 kV and 300 kV. Thus, it is

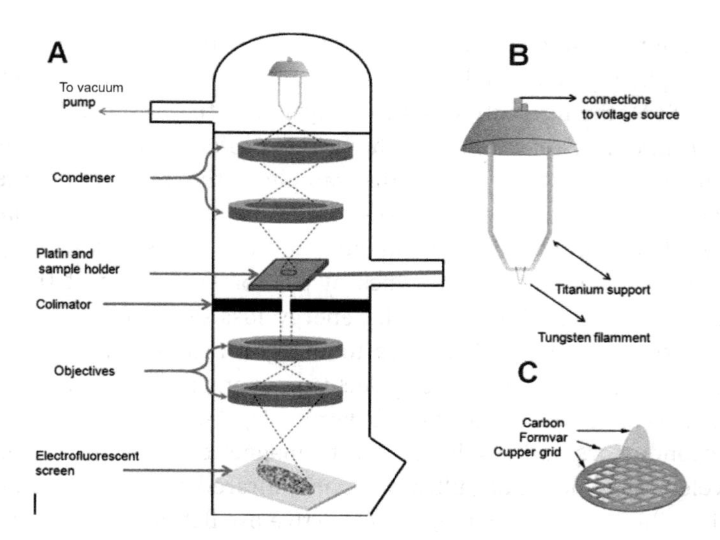

Figure 5.1 Schematic for the operation of a TEM: (A) outlining the basic components of a TEM, (B) filament and support, and (C) shows a carrier grid.

possible to define very important features in structure and material composition, level of crystal lattice dimensions, and even the provision of crystal unit cells. In this way, it is possible to determine the crystal structures of a sample based on the information provided for the beam. Diffraction patterns that are characteristic for certain structures are stored in database [85, 86].

5.4.2 *Scanning Electron Microscopy*

Scanning electron microscopy (SEM) is one of the most versatile techniques used in the study and analysis of the microstructural characteristics of solid objects. This technique allows us to observe samples in the fields of materials science and biological materials. The most important feature of SEM is that it is possible to observe samples in three dimensions, opposite to TEM in which the samples are observed in two dimensions, which represents a loss in information related to thickness. The versatility of SEM regarding transmission is largely derived from the variety of interactions that undergo the electron beam in the specimen preparation.

Interactions can give information about material composition, topography, crystallography, electrical potential, local magnetic field, etc. In SEM, the interaction of the sample with the beam can be of elastic and inelastic scattering. The first one affects the trajectories of the electron beam within the sample without changing the kinetic energy of the electrons. The elastic scattering is responsible for electron backscatter phenomenon, and the inelastic scattering events give rise to different types of phenomena. Some of these signals can be resulting from the energy loss or energy transfer within the specimen. This leads to the generation of secondary electrons, which could be Auger electrons, characteristic X-rays and bremsstrahlung (continuous or background), electron-hole pairs in semiconductors and insulators, electromagnetic radiation of long wavelength in the visible, ultraviolet, and infrared radiation spectra, lattice vibrations (phonons), and collective oscillations of electrons in metals (plasmons). The two types of microscopies (TEM and SEM) should work under a vacuum system that enables the electron beam to travel through the column. The vacuum system is generally similar in both systems. The necessary vacuum level depends on the equipment requirements and techniques.

For observation, the sample is placed in a small space, which is evacuated after the door is closed. The door has three levers that the operator uses to raise and lower the sample, rotate the sample, and zoom in or out. A thin electron beam is produced on top of the microscope by heating a metal filament (10–30 kV). The primary electron beam follows a path through the empty column of the microscope, with the purpose of avoiding the scattering of electrons. The path of the electron beam is quickly modified by a set of deflection coils, which are able to scan the sample point by point and along parallel lines. Subsequently, the diameter of the electron beam can be modified by passing it through the objective lens to control the amount of electrons that will impact the sample. When the primary electrons strike the specimen, secondary electrons are emitted from the specimen itself. These secondary electrons are attracted by a collector and are accelerated and directed to the scintillator, where the kinetic energy is converted into points with higher or lower brightness. This light is directed to an amplifier where it is converted into an electrical signal, which passes an observation screen where the image is formed, line by line and point by point. The circuits that sweep coils, which force the beam to sweep the sample, are the same that drive the collection of electrons that produce the image.

5.4.3 *Scanning Probe Microscopy*

Scanning probe microscopy (SPM) is a family of microscopies where a sharp probe scans the surface of a sample by monitoring the interactions occurring between the tip and the sample. It is an imaging tool with a wide dynamic range, spanning the realms of light and electron microscopes. It is also considered a profiler with a resolution in three dimensions.

The applications are very different: measures of properties such as surface conductivity, static charge distribution, localized friction, magnetic fields, and elastic modulation. The two main forms of SPM are tunneling microscopy, developed by Binning and Roher at IBM Laboratories (Switzerland), a discovery for which they received the Nobel Prize in Physics in 1986, and atomic force microscopy (AFM). In this, three main modes, (i) contact mode, (ii) non-contact mode, and (iii) tapping mode, are distinguished.

All SPM microscopy techniques have, in common, a tip (cantilever), a nano-displacement system, a device for tip/sample approximation, and a computer set that allows the control and data collection. The movements are performed by piezoelectric ceramics to ensure the movement of the tip or the sample in all three axes with high precision.

Any interaction strength that can be measured between the tip and the sample maybe a sort of SPM. The resolution of each type of microscopy depends, ultimately, on the dependence of the interaction force measured and the tip-to-sample distance. During scanning over the sample, an image is created, which gives the variations of forces depending on the tip position on the surface. Generally, the measurement obtained by an image contrasting color is represented. The general characteristics of the SPM techniques are displacements of up to 150 microns in the plane, and 10 to 15 microns in height. A resolution up to 0.01 Å can be achieved, which is the theoretical resolution of piezoelectric ceramics. It is also possible to work in a widely varying ways: air, controlled vacuum and ultra-high vacuum atmosphere, high/low temperatures, liquids, etc.

In particular, an atomic force microscope is a mechanic-optical instrument capable of detecting forces of nanonewtons. When analyzing a sample, an atomic force microscope is able to continuously record the height above the surface of a probe tip or crystal pyramid shape. The probe is coupled to the effect of the forces of a highly sensitive cantilever, a microscopic bar of only about 200 microns in length [87].

The atomic force can be detected when the tip is very close to the sample surface. Then it is possible to register the small flex of the cantilever by using a laser beam reflected on its back. An auxiliary piezoelectric system displaces the sample in three dimensions, while the cantilever is deflected through the surface topology of the sample. All movements are controlled by a computer. The resolution of the instrument is less than 1 nm, which allows to distinguish details on the sample surface at a magnification of several million times.

5.4.4 *X-Ray Diffraction*

The English physicists Sir W. H. Bragg and his son Sir W. L. Bragg (1913) gave an explanation of the X-ray diffraction phenomenon of powdered crystalline samples generated by an X-ray beam. Polycrystalline samples were first revealed in Germany by P. Scherrer and P. Debye in 1916. They state that every crystalline substance gives a specific pattern and that in a mixture of substances, each produces its pattern independent of the others. The phenomenon of X-ray diffraction is essentially a process of producing a constructive wave interference that produces X-rays in particular directions of space, which interact with a material exhibiting periodic and regular structures, such as a glass or a liquid crystal. This phenomenon occurs because the wavelengths delivered by the X-rays are comparable with the interatomic distance of a crystal. Then, when an X-ray beam strikes an atomic plane, it will be diffracted, generating a phase angle at a defined wavelength. If the plane has a certain periodicity, the diffracted waves in the same specific angle will possess a similar phase, generating a resultant interference. This will be proportional to the number of planes present that generate the diffracted wave, thus allowing identification of the structure of the studied material. Today this technique is a standard tool in the determination of structures of new materials.

The identification of crystalline phases by the powder method is one of the most important fields for substance-identification application. The identification of a crystal phase by this method is based on the fact that each substance in a crystalline state has a characteristic X-ray diagram. These diagrams are collected in databases of the Joint Committee on Powder Diffraction Standards and grouped index of organic, inorganic, and mineral compounds. However, the description of the structural characteristics obtained by XRD about nanomaterials is not just limited to the identification of crystal structures. It is also possible to obtain information on the size by using the Scherrer formula. So the crystallite size is obtained by measuring the extension of the X-ray diffraction of a flat reflection particularly within the crystal unit cell. This is due to the fact that the periodicity of the crystallite domains in phase reinforces the X-ray diffraction, resulting in a high and narrow peak. If the crystals are free from defects and periodically arranged, the X-ray is diffracted at the same angle. If the crystals are arranged randomly, or have very low degree of frequency, the peaks are wider.

On the other hand, small angle X-ray scattering (SAXS) is an analytical technique that can be used for structural characterization of materials in the range of nanometers. From the intensity distribution at a very low angle, it is possible to obtain information on size for the particle size distribution, particle shape, and internal structure. This technique is used in particles with a size between 0.5 nm and 50 nm of materials such as liquid crystals, polymer films, microemulsions, catalysts, proteins, nanoparticles, and viruses.

The interpretation of the SAXS data can be really complex. In the most favorable cases, it is possible to use direct methods in which the data, corrected with the background, are used without manipulation. It is also possible to apply the Fourier transform to real space information in an analogous manner to TEM.

Finally, the nanoscale interactions between the environment and the surface of the material become more important at the analytical level. Therefore, those symmetry planes prohibited according to crystallographic laws may appear reflected with some intensity due to the large amount of surface present in the nanometric samples. These now observable planes may provide important information on the surface and the interaction of this with its immediate environment.

5.4.5 *X-Ray Photoelectron Spectroscopy*

In 1887, Heinrich Rudolf Hertz discovered the photoelectric effect. Albert Einstein explained the effect in 1905, which earned him the Nobel Prize in Physics in 1921. In 1907, P. D. Innes experimented with a Röntgen tube coupled with Helmholtz coils (electron energy analyzer) and recorded, in photographic plates, electron bands emitted depending on their speed. Then, when electrons are irradiated with light of a certain wavelength, they can be emitted to specific speeds depending on their atomic binding forces. When selecting the emitted electrons with high energy light (Fig. 5.2), such as X-rays, each atom generates a characteristic pattern of signals depending on its oxidation state and present amount.

X-ray photoelectron spectroscopy (XPS) is a low-resolution quantitative method, which was demonstrated by Dr. Siegbahn, and he was awarded the Nobel Prize in 1981 for it. He built the first

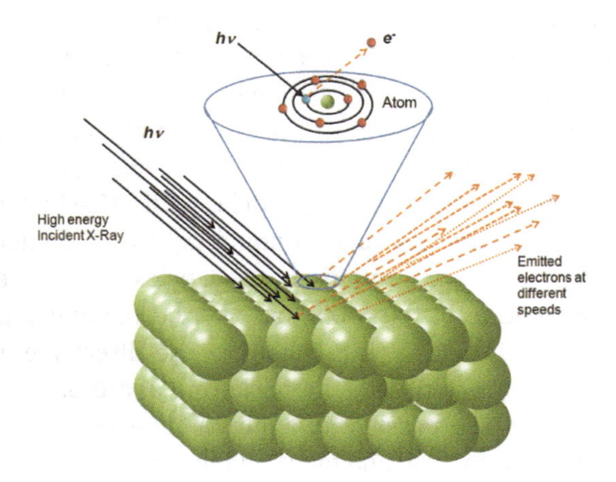

Figure 5.2 Basic scheme of the basis of the technique of electron spectroscopy for chemical analysis (XPS-ESCA), and equipment used for the measurement of samples.

device for electron spectroscopy for chemical analysis (ESCA). He observed that the signals correspond to the specific emitted electron configuration for each element. To generate the atomic percentage values, each obtained XPS signal must be corrected by dividing the signal intensity (number of detected electrons) by a "relative sensitivity factor" and normalized over the entire detected elements.

5.4.6 *Dynamic Light Scattering*

Dynamic light scattering (DLS) is a technique used to measure particle size, typically in the sub-micron region. DLS measures the Brownian movements of the particles, which are dependent of their sizes. Brownian motion is the random movement of the particles due to the shocks given by solvent molecules surrounding them. Normally, DLS involves measuring particles suspended in a liquid.

The larger particles show a slow Brownian motion, while smaller particles collide with molecules of the solvent and, therefore, move faster. For this reason, it is necessary to know accurately the temperature of the suspension, which in turn should remain stable during measurement. This will prevent convection currents in the sample from generating random movements that lead to an incorrect interpretation of the size. The velocity of the Brownian motion is defined by a translational diffusion coefficient (D) and, therefore, it is also necessary to know the viscosity of the sample. Then the size of a particle is calculated from the translational diffusion coefficient using the Stokes–Einstein equation $dH = KT/3\pi\eta D$, where D is the translational diffusion coefficient, which is measured by the equipment; K is the Boltzmann constant; T is the temperature in Kelvin; and η is the viscosity at the temperature T of the sample.

It is necessary to consider that the measured diameter by DLS is a value referred in how the particle diffuses within a fluid and, therefore, is known as a hydrodynamic diameter (dH). The diameter obtained with this technique is the diameter of a sphere having the same translational diffusion coefficient of the particle. The translational diffusion coefficient depends not only on particle size but also on any surface structure, as well as the concentration and type of ions in the medium.

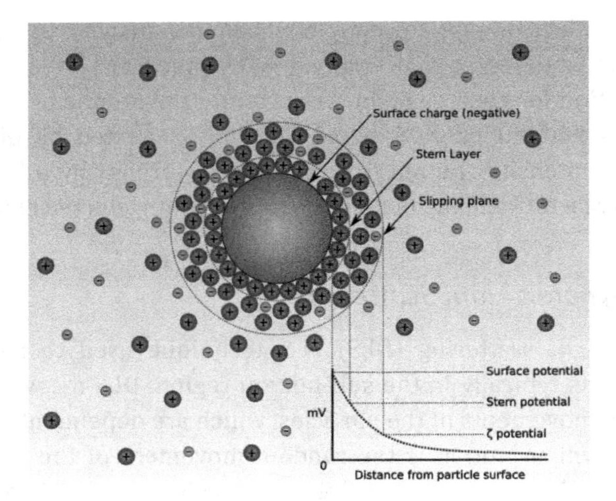

Figure 5.3 Diagram showing the ionic concentration and potential difference as a function of distance from the charged surface of a particle suspended in a dispersion medium [88].

5.4.7 *Zeta Potential*

The nanoparticle suspension may be considered an intrinsic colloid particle charged in a colloidal suspension, plus all ions adsorbed and stabilized from what is known as the electrical double layer. The thickness of this double layer is typically the Debye length ($k - 1$). The charges that remain within this length do not exert effective force, and a shielding phenomenon is produced. Modifying the surface charge distribution of the surrounding ions results in an area around the particle that is electrically different from the bulk of the solution. This area can be divided into two parts: the inner layer adhered strongly to the particle, which will move along with the particle, and the so-called stern layer, and an outer region, which is diffused as soon as the distance from the surface increases. In this region, there is a conceptual boundary beyond which the ions do not move along with the particle. This boundary is known as the sliding plane or hydrodynamic cutting plane and corresponds to the end of the electrical double layer (Fig. 5.3).

From the above we can say that the zeta potential (ζ) is the electric potential at the shear plane, which is usually expressed in millivolts (mV). It corresponds to a measurable magnitude of the electrical repulsion or attraction between the particles that compose the colloid. Its measurement confirms the mechanism of stabilization of a colloid. A zeta potential of great magnitude is correlated with good colloidal stability; particles with high zeta potential, either positive or negative, will repel each other. However, it is necessary to consider that colloids could be stabilized by steric effects that are independent of the charge.

The measurement of zeta potential can be performed in two ways: by electrophoresis or by electroacoustic methods using ultrasound. From the first method, ζ is calculated by electrophoretic mobility, which is determined experimentally. The oldest measurement equipment had a magnifying glass with graduated scale distances that scale the distance traveled by the particles. Using a timer, it is possible to determine the time and, therefore, the speed. Then the zeta potential could be calculated, knowing the viscosity and permittivity of the medium at the temperature of the experiment. Most modern computers use more sophisticated techniques to measure the velocity of the particles, which allows to measure the zeta potential of smaller particles (nanoparticles) colloid. Among these techniques are laser Doppler velocimetry and phase analysis of the scattered light.

Acknowledgments

A. Meneses-Franco acknowledges Conicyt scholarship for doctoral studies; E. A. Soto-Bustamante thanks Fondecyt Project 1130187; and M. Kogan-Bocian thanks Fondap project 15130011.

References

1. Koo O.M., Rubinstein I., and Onyuksel H. (2005). Role of nanotechnology in targeted drug delivery and imaging: A concise review, *Nanomedicine: Nanotechnology, Biology, and Medicine*, **1**, pp. 193–212.
2. Hyuk Suh W., Suslick K.S., Stucky G.D., and Suh Y. (2009). Nanotechnology, nanotoxicology, and neuroscience, *Progress in Neurobiology*, **87**, pp. 133–170.

3. Chung C.K., Fung P.K., Hong Y.Z., Ju M.S., Lin C.C.K., and Wu T.C. (2006). A novel fabrication of ionic polymer-metal composites (IPMC) actuator with silver nano-powders, *Sensors and Actuators B,* **117**, pp. 367–375.
4. Wilson S.A. (2007). New materials for micro-scale sensors and actuators an engineering review, *Materials Science and Engineering R,* **56**, pp. 1–129.
5. Fang X.Y., Tan O.K., Wei Q., Yao M.W., and Tjin S.C. (2006). Dielectric film for biosensor application, *Sensors and Actuators B,* **119**, pp. 78–83.
6. Bennewitz M.F. and Saltzman W.M. (2009). Nanotechnology for delivery of drugs to the brain for epilepsy, *Neurotherapeutics: The Journal of the American Society for Experimental NeuroTherapeutics,* **6**, pp. 323–336.
7. Edelstein A.S and Cammaratra R.C (1996). *Nanomaterials: Synthesis, Properties and Applications,* 2nd edn, Taylor & Francis, New York.
8. Tavakoli A., Sohrabi M., and Kargari A. (2007). A review of methods for synthesis of nanostructured metals with emphasis on iron compounds, *Chemical Papers,* **61**, pp. 151–170.
9. Ashkenas H. and Sherman F.S. (1966). *Rarified Gas Dynamics,* 1st edn (De Leewuw, J.H., ed), Academic Press, New York.
10. Richards R., Li W., Decker S., Davidson C., Koper O., Zaikovski V., Volodin A., Rieker T., and Klabunde K.J. (2000). Consolidation of metal oxide nanocrystals. Reactive pellets with controllable pore structure that represent a new family of porous, inorganic materials, *Journal of the American Chemical Society,* **122**, pp. 4921–4925.
11. Wagner G.W., Koper O.B., Lucas E., Decker S., and Klabunde K.J. (2000). Reactions of VX, GD and HD with nanosize CaO: Autocatalytic dehydrohalogenation of HD, *Journal of Physical Chemistry B,* **104**, pp. 5118–5123.
12. Yamashita M. and Fenn J.B. (1984). Electrospray ion source. Another variation on the free-jet theme, *Journal of Physical Chemistry,* **88**, pp. 4451–4459.
13. Brechignac C., Houdy P., and Lahmani M. (2007). *Nanomaterials and Nanochemistry,* Springer Verlag, Berlin Heidelberg.
14. Sergeev G.B. and Klabunde K.J. (2013). Size Effects in Nanochemistry, in *Nanochemistry* (Sergeev G.B. and Klabunde K.J, eds), 2nd edn, Springer Verlag, Berlin Heidelberg, pp. 275–297.
15. Schmidt H. (2001). Nanoparticles by chemical synthesis, processing to materials and innovative applications, *Applied Organometalic Chemistry,* **15**, pp. 331–343.

16. Xiangqian L., Huizhong X., Zhe-Sheng C., and Guofang C. (2011). Biosynthesis of nanoparticles by microorganisms and their applications, *Journal of Nanomaterials*, Article ID 270974, doi:10.1155/2011/270974.
17. Mohanpuria P., Rana K.N., and Yadav S.K. (2008). Biosynthesis of nanoparticles: Technological concepts and future applications, *Journal of Nanoparticle Research*, **10**, pp. 507–517.
18. Xia H. and Yang G. (2012). Facile synthesis of inorganic nanoparticles by a precipitation method in molten ε-caprolactam solvent, *Journal of Materials Chemistry*, **22**, pp. 18664–18670.
19. Martínez-Miranda L.J., Traister K.M, Meléndez-Rodríguez I., and Salamanca-Riba L. (2010). Liquid crystal-ZnO nanoparticle photovoltaics: Role of nanoparticles in ordering the liquid crystal, *Applied Physics Letters*, **97**, pp. 223301–223303.
20. Pottier A., Cassaignon S., Chanéac C., Villain F., Tronca E., and Jolivet J-P. (2003). Size tailoring of TiO_2 anatase nanoparticles in aqueous medium and synthesis of nanocompósitos: Characterization by Raman spectroscopy, *Journal of Materials Chemistry*, **13**, pp. 877–882.
21. Hong R., Pan T., and Qian J., Li H. (2006). Synthesis and surface modification of ZnO nanoparticles, *Chemical Engineering Journal*, **119**, pp. 71–81.
22. Jiang L., Sun G., Zhou Z., Sun S., Wang Q., Yan S., Li H., Tian J., Guo J., Zhou B., and Xin Q. (2005). Size-controllable synthesis of monodispersed SnO_2 nanoparticles and application in electrocatalysts, *Journal of Physical Chemistry B*, **109**, pp. 8774–8778.
23. Si S., Li C., Wang X., Yu D., Peng Q., and Li Y. (2005). Magnetic monodisperse Fe_3O_4 nanoparticles, *Crystal Growth & Design*, **5**, pp. 391–393.
24. Lyon J.L., Fleming D.A., Stone M.B., Schiffer P., and Williams M.E. (2004). Synthesis of Fe oxide core/Au shell nanoparticles by iterative hydroxylamine seeding, *Nano Letters*, **4**, pp. 719–723.
25. Tang Y., Li Z., He N., Zhang L., Ma C., Li X., Li C., Wang Z., Deng Y., and He L. (2013). Preparation of functional magnetic nanoparticles mediated with PEG-4000 and application in *Pseudomonas aeruginosa* rapid detection, *Journal of Biomedical Nanotechnology*, **9**, pp. 312–317.
26. Hazra S. and Ghosh N.N. (2014). Preparation of nanoferrites and their applications, *Journal of Nanoscience and Nanotechnology*, **14**, pp. 1983–2000.
27. Lima E., Winkler E.L., Tobia D., Troiani H.E., Zysler R.D., Agostinelli E., and Fiorani D. (2012). Bimagnetic CoO core/$CoFe_2O_4$ shell nanoparti-

cles: Synthesis and magnetic properties, *Chemistry of Materials*, **24**, pp. 512–516.

28. Mandal M., Kundu S., Ghosh S.K., Panigrahi S., Sau T.K., Yusuf S.M., and Pal T. (2005). Magnetite nanoparticles with tunable gold or silver shell, *Journal of Colloid and Interface Science*, **286**, pp. 187–194.
29. Kostura L., Kraitchman D.L., Mackay A.M., Pittenger M.F., and Bulte J.W.M. (2004). Feridex labeling of mesenchymal stem cells inhibits chondrogenesis but not adipogenesis or osteogenesis, *NMR in Biomedicine*, **17**, pp. 513–517.
30. Reimer P. and Balzer T. (2003). Ferucarbotran (Resovist): A new clinically approved RES-specific contrast agent for contrast-enhanced MRI of the liver: Properties, clinical development, and applications, *European Radiology*, **13**, pp. 1266–1276.
31. Skotland T., Sontum P.C., and Oulie I. (2002). In vitro stability analyses as a model for metabolism of ferromagnetic particles (Clariscan™), a contrast agent for magnetic resonance imaging, *Journal of Pharmaceutical and Biomedical Analysis*, **28**, pp. 323–329.
32. Gerion D., Pinaud F., Williams S.C., Parak W.J., Zanchet D., Weiss S., and Alivisatos A.P. (2001). Synthesis and properties of biocompatible water-soluble silica-coated CdSe/ZnS semiconductor quantum dots, *Journal of Physical Chemistry B*, **105**, pp. 8861–8871.
33. Dabbousi B.O., Rodriguez-Viejo J., Mikulec F.V., Heine J.R., Mattoussi H., Ober R., Jensen K.F., and Bawendi M.G. (1997). (CdSe)ZnS core–shell quantum dots: Synthesis and characterization of a size series of highly luminescent nanocrystallites, *Journal of Physical Chemistry B*, **101**, pp. 9463–9475.
34. Vlasov Y.A., Yao N., and Norris D.J. (1999). Synthesis of photonic crystals for optical wavelengths from semiconductor quantum dots, *Advanced Materials*, **11**, pp. 165–169.
35. Liu Y-F. and Yu J.S. (2010). In situ synthesis of highly luminescent glutathione-capped CdTe/ZnS quantum dots with biocompatibility, *Journal of Colloid and Interface Science*, **351**, pp. 1–9.
36. Li Y., Zhu H., and Hou C. (2013). Hydrothermal & solvothermal synthesis of nanoscale magnetic materials, *Progress in Chemistry*, **25**, pp. 276–287.
37. Byrappa K. and Adschiri T. (2007). Hydrothermal technology for nanotechnology, *Progress in Crystal Growth and Characterization of Materials*, **53**, pp. 117–166.
38. Crooks R.M., Zhao M., Sun L., Chechik V., and Yeung L.K. (2001). Dendrimer-encapsulated metal nanoparticles: Synthesis, characteriza-

tion and applications to catalysis, *Accounts of Chemical Research*, **34**, pp. 181–190.

39. Wiley B., Sun Y., Mayers B., and Xia Y. (2005). Shape-controlled synthesis of metal nanostructures: The case of silver, *Chemistry: A European Journal*, **11**, pp. 454–463.
40. Zhu L-P., Xiao H-M., Zhang W-D., Yang Y., and Fu S-Y. (2008). Synthesis and characterization of novel three-dimensional metallic Co dendritic superstructures by a simple hydrothermal reduction route, *Crystal Growth and Design*, **8**, pp. 1113–1118.
41. Mousavand T., Takami S., Umetsu M., Ohara S., and Adschiri T. (2006). Supercritical hydrothermal synthesis of organic-inorganic hybrid nanoparticles, *Journal of Materials Science*, **41**, pp. 1445–1448.
42. Yu J. and Yu X. (2008). Hydrothermal synthesis and photocatalytic activity of zinc oxide hollow spheres, *Environmental Science and Technology*, **42**, pp. 4902–4907.
43. Cozzoli P.D., Kornowski A., and Weller H. (2003). Low-temperature synthesis of soluble and processable organic-capped anatase TiO_2 nanorods, *Journal of the American Chemical Society*, **125**, pp. 14539–14548.
44. Zhou J., Liu Z., and Li F. (2012). Upconversion nanophosphors for small-animal imaging, *Chemical Society Reviews*, **41**, pp. 1323–1349.
45. Huang Y., You H., Jia G., Song Y., Zheng Y., Yang M., Liu K., and Guo N. (2010). Hydrothermal synthesis, cubic structure, and luminescence properties of BaYF5:RE (RE = Eu, Ce, Tb) nanocrystals, *Journal of Physical Chemistry C*, **114**, pp. 18051–18058.
46. Li C., Xu Z., Yang D., Cheng Z., Hou Z., Ma P., Lian H., and Lin J. (2012). Well-dispersed KRE3F10 (RE = Sm–Lu, Y) nanocrystals: Solvothermal synthesis and luminescence properties, *CrystEngComm*, **14**, pp. 670–678.
47. Qu X., Yang H.K., Pan G., Chung J.W., Moon B.K., Choi B.C., and Jeong J.H. (2011). Controlled fabrication and shape-dependent luminescence properties of hexagonal $NaCeF_4$, NaCeF4:Tb3+Nanorods via polyol-mediated solvothermal route, *Inorganic Chemistry*, **50**, pp. 3387–3391.
48. Chen J. and Whittingham M.S. (2006). Hydrothermal synthesis of lithium iron phosphate, *Electrochemistry Communications*, **8**, pp. 855–858.
49. Gogotsia Y., Libera J.A., and Yoshimura M. (2000). Hydrothermal synthesis of multiwall carbon nanotubes, *Journal of Materials Research*, **15**, pp. 2591–2594.

50. Kleitz F., Choia S.H., and Ryoo R. (2003). Cubic Ia3d large mesoporous silica: Synthesis and replication to platinum nanowires, carbon nanorods and carbon nanotubes, *Chemical Communications*, **17**, pp. 2136–2137.
51. Calderon Moreno J.M. and Yoshimura M. (2001). Hydrothermal processing of high-quality multiwall nanotubes from amorphous carbon, *Journal of the American Chemical Society*, **123**, pp. 741–742.
52. Brinker C.J. and Scherer G.W. (1990). *Sol-gel Science: The Physics and Chemistry of Sol-gel Processing*, Gulf Professional Publishing, London, U.K.
53. Hench L.L. and West J.K. (1990). The sol–gel process, *Chemical Reviews*, **90**, pp. 33–72.
54. Corriu R. and Trong A.N. (2009). *Molecular Chemistry of Sol-Gel Derived Nanomaterials*, 1st edn, John Wiley and Sons, West Sussex, U.K.
55. Niederberger M. and Pinna N. (2009). *Metal Oxide Nanoparticles in Organic Solvents: Synthesis, Formation, Assembly and Application*, Springer-Verlag, London, U.K.
56. Niederberger M. (2007). Nonaqueous sol-gel routes to metal oxide nanoparticles, *Accounts of Chemical Research*, **40**, pp. 793–800.
57. Aruna S.T. and Mukasyan A.S. (2008). Combustion synthesis and nanomaterials, *Current Opinion in Solid State and Materials Science*, **12**, pp. 44–50.
58. Wang X., Zhao C., Wang Z., Wu F., and Zhao M. (1994). Synthesis of $BaTiO_3$ nanocrystals by stearic acid gel method, *Journal of Alloys and Compounds*, **204**, pp. 33–36.
59. Meneses-Franco A., Trujillo-Rojo V.H., and Soto-Bustamante E.A. (2010). Synthesis and characterization of pyroelectric nanocomposite, formed of $BaTiO_3$ nanoparticles and a smectic liquid-crystal matrix, *Phase Transitions*, **83**, pp. 1037–1047.
60. Meldrum A., Boatner L.A., Weber W.J., and Ewing R.C. (2002). Amorphization and recrystallization of the ABO_3 oxides, *Journal of Nuclear Materials*, **300**, pp. 242–254.
61. Wang L., Liu L., Xue D., Kanga H., and Liu C. (2007). Wet routes of high purity $BaTiO_3$ nanopowders, *Journal of Alloys and Compounds*, **440**, pp. 78–83.
62. Gash A.E., Tillotson T.M., Satcher Jr. J.H., Poco J.F., Hrubesh L.W., and Simpson R.L. (2001). Use of epoxides in the sol–gel synthesis of porous iron (III) oxide monoliths from Fe (III) salts, *Chemistry of Materials*, **13**, pp. 999–1007.

63. Samira B., Chandrappa K.G., and Sharifah B.A.H. (2013). Generation of hematite nanoparticles via sol-gel method, *Research Journal of Chemical Sciences*, **3**, pp. 62–68.
64. Lu Y., Yin Y., Mayers B.T., and Xia Y. (2002). Modifying the surface properties of superparamagnetic iron oxide nanoparticles through a sol–gel approach, *Nano Letters*, **2**, pp. 183–186.
65. Hamidia M., Azadia A., and Rafiei P. (2008). Hydrogel nanoparticles in drug delivery, *Advanced Drug Delivery Reviews*, **60**, pp. 1638–1649.
66. National Nanotechnology Initiative, www.nano.gov, on line, The NNI website 2014.
67. Mohanpuria P., Rana N.K., and Yadav S.K. (2008). Biosynthesis of nanoparticles: Technological concepts and future applications, *Journal of Nanoparticle Research.*, **10**, pp. 507–517.
68. Li X., Xu H., Chen Z-S., and Chen G. (2011). Review article: Biosynthesis of nanoparticles by microorganisms and their applications, *Journal of Nanomaterials*, Article ID 270974.
69. Ahmad A., Senapati S., Khan M.I., Kumar R., and Sastry M. (2003). Extracellular biosynthesis of monodisperse gold nanoparticles by a novel extremophilic actinomycete, *Thermomonospora* sp., *Langmuir*, **19**, pp. 3550–3553.
70. Lengke M. F., Fleet M.E., and Southam G. (2006). Morphology of gold nanoparticles synthesized by filamentous cyanobacteria from gold(I)-thiosulfate and gold(III)-chloride complexes, *Langmuir*, **22**, pp. 2780–2787.
71. Husseiny M.I., El-Aziz M. A., Badr Y., and Mahmoud M. A. (2007). Biosynthesis of gold nanoparticles using *Pseudomonas aeruginosa*, *Spectrochimica Acta Part A*, **67**, pp. 1003–1006.
72. Klaus T., Joerger R., Olsson E., and Granqvist C.G. (1999). Silver based crystalline nanoparticles, microbially fabricated, *Proceedings of the National Academy of Sciences USA*, **96**, pp. 13611–13614.
73. Klaus T., Joerger R., Olsson E., and Granqvist C.G. (2001). Bacteria as workers in the living factory: Metal-accumulating bacteria and their potential for materials science. *Trends in Biotechnology*, **19**, pp. 15–20.
74. Blakemore R. (1975). Magnetotactic bacteria, *Science*, **190**, pp. 377–379.
75. Roh Y., Lauf R.J., McMillan A.D., Zhang C., Rawn C.J., Bai J., and Phelps T.J. (2001). Microbial synthesis and the characterization of metal-substituted magnetites, *Solid State Communications*, **118**, pp. 529–534.
76. Dameron C.T., Reeser R.N., Mehra R.K., Kortan A.R., Carroll P.J., Steigerwaldm M.L., Brus L.E., and Winge D.R. (1989). Biosynthesis of

cadmium sulphide quantum semiconductor crystallites. *Nature*, **338**, pp. 596–597.

77. Kumar S.A., Ansary A.A., Abroad A., and Khan M.I. (2007). Extracellular biosynthesis of CdSe quantum dots by the fungus, *Fusarium oxysporum*, *Journal of Biomedical Nanotechnology*, **3**, pp. 190–194.
78. Bansal V., Rautaray D., and Bharde A. (2005). Fungus-mediated biosynthesis of silica and titania particles, *Journal of Materials Chemistry*, **15**, pp. 2583–2589.
79. Bansal V., Poddar P., Ahmad A., and Sastry M. (2006). Room-temperature biosynthesis of ferroelectric barium titanate nanoparticles, *Journal of the American Chemical Society*, **128**, pp. 11958–11963.
80. Bansal V., Rautaray D., Ahmad A., and Sastry M. (2004). Biosynthesis of zirconia nanoparticles using the fungus *Fusarium oxysporum*, *Journal of Materials Chemistry*, **14**, pp. 3303–3305.
81. Senapati S., Mandal D., and Ahmad A. (2004). Fungus-mediated synthesis of silver nanoparticles: A novel biological approach, *Indian Journal of Physics A*, **78A**, 101–105.
82. Armendariz V., Herrera I., Peralta-Videa J.R., Jose-Yacaman M., Troiani H., Santiago P., and Gardea-Torresdey J.L. (2004). Size controlled gold nanoparticle formation by Avena sativa biomass: Use of plants in nanobiotechnology, *Journal of Nanoparticles*, **6**, pp. 377–382.
83. Shankar S.S., Ahmad A., Pasrichaa R., and Sastry M. (2003). Bioreduction of chloroaurate ions by geranium leaves and its endophytic fungus yields gold nanoparticles of different shapes. *Journal of Materials Chemistry*, **13**, pp. 1822–1826.
84. Schabes-Retchkiman P.S., Canizal G., Herrera-Becerra R., Zorrilla C., Liu H.B., and Ascencio J.A. (2006). Biosynthesis and characterization of Ti/Ni bimetallic nanoparticles. *Optical Materials*, **29**, pp. 95–99.
85. Freundlich M.M. (1963). Origin of the electron microscope, *Science: New Series*, **142**, pp. 185–188.
86. Williams D.B. and Carter C.B. (2009). *Transmission Electron Microscopy: A Textbook for Materials Science*, Springer, Springstreet, NY, USA.
87. Neuman K.C. and Nagy A. (2008). Single-molecule force spectroscopy: Optical tweezers, magnetic tweezers and atomic force microscopy, *Nature Methods*, **5**, pp. 491–505.
88. Russel W.B., Saville D.A., and Schowalter W.R. (1992). *Colloidal Dispersions*, Cambridge University Press, ISBN 0-521-42600-6.

Chapter 6

Plasmon–Gain Interplay in Metastructures

Antonio De Luca,[a] Melissa Infusino,[b] Alessandro Veltri,[b] Kandammathe V. Sreekanth,[c] Roberto Bartolino,[a,d] and Giuseppe Strangi[a,c]

[a]*Department of Physics and CNR-IPCF UOS di Cosenza, University of Calabria, 87036 Rende, Italy*
[b]*Colegio de Ciencias e Ingeniería, Universidad San Francisco de Quito, Quito, Ecuador*
[c]*Department of Physics, Case Western Reserve University, 44106-7079 Cleveland, USA*
[d]*Centro Linceo dell'Accademia Nazionale dei Lincei, Via Lungara, Rome, Italy*
gxs284@case.edu

6.1 Introduction: Background and Significance

The aim of this chapter is a profound discussion of the interplay between gain media and plasmon elements in metamaterials for visible light. The exciton–plasmon dynamics arising by specific coupling configurations has been the core of scientific discussions and experimental studies to explain extraordinary physical processes. In fact, the coupling plasmon–gain has been proposed as a challenging solution to tackle and solve the unavoidable issue of optical losses in metal-based nanostructures with plasmonic resonances at optical

Active Plasmonic Nanomaterials
Edited by Luciano De Sio

ISBN 978-981-4613-00-2 (Hardcover), 978-981-4613-01-9 (eBook)
www.panstanford.com

frequencies. The fascinating ability of metal nanostructures to localize light at scales much shorter than visible wavelengths is accompanied by enormous ohmic losses, with direct consequences as the remarkable augment of the extinction cross section of the material. This implies that extraordinary physical properties related to light localization effects at the nanoscale cannot be harnessed to design optical materials because of the strong radiation damping. The idea to bring gain molecules in close proximity to metallo-dielectric nanostructures is based on coherent effects of excitation energy transfer between resonant bands of the two materials. Radiation-less transitions from exciton levels of chromophores to plasmon states of metal nanoparticles (NPs) are responsible of the modification of the energy of their quasi-static field, that in turn modify the complex dielectric function of metastructures. To increase the efficiency of the resonant energy transfer, several coupling parameters have to be accounted both chemically and physically [Fang et al. (2009)]. This would enable promising new applications of these materials in fields such as materials science [Imahori and Fukuzumi (2001)], biophysics [Powell et al. (1997)], molecular electronics, and fluorescence-spectral engineering based on surface-enhancement effects [Lakowicz (2001)]. Compensation of the strong losses caused by metal absorption would permit us to operate at optical frequencies, opening the possibility to investigate phenomena such as perfect lenses [Pendry (2000)], cloaking [Schurig et al. (2006); Cai et al. (2008); Veltri (2009)], and others not yet conceived. Recent experimental works performed on plasmonic structures with gain units dissolved in solution [Noginov et al. (2006); De Luca et al. (2011, 2013); Strangi et al. (2011)] showed that the presence of fluorescent molecules in a mixture may modify the scattering intensity as a function of the gain owing to the enhancement of the quality factor of surface plasmon resonances (SPRs). It is well known that relevant modifications of the fluorescence of dye molecules placed in close proximity to metal NPs are due to mutual interactions with NPs surface plasmons, including resonant energy transfer (RET) [Gersten and Nitzan (1981); Das and Puri (2002); Weitz et al. (1983)]. A localized surface plasmon represents a collective oscillation of electron charges in metallic NPs, whose resonance frequency is sensitive to dielectric changes of the environment, as well as to the size and shape of the NP. A phenomenon relevant to localized surface plasmons is a surface plasmon polariton (SPP), that is, a surface electromagnetic wave propagating parallel to the interface between two media possessing permittivities with opposite signs, such as a metal and a dielectric. In both cases, oscillations are excited by light, exhibiting enhanced near-field amplitude at the resonance wavelength. Localized surface plasmons have been found on rough surfaces [Ritchie (1973); Moskovits (1985)], in engineered nanostructures [Quinten et al. (1998); Averitt et al. (1999); Brongersma et al. (2000); Mock et al. (2002)], as well as in clusters and aggregates of NPs [Kreibig and Vollmer (1995); Su et al. (2003); Quinten (1999)]. In 1989, Sudarkin and Demkovich suggested to increase the propagation length of an SPP by utilizing the population inversion created in the dielectric medium adjacent to the metallic film. Recently, gain-assisted propagation of SPPs at the interface between a metal and a dielectric with optical gain has been analyzed theoretically [Nezhad et al. (2004); Avrutsky (2004)]. The enhancement of two orders of magnitude of the SPP at the interface between the silver film and the dielectric medium with optical gain (laser dye) has been demonstrated by Seidel et al. (2005), and described by Noginov for Ag aggregates in solution [Noginov et al. (2006)]. Furthermore, a relevant phenomenon of surface plasmon amplification by stimulated emission of radiation (SPASER), based on Förster-like energy transfer from excited molecules to resonating metallic nanostructures introduced by Stockman et al. in 2003 [Bergman and Stockman (2003); Stockman (2008)], has been theoretically analyzed by Zheludev et al. (2008) and experimentally demonstrated by Noginov et al. (2009). At the same time, a theoretical self-consistent calculation on gain-assisted metamaterials was proposed in 2009 by Fang et al. [Noginov et al. (2006)], showing that 2D dispersive metamaterial losses can be compensated ($\mathrm{Im}(\varepsilon) = 0$), whereas both positive and negative values of $\mathrm{Re}(\varepsilon)$ can be obtained. Nevertheless, most of these studies have been performed in "gain-assisted" systems, where striking plasmonic properties have been evidenced, but the unavoidable problem of absorptive losses still requires many technical and scientific problems to be solved. The research strategies reported in this chapter deal with cross-disciplinary approaches that involve design and tailoring of electromagnetic properties, materials preparation, advanced experimental studies, and theoretical modeling. In particular, in Section 6.2 theory and modeling of the plasmon–gain interplay in metastructures is discussed by means of the semiclassical approach, where different systems are modeled as function of the gain. Materials functionalization and experimental investigations are reported in Section 6.3, time-resolved and transient absorption spectroscopy, spectrophotometry and spectroscopic ellipsometry are only a few of the experimental techniques utilized to study how plasmon–gain dynamics can be directed to mitigate optical losses across scales.

6.2 Plasmon–Gain Interplay: Theory and Modeling

Following the work presented by Lawandy (2004), it has been recently conjectured through a purely classical, yet geometrically sound, approach that a single-metallic NP immersed in a gain medium may produce amplification and distortions in some regions of the plasmonic spectrum [Veltri and Aradian (2012)].

In an even more recent paper [Veltri et al. (submitted)], it has been shown that below a singular point (i.e., when the gain level is lower than some threshold value at which the sign of the imaginary polarizability will become negative), the quasi-static classical approach provides an accurate description of the interplay between gain and plasmonic nano-objects and allows to predict where amplifying regimes will take place as well as to correctly estimate the gain level needed to achieve amplification.

Taking into account that we can use this simple approach to model more and more complex nanostructures. In the following subsections, we will present a few examples of how this can be applied to the nano- and mesoscale: a single nanosphere in a gain-assisted medium, a core–shell structure where the gain is embedded in the shell and a plasmonic mesocapsule.

6.2.1 *Single Spherical Nanoparticle*

We consider a single spherical NP made of a metal with a relative permittivity ε_1 depending on the frequency ω: $\varepsilon_1(\omega) = \varepsilon_1'(\omega) + i\varepsilon_1''(\omega)$, with $\varepsilon_1'(\omega) < 0$ (for ω below plasma frequency) and $\varepsilon_1''(\omega) > 0$ (representing ohmic losses). In the following discussion, actual values of ε_1 for gold and silver have been interpolated [Johnson and Christy (1972)]. The nanosphere is immersed in a host medium, e.g., a solution of dye molecules or quantum dots (QDs), with relative permittivity $\varepsilon_2 = \varepsilon_2' + i\varepsilon_2''$. We consider the polarizability α of the NP, which relates the total dipole moment $\mathbf{p}$ to the local electric field $\mathbf{E}_{\text{loc}}$ as $\mathbf{p} = \alpha\mathbf{E}_{\text{loc}}$. The polarizability is classically given as [Craig F. Bohren (1998)]:

$$\alpha(\omega) = 4\pi r^3 \frac{\varepsilon_1(\omega) - \varepsilon_2(\omega)}{\varepsilon_1(\omega) + 2\varepsilon_2(\omega)}, \tag{6.1}$$

In the common case of a sphere immersed in a passive, dielectric host medium with negligible losses ($\varepsilon_2' > 0, \varepsilon_2'' = 0$), Eq. 6.1 predicts the appearance of the localized surface plasmon resonance centered around the frequency ω_0 defined by $\varepsilon_1'(\omega_0) + 2\varepsilon_2(\omega_0) = 0$. The real part $\alpha'(\omega)$ assumes a classical ripple-like "up-down" lineshape, while the imaginary part has a bell-like Lorentzian lineshape. This behavior is very clearly exhibited in silver NPs (see left of Fig. 6.1a), while the response for Au NPs shows some distortions under the effect of a higher level of interband losses (see right of Fig. 6.1a).

For many applications and devices based on plasmonic resonators, one is interested in working in the regions where the α' (real) values are largest (positive or negative), that is, on the wings of the resonance; however, these are inevitably associated with significant α'' losses. The ideal situation looked for is that of singular "perfect plasmons" arising in lossless metals: When $\varepsilon_1''(\omega) = 0$, the real response becomes a singular function $\alpha' \sim 1/(\omega - \omega_0)$. Meanwhile, the width of the Lorentzian shrinks to a Dirac peak, centered at ω_0, which means that losses essentially vanish over the whole spectrum except at the resonance frequency.

The question then arises whether it is possible to approximate this behavior in any way by compensating the losses in real metals with a surrounding gain medium. Lawandy (2004), also considering Eq. 6.1, has found the conditions for singularity to be recovered,

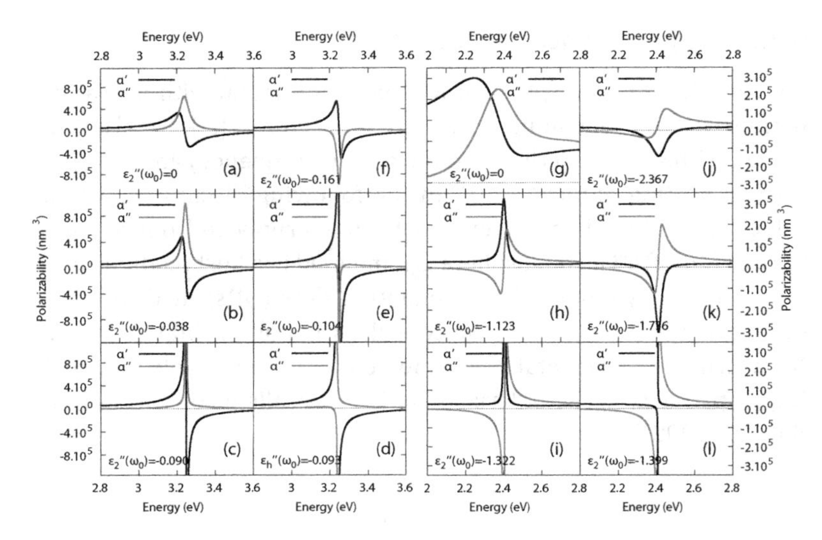

Figure 6.1 Evolution of the plasmon resonance of a 10 nm silver NP as gain is increased from (a/g) to (f/j), in the surrounding medium, before and after the singular plasmon value (d/l). Parameters: $\varepsilon_\mathrm{b} = 1.769$ (water) *Silver* (Left): $\omega_0 = \omega_\mathrm{g} = 3.24$ eV, $\Delta = 0.15$. *Gold* (Right): $\omega_0 = \omega_\mathrm{g} = 2.41$ eV, $\Delta = 0.35$.

but to the best of our knowledge, the exact spectral behavior of the plasmonic response at singularity has never been studied in detail, nor at lower or higher gain levels. We here show that very interesting new plasmonic behaviors can arise. We do not discuss here the question of the stability of the found solutions, and we present results of the steady state, linear solutions, which however have been proven to be valid until a threshold value for the gain is reached [Veltri et al. (submitted)]. We, therefore, take interest in the behavior of $\alpha(\omega)$ in the presence of an amplifying medium surrounding the NP, with $\varepsilon_2''(\omega) < 0$. Without loss of generality, a single Lorentzian emission lineshape is assumed for the gain host:

$$\varepsilon_2(\omega) = \varepsilon_\mathrm{b} - \frac{\varepsilon_2''(\omega_\mathrm{g})\Delta}{2(\omega - \omega_\mathrm{g}) + i\Delta}. \tag{6.2}$$

This permittivity corresponds to the steady-state solution to the optical Bloch equations for a two-level system, where ε_b is the real, positive permittivity of the background medium in which the gain elements are dispersed, and

$$\varepsilon_2''(\omega_g) = -4\pi \frac{n\mu^2}{3\hbar\Delta} \quad (6.3)$$

where Eq. 6.3 denotes the maximum value of $\varepsilon_2''(\omega)$ and sets the global level of gain in the medium, Δ is related to the emission linewidth and ω_g is the emission central frequency, and μ is the transition dipole moment of a single gain element. The gain molecules or nanocrystals are externally pumped at some (absorption) frequency considerably shifted from the plasmon resonance at ω_0.

In the following discussion, we are going to present results assuming that the dye emission frequency is centered on the plasmon resonance, that is, $\omega_g = \omega_0$. This assumption, here made for the sake of simplicity, can be relieved: Our calculations show that the same results are obtained with uncentered dyes, with the only differences that higher global gain levels are required, and that the plasmon resonance will progressively drift from its intrinsic (zero-gain) frequency in the direction of the dye emission frequency as the gain level in the system is increased.

Singular Plasmons Singular plasmonic behavior can be retrieved by completely cancelling the denominator in Eq. 6.1 at the plasmon frequency ω_0 [Lawandy (2004)], that is, by having not only $\varepsilon_1'(\omega_0) = -2\varepsilon_2'(\omega_0)$, but also $\varepsilon_1''(\omega_0) = -2\varepsilon_2''(\omega_0)$ through proper adjustment of the amount of gain.

The corresponding singular curves $\alpha'(\omega)$ and $\alpha''(\omega)$ for 10 nm silver and Au NPs are plotted numerically in Fig. 6.1d left and right, and deserve several comments: (i) as expected, the resonances are very sharp, but *both* α' and α'' have a $1/(\omega - \omega_0)$ behavior, which means that gain-assisted plasmons—even singular—have a spectrally spread, nonvanishing imaginary response. This is a crucial difference to be hoped for, perfect plasmons stemming from the fact that in the denominator of Eq. 6.1, losses can be compensated at only one frequency ($\omega = \omega_0$) rather than over the whole spectrum. This asymptotic imaginary response depends crucially on the intrinsic metal properties, and, in this sense, the compensation of losses is more efficient in silver than in gold, which has more significant spreading; (ii) as ω is increased, the particle's behavior transits from an active state ($\alpha'' < 0$) before the resonance, to a passive, absorptive behavior beyond it ($\alpha'' > 0$); therefore, for singular plasmons, spaser-like or emissive situations should be looked only for $\omega < \omega_0$. If one is interested in such situations, gold may be more appropriate, since the negative imaginary part α'' is stronger than in silver; (iii) the level of gain required to retrieve singularity is much larger for gold [$\varepsilon_2''(\omega_0) = -1.399$ in Fig. 6.1j, due to higher interband losses, than for silver [$\varepsilon_2''(\omega_0) = -0.093$ in Fig. 6.1d].

Low-loss Metal Behavior We now turn to the particle response before and after the singular point. The behavior is significantly different, depending on the nature of the metal, silver or gold. Looking at the situation for silver (Fig. 6.1 left), we see that it simply corresponds to that of a plasmon of increasing quality and amplitude as $\varepsilon_2''(\omega_0)$ increases from zero (Fig. 6.1b,c), until the singular point is reached (Fig. 6.1d). As $\varepsilon_2''(\omega_0)$ is increased after the singular point, the plasmon gradually degrades due to excess gain, since the denominator in Eq. (6.1) acquires a growing (negative) imaginary part. Note, however, that the imaginary part of α is now negative in the resonant region, meaning a state where the particle acts as a net emitter of light (Fig. 6.1e,f left). Very close to the singular point, the emission line can be very sharp (Fig. 6.1e); this could probably yield appropriate conditions for the appearance of spasing [Noginov et al. (2009); Stockman (2008)].

High-loss Metal Behavior The situation for gold is strikingly different, with richer behavior (Fig. 6.1 right). In the absence of gain [$\varepsilon_2''(\omega_0) = 0$], the plasmon resonance for gold, compared to that of silver, is distorted and not very pronounced, due to high interband losses. As gain is added toward the singular point, the resonance takes on sharper features (increased quality factor) but is also increasingly distorted. Indeed, for a large range of gain values (Fig. 6.1h,i), one observes a most interesting situation arising, where the *real* part $\alpha'(\omega)$ has a bell-like shape (whereas this is what is usually seen for imaginary response, as in silver for example), and conversely the *imaginary* part has now the ripple-like shape (normally expected for real response). We call this original behavior "conjugate plasmon," since usual real and imaginary parts are swapped. This new behavior shows one particularly attractive property: at the plasmon frequency, where the real response is maximal, losses are also close to zero; this is in fact much more favorable for practical applications than the situation of usual plasmons, where the extrema of the real response, located on the wings of the resonance, come with non-negligible losses, which degrade plasmonic system performance. Increasing the gain level past the singular point (Fig. 6.1l) reveals a symmetrical situation where conjugate plasmons also appear. While conjugate plasmons obtained before the singular point had a positive real part, here, they display a negative real part (Fig. 6.1k,l right). This type of responses could, therefore, be seen as even more interesting than their positive counterparts, if one is interested in obtaining artificial, low-loss media with so-called "negative" properties. We again performed an analytical study of the set of behaviors displayed in Fig. 6.1 right, confirming that the appearance of such conjugate plasmons in gold (negligible for silver) is directly due to the higher intrinsic loss level (interband transition).

6.2.2 *Core–Shell Spherical Nanoparticles*

The approach presented in the previous section can be correctly applied to model experimental measures in which the gain level is lower than the one required for the singular behavior. Here we use the quasi-static expression for the polarizability of a core–shell particle of internal radius r_1 and external radius r_2, having ε_1, ε_2, and ε_3 as the complex permittivity of the core, the shell, and the external medium, respectively [Craig F. Bohren (1998)], which can be written as:

$$\alpha = 4\pi r_2^3 \frac{(\varepsilon_2 - \varepsilon_3)(\varepsilon_1 + 2\varepsilon_2) + \rho^3(\varepsilon_1 - \varepsilon_2)(\varepsilon_3 + 2\varepsilon_2)}{(\varepsilon_2 + 2\varepsilon_3)(\varepsilon_1 + 2\varepsilon_2) + 2\rho^3(\varepsilon_2 - \varepsilon_3)(\varepsilon_1 - \varepsilon_2)} \quad (6.4)$$

where $\rho = r_1/r_2$. Again we used a spline interpolation of the Johnson and Christy (1972) dataset for gold to describe $\varepsilon_1(\omega)$ and expression 6.2 for $\varepsilon_2(\omega)$. Now ε_b is the permittivity of the host in which the gain is dissolved, Δ is the bandwidth, and ω_0 is the center of the line for the used dye. Once again $\varepsilon_2''(\omega_g)$ sets the level of gain and can be related to the density of fluorophores via expression 6.3. For the sake of simplicity, we used a dye having the emission band centered in the plasmonic resonance, with a bandwidth comparable with the one of the plasmon without gain. For ε_b and for ε_3, we used the values of silica and ethanol ($\varepsilon_b = 2.1316$, $\varepsilon_3 = 1.8496$). Using Eq. 6.4, it is possible to calculate the absorption cross section of our NP as:

$$\sigma = \frac{2\pi\sqrt{\varepsilon_3}\alpha''}{\lambda}, \tag{6.5}$$

from which we can finally calculate the absorbance of the system:

$$A = \frac{\sigma n_{NP} L}{\ln(10)}, \tag{6.6}$$

where n_{NP} is the density of NPs ($n = \rho / W_c$, where W_c is the weight of the gold core, given by $W_c = \frac{4\pi}{3} r_1^3 \rho_0$, where $\rho = 0.38$ g/L and $\rho_0 = 0.193$ g/L) and $L = 1$ mm, the thickness of the sample. By considering values of $r_1 = 0.6$ nm and $r_2 = 0.4$ nm, the simplified model predicts an absorbance behavior for the system in absence of gain [$\varepsilon_2''(\omega_0) = 0$)] that is reported in Fig. 6.2 (black solid line). By adding a certain amount of gain into the shell [$\varepsilon_2''(\omega_0) = -0.5$)], the model predicts the modification of the absorbance curve (green solid line in Fig. 6.2) in a way that follows the same behavior of

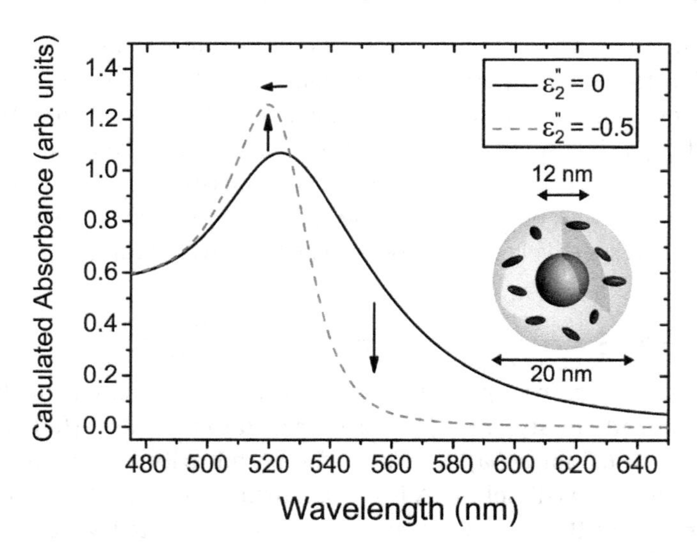

Figure 6.2 Calculated absorbance of a gold core/silica shell (12 nm/4 nm) with dye-encapsulated into the shell. $\varepsilon_2''(\omega_0) = 0$ means zero gain, whereas $\varepsilon_2''(\omega_0) = -0.5$ means a certain gain considered into the silica shell.

the experimental data reported in Section 6.3.1, including the slight maximum blue shift. The calculated absorbance reported in Fig. 6.2 is relative to a gain-encapsulated silica shell of thickness 4 nm. This choice is due to the fact that only fluorophores in close proximity result strongly coupled with plasmon gold core, thus promoting a remarkable resonant energy transfer process.

6.2.3 *Plasmonic Mesocapsules*

In order to describe the mitigation of losses in these systems, we calculate the absorption cross section using expression 6.5.

The polarizability of the mesocapsule $\alpha(\omega)$ is modeled as a core-shell in which the shell permittivity ε_2 is calculated via the Maxwell–Garnett mixing rule (Fig. 6.3):

$$\varepsilon_2 = \varepsilon_{22}\frac{\varepsilon_{21}(1+2f) - \varepsilon_{22}(2f-2)}{\varepsilon_{22}(2+f) + \varepsilon_{21}(1-f)}, \tag{6.7}$$

where ε_{21} is the permittivity of the gold inclusions, $\varepsilon_{22} = 2.10143$ is the one of the silica matrix, and f is the volume fraction of

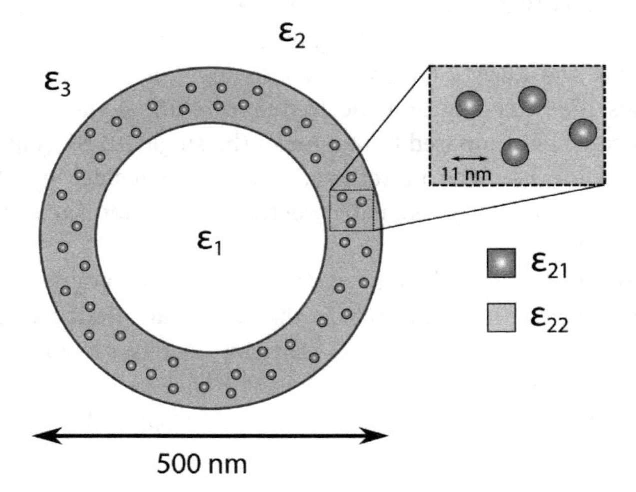

Figure 6.3 We model the mesocapsule as a core–shell particle in which the shell permittivity ε_2 is calculated via Maxwell–Garnett mixing rule. Here ε_{21} is the permittivity of the metal inclusions, ε_{22} is the permittivity of the silica structure, ε_1 is the permittivity in the hollow core, and ε_3 is the permittivity of the host.

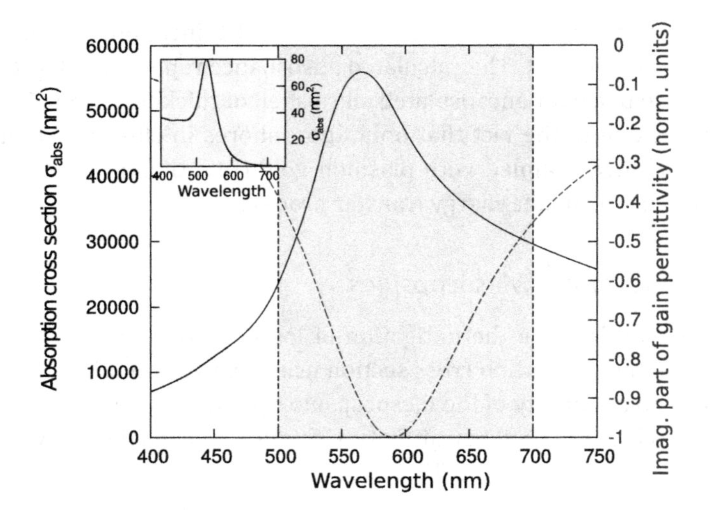

Figure 6.4 Calculated absorption cross section (black continuous line) and imaginary part of gain permittivity (red dashed line); in the inset we show the absorption cross section of a single-metal NP.

the inclusions (we used $f = 0.3$). Once again, to give a realistic description of ε_{21} at the optical frequencies, we interpolated the data of Johnson and Christy for gold permittivity. This model correctly describes the widening and the frequency shift of the collective resonance when compared to the one of the single Au NP (Fig. 6.4) and underline how the mesocapsules can be considered a loss-compensation model system in between a single nanoresonator and a bulk material.

Mesocapsules are objects whose size is comparable to the incident wavelength; for this reason the quasi-static limit cannot be applied and the polarizability $\alpha(\omega)$ has to be calculated using the Mie solution, which for a core–shell particle with core radius r_1 and shell radius r_2, made of three uniform materials with permittivity ε_1, ε_2, and ε_3, is:

$$\alpha = \frac{3i}{4}\frac{\lambda^3 \varepsilon_3}{\pi^2} a_1 \tag{6.8}$$

with a_1 being the first Mie coefficient for a core–shell spherical particle:

$$A_1 = \frac{m_2\psi_1(m_2x)\psi_1'(m_1x) - m_1\psi_1'(m_2x)\psi_1(m_1x)}{m_2\chi_1(m_2x)\psi_1'(m_1x) - m_1\chi_1'(m_2x)\psi_1(m_1x)} \tag{6.9}$$

$$a_1 = \frac{\psi_1(y)\left[\psi_1'(m_2y) - A_1\chi_1'(m_2y)\right] - m_2\psi_1'(y)\left[\psi_1(m_2y) - A_1\chi_1(m_2y)\right]}{\zeta_1(y)\left[\psi_1'(m_2y) - A_1\chi_1'(m_2y)\right] - m_2\zeta_1'(y)\left[\psi_1(m_2y) - A_1\chi_1(m_2y)\right]} \tag{6.10}$$

here $m_1 = n_1/n_3$, $m_2 = n_2/n_3$, where n_1 is the index of the sphere ($n_1 = \sqrt{\varepsilon_1}$), n_2 the index of the shell ($n_2 = \sqrt{\varepsilon_2}$), and n_3 the index of the external medium ($n_3 = \sqrt{\varepsilon_3}$) while $x = 2\pi\lambda/r_1$ and $y = 2\pi\lambda/r_2$; ψ, χ and ζ are the Riccati-Bessel functions.

The result of this calculation, in this case the mesocapsules are diluted in ethanol ($\varepsilon_1 = \varepsilon_3 = 1.8496$), is presented in Fig. 2 of original work.

Effect of Gain Here we used expression 6.2 to calculate both ε_1 and ε_3:

$$\varepsilon_{1,3}(\omega) = \varepsilon_b - \frac{\varepsilon_{1,3}''(\omega_g)\Delta}{2(\omega - \omega_g) + i\Delta}, \tag{6.11}$$

where ε_b is the real, positive permittivity of the background medium in which the gain elements are dispersed, $\varepsilon_{1,3}''(\omega_g)$ denotes the maximum value of $\varepsilon_{1,3}''(\omega)$ and sets the global level of gain in the medium, Δ is related to the emission linewidth, and ω_g is the emission central frequency. In order to be closer to experiments, we choose $\hbar\omega_g = 2.19441$ eV (corresponding to $\lambda_g = 590$ nm, that is, the wavelength corresponding to the fluorescence maximum when R6G concentration is 1.2 mg mL^{-1}) and $\hbar\Delta = 0.6$ eV.

To highlight the effect of the global level of gain, we define:

$$\Delta\sigma_{abs}^{\%} = \frac{\sigma_{abs}(\varepsilon_3''(\omega_g)) - \sigma_{abs}^0}{\sigma_{abs}^0}, \tag{6.12}$$

where σ_{abs}^0 is the absorption cross section of the mesocapsule when no gain is added to the system. The behavior of $\Delta\sigma_{abs}^{\%}$ has been reported in the paper, where we presented two characterizations: one assuming the gain to be only outside the gold-rich inner shell (Fig. 6.14a) and another assuming the gain to be also in the hollow core (Fig. 6.14b).

This result is not surprising because in the hollow core of the mesocapsule, the electric field is uniform and, consequently, it is

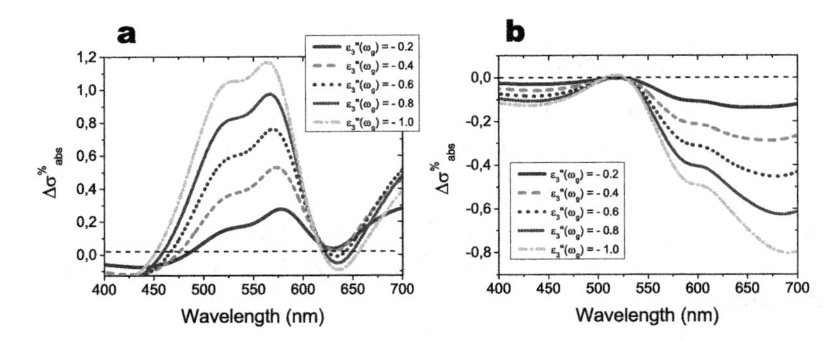

Figure 6.5 (a) Calculated behavior for $\Delta\sigma_{\text{abs}}^{\%}$ for different values, levels of gain ($\epsilon_3''(\omega_g)$ from -0.2 to -1.0). (c) Gain elements are assumed to be only outside the core of mesocapsules; (b) they are assumed to be both outside and in the core of mesocapsules.

easier for the dipole moments of gain elements to be aligned with it, making more efficient the energy transfer needed to realize an effective loss mitigation. From these results one can infer that, in the experiments, a not negligible amount of gain molecules infiltrated the hollow core of mesocapsules, showing how important it is to strategically position the gain in dependence of the geometry of the plasmonic structure.

6.2.4 *Effective Medium Theory*

In this section, we discuss the effective medium theory (EMT) to study the wave propagation of light in one-dimensional metal/dielectric multilayer structure [hyperbolic metamaterials (HMMs)]. Such indefinite hyperbolic media possess a diagonal form of the permittivity tensor ($\varepsilon = diag(\varepsilon_x, \varepsilon_y, \varepsilon_z)$) in which diagonal elements have different signs ($\varepsilon_x = \varepsilon_x$ and $\varepsilon_x \varepsilon_z < 0$) that lead to a hyperbolic dispersion, that is, $\omega^2/c^2 = (k_x^2 + k_y^2)/\varepsilon_z + k_z^2/\varepsilon_x$. It shows that HMMs have dielectric properties ($\varepsilon > 0$) in one direction and metallic properties in another direction ($\varepsilon < 0$). Interestingly, HMM supports high wavevector propagating waves (high-k modes) due to the hyperbolic dispersion. Note that these high wavevector propagating waves have an effective wavelength inside the medium, which is much smaller than the free space wavelength. However, a conventional isotropic medium is associated with a closed spherical dispersion and the related waves are evanescent. In order to achieve homogeneity and hence EMT to be valid, the unit cell of HMM must be much smaller ($\lambda/10$) than the excitation wavelength. Since HMM is an anisotropic medium with uniaxial dielectric tensor components such as $\varepsilon_{xx} = \varepsilon_{yy} = \varepsilon_{\|}$ and $\varepsilon_{zz} = \varepsilon_{\perp}$, which are approximated using EMT as follows:

$$\varepsilon_{\|} = \frac{t_m\varepsilon_m + t_d\varepsilon_d}{t_m + t_d} \tag{6.13}$$

$$\varepsilon_{\perp} = \frac{\varepsilon_m\varepsilon_d(t_m + t_d)}{t_m\varepsilon_d + t_d\varepsilon_m}. \tag{6.14}$$

In Eqs. 6.13 and 6.14, (t_d, ε_d) and (t_m, ε_m) are the thickness and dielectric permittivity of dielectric and metal, respectively. Here, we study the wave propagation in Au/TiO_2 multilayer structure using EMT. In the calculation, the optical constants of Au are calculated based on Drude free-electron theory, $\varepsilon_m = (\omega_p^2/\omega(\omega + i/\tau))$, where ω_p is the plasma frequency of Au, ω is the excitation frequency, and τ is the relaxation time. The dielectric constant of TiO_2 is set to be 7.3. The thicknesses of Au and TiO_2 are assumed to be 16 nm and 32 nm, respectively. The simulated result is shown in Fig. 6.6. According to

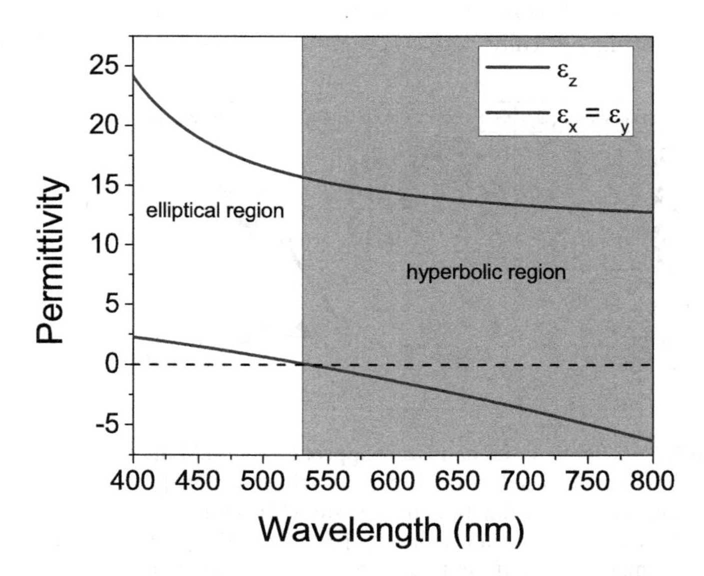

Figure 6.6 Variation of effective permittivity with wavelength.

Fig. 6.6, the hyperbolic dispersion is obtained for certain wavelength in which values of the real part of $\varepsilon_{\parallel}$ are negative. However, the values of the real part of $\varepsilon_{\parallel}$ result positive in the elliptical region. The critical wavelength, that is, the wavelength at which transition occurs from elliptical to hyperbolic dispersion is 530 nm. The tuning of the critical wavelength is possible by varying the periodicity of the HMM. The experimental validation of this simulation result is shown in the following sections.

6.3 Systems and Approaches Across Scales

In this section, we will present experimental studies aimed to demonstrate effective chemical and physical approaches to mitigate the absorptive loss effect in different systems, depending on the scale range. We will start from the nanoscale, which includes solutions of "gain-assisted" (G_A) and "gain-functionalized" (G_F) core–shell metal nanospheres, selected as metamaterial building blocks. In the gain-assisted system, fluorescent guest molecules are dissolved in solution with the plasmonic NPs, whereas NP gain functionalization comprises "smart nanostructures" obtained by incorporating optically active components (QDs or organic dyes)

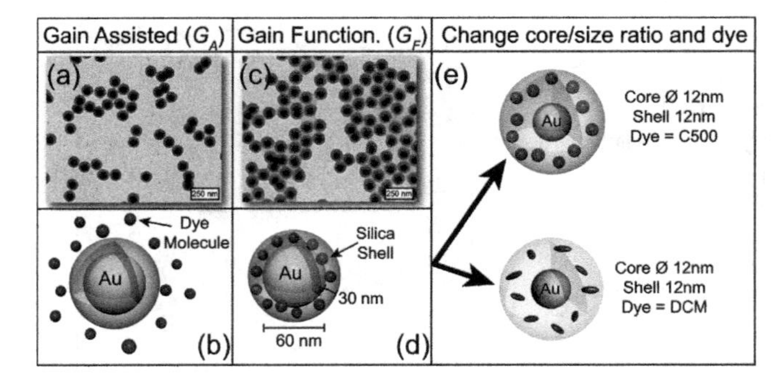

Figure 6.7 Scheme of the nanoscale approach. (a) TEM image of the gold-core/silica-shell NP used to obtain a gain-assisted system [sketch in (b)]. (c) TEM image of the gain-functionalized system [Au@(SiO_2+ R6G)], with R6G incorporated within the silica shell. (d) Sketch of the gain-funtctionalized (G_F) system. (e) Two different G_F systems, with a modified core–shell size ratio (1:1) and different dyes encapsulated into the silica shell.

within the silica shell (see Fig. 6.7). Then we will handle two novel different approaches at meso- and macroscale, showing that transferring the attention to bulk systems, the plasmon–exciton coupling assumes new characteristics, and collective resonances and plasmon hybridization processes have to be accounted. RET processes are at the basis of plasmon–gain interplay in metastructure, as well as optical loss-compensation mechanisms and require a series of physical and chemical conditions that have to be strictly satisfied. Spectral overlapping between plasmonic resonances and gain emission band, dye molecules concentration, NPs-dye molecules interdistance, core–shell size ratios. Material parameters have to be accounted to optimize dipolar, as well as multipolar, interactions that are responsible for nonradiative resonant transfer of the excitation energy from gain units to plasmonic nano-objects.

6.3.1 *Nanoscale Approach*

The approach at nanoscale has been performed by investigating primarily the photo-physical properties of dye-NPs solutions, to evaluate the main physical parameters governing the RET mechanisms. Subsequently, gain-functionalized composites have been considered, whose electromagnetic properties were modeled taking into account the presence of gain elements that can shift plasmonic resonances modifying the frequency response of the material. In particular, the effective rate of (nonradiative) excitation energy transfer from fluorescent dye molecules to surface plasmon modes of functionalized gold-core/silica-shell NPs has been investigated by exploiting a surface coating approach (see Fig. 6.7). The predicted behavior of the nonradiative transfer rate governing the Coulombic interactions [Förster (1948)] and a nontrivial NP size dependence represent key parameters to properly include gain units and provide effective gain to plasmonic elements (inset of Fig. 6.8 shows an RET process from gain to plasmonic particle). Therefore, a multipronged synthetic strategy was followed to bring the required functionalities at the plasmonic NPs acting as low-loss nanoresonators. The first step involved the dissolution of organic dyes in the NP dispersion to optimize compatibility, spectral overlap, and nonradiative transfer rate [gain-assisted approach, (a) and (b) in Fig. 6.7]; then this system

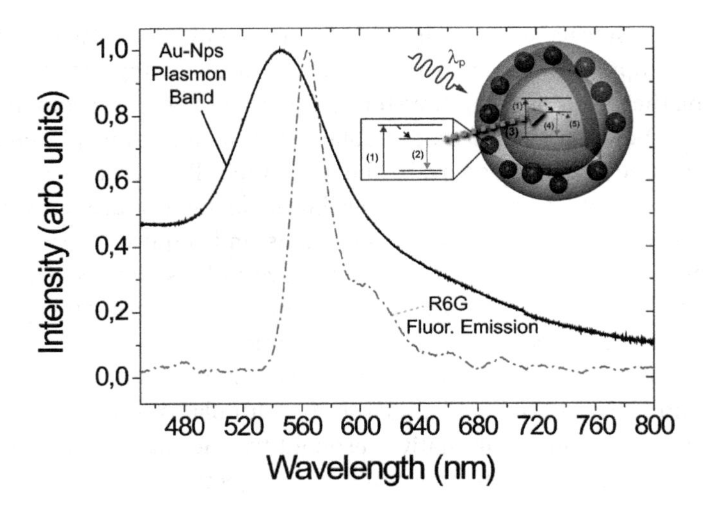

Figure 6.8 Emission (dash dot-dot) spectrum of rhodamine 6G dye in ethanol and plasmon band of G_A and G_F Au NPs. Inset: Sketch of the nonradiative RET process [green dashed line (3)] occurring from fluorescent dye molecules (red spheres, inside or outside the silica shell) to surface plasmon modes of properly functionalized gold cores (yellow sphere). *p* is the pump wavelength, (1) is excitation; (2) dye radiative de-excitation; (3) nonradiative energy transfer; (4) plasmon state de-excitation; (5) thermal de-excitation (electron–phonon coupling).

was compared to the G_F system where the selected fluorescent molecules were encapsulated within the silica shell surrounding the noble metal NPs. More in detail, the first gain-functionalized system consists of a gold core (diameter $d \sim 60$ nm) coated with a silica shell (30 nm thick) containing organic dye molecules (Fig. 6.7) [Doussineau et al. (2009); Fernández-López et al. (2009); Rodríguez-Fernández et al. (2006); Tovmachenko et al. (2006); Draine (2000)]. The photostable organic dye [rhodamine 6G (R6G)] was optically excited by laser pulses at 355 nm, showing that its gain curve overlaps the plasmon band of Au NPs in both systems: "gain-assisted" (G_A) and "gain-functionalized" (G_F) (Fig. 6.8). This represents another key parameter for an effective plasmon–exciton coupling. In case of (G_A) NPs, Au-core/silica-shell nanospheres (core diameter 60 nm, silica shell 30 nm) were dispersed in an ethanol solution of R6G (0.01 wt%). Fluorescence spectroscopy allowed us to gain understanding about the coupling between active dyes and plasmonic nanospheres. As observed by Dulkeith et al. (2002), if a donor fluorescent molecule is maintained in the vicinity of a metal NP, a drastic quenching of fluorescence is expected. This effect was predicted by Gersten and Nitzan (1981) by considering a remarkable enhancement of nonradiative resonant energy transfer between donor fluorophore and plasmonic acceptor. Figure 6.9a shows the fluorescence quenching observed in Au@SiO_2/R6G system with respect to the pure R6G ethanol-based solution (same concentration, 0.01 wt%), when optically pumped with large spotted pulse trains of a tripled Nd:Yag laser (λ = 355 nm) at the same pump energy.

By plotting the maximum fluorescence emission as a function of pump energy for both solutions (Fig. 6.9b), a different rate was observed. This can be the consequence of the resonant nonradiative energy transfer process occurring between gain molecules and metal units present in the surrounding volume, which causes the decrease in the radiative rate. Indeed, upon studying fluorescence changes in these gain-assisted systems, only preliminary indications of excitation energy transfer processes can be obtained. These observations are necessary but not sufficient to demonstrate mitigation of absorptive losses. Hence, time-resolved fluorescence spectroscopy along with pump-probe experiments have been carried out both on gain-assisted and gain-functionalized systems, to perform a comparative analysis of gain-induced optical loss modifications.

The step forward is represented by the functionalization of core–shell NPs by encapsulating properly selected fluorescent guest molecules within the silica shell to induce effective resonant energy transfer in dye-doped metastructures. In contrast with the gain-assisted system, this approach allows a fine control of key parameters such as dye-metal core interdistance and dye concentration, offering the advantage of clearing inactive fluorescent molecules from solution, and allowing to maintain almost unmodified the energy density of the optical pump. In this framework, material parameters of core–shell NPs have not been modified, the difference only being the encapsulation of R6G dye molecules within the shell. This process allows us to bring gain to single Au NPs,

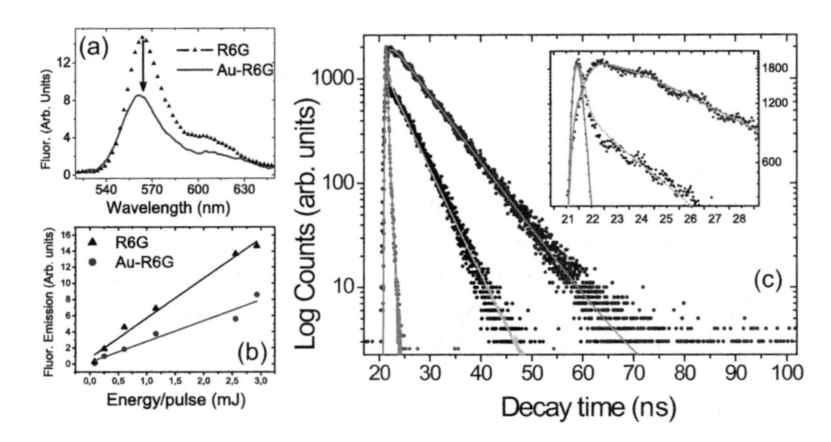

Figure 6.9 (a) Fluorescence quenching observed in G_A (black triangular dots) with respect to the R6G ethanol-based solution (solid blue line) under the same pump energy value. (b) Fluorescence emission maxima of the gain-assisted sample (blue circles) with respect to the R6G ethanol-based solution (black triangles) as a function of pump energy (λ@355 nm). (c) Time-resolved fluorescence intensity decays: G_A and G_F systems are compared to pure R6G dye solution. Long-living emission of pure R6G characterized by a time constant $\tau_{\text{Fluo}} = 5.4$ ns (dark red dots). Two components are identified in the decay dynamics of the G_A system: $\tau_{A1} \sim 190$ ps and $\tau_{A2} \sim 6.0$ ns (black dots). A short-living emission is identified for G_F ($\tau_F \sim 120$ ps), clearly indicating a strong dye-NP coupling. Inset: Zoom of the time-resolved fluorescence decays. The double decay time in the case of G_A system is more evident.

which can overcome direct energy feeding through nonradiative mechanisms. Fluorescence lifetime measurements demonstrate quenching behavior, consistent with a small separation distance from dyes to the surface of the NP, because of their strong resonant coupling. Figure 6.9c reports the time-correlated single-photon counting (TCSPC) data at 560 nm for G_A and G_F systems with respect to the pure R6G dye solution, when irradiated with a 265 nm NanoLed pulsed laser diode. The time-resolved fluorescence intensity decay of the ethanol solution of pure R6G molecules is fitted as a single exponential function in Fig. 6.9c (dark red dots and green line fit), giving a time constant of $\tau_{\text{Fluo}} = R_{\text{Fluo}}^{-1} = 5.4ns$ ($\chi^2 = 0.978$). From the TCSPC data of the gain-assisted system (black dots and cyan line fit), two components can be identified in the decay dynamics. A fast decay (of $\tau_{A1} \sim 190\,ps$; $\chi^2 = 0.978$, inset of Fig. 6.9c) is accompanied by a long-living emission where the decay kinetics resembles the fluorescence decay for pure R6G dye molecules. The first decay time is attributed to the fraction of dye molecules decorating the silica shell that experiences the resonant energy transfer process; the long-living emission is related to the fraction of unbound dye molecules (the largest fraction) present in solution but that are not coupled to plasmonic NPs. Strikingly, the time-resolved spectrum of the G_F system (pink dots) shows only a short-living intensity emission decay, fitted as a single exponential with a time constant $\tau_{A1} \sim 120\,ps$ ($\chi^2 = 1.058$). Accordingly, this decay time can be attributed to the encapsulated fluorescent molecules resonantly coupled to the plasmonic gold core. It is worth noting that in G_F system, we have not measured long-living emission due to unbound R6G molecules. Thus, the identification of a single exponential decay of the short-living encapsulated dye emission indicates a complete and effective dye-NP coupling, which is manifested as a significant reduction of the radiative rate. To exclude a possible interference caused by the presence of Au@SiO_2 particles, TCSPC data have been collected from a solution containing the Au@SiO_2 particles only; the results [not reported here, see De Luca et al. (2011)] obtained in the same experimental conditions as used for the G_F system are clearly different, confirming that the origin of the short decay is due to the dye-NP coupling. All these results elucidate the nonradiative energy transfer rate as a consequence of strong coupling for the G_F system. According to Beer–Lambert–Bouguer law, by measuring simultaneously Rayleigh scattering and transmission, either in the absence or in the presence of gain, we should be able to understand if the absorptive power of the material is affected by excitation energy transfer. Thus, modifications of Rayleigh scattering and transmitted intensity of a constant probe beam (λ = 532 nm) have been monitored as a function of the pump energy for both systems (excitation @ 355 nm, see pump laser in the setup of Fig. 6.10). As first proposed by Lawandy (2004), the localized surface plasmon resonance in metallic nanospheres is predicted to exhibit a singularity when the surrounding dielectric medium has a critical value of optical gain. This can be evidenced by an increase

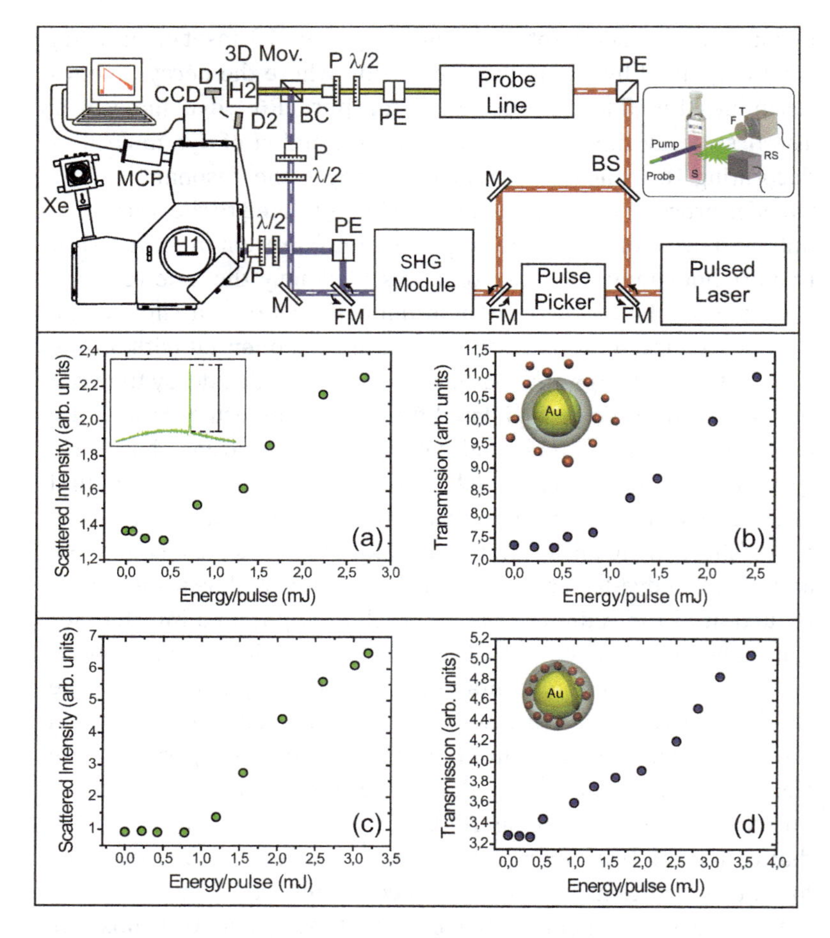

Figure 6.10 Multipronged spectroscopic setup, including pulsed laser at different wavelengths and repetition rates, and different probe lines, including a supercontinuum generation. Comparison of scattering and transmission enhancement signals between G_A (a and b) and G_F (c and d) samples. Enhancement of surface plasmon evidenced by an increase in normalized Rayleigh scattering signal of probe beam as a function of pump energy (355 nm) for (a) Au@SiO_2/R6G and (c) Au@(SiO_2+ R6G) samples. Normalized transmission increase of a probe beam at 532 nm as a function of pump energy for (b) Au@SiO_2/R6G and (d) Au@(SiO_2+ R6G) samples. Comparable threshold values for both transmission and scattering signals have been observed in the two samples.

in the Rayleigh scattering signal because of the enhancement of the local field surrounding the NP. Hence, a pump-probe Rayleigh scattering experiment enables to observe the enhancement of the surface plasmon resonance due to the gain material present in the solution or chemically encapsulated in the silica shell surrounding the gold core. Pump-probe Rayleigh scattering experiments were performed by a co-linear launch of a probe beam in a small portion of the volume excited by a wider pump spot at $\lambda = 355$ nm.

The scattered probe light, acquired by means of the optical fiber of a high-resolution spectrometer, was observed in the spectrum as a relatively narrow line centered at 532 nm, on the side of a much broader emission band relative to the dye fluorescence (see inset of Fig. 6.10a). Thus, it could be easily separated from the dye emission. The difference between the maximum of the scattered beam and the corresponding value in the fluorescence emission spectrum is plotted as a function of the pump energy. Figure 6.10a shows the enhancement of Rayleigh scattering signal as a function of excitation energy in system G_A. The super-linear increase in the scattered signal above a certain threshold value of gain is a demonstration of the enhancement of the quality factor of surface plasmon resonance mediated by resonant energy transfer processes between active elements and gold cores within the composite NPs. The presence of a threshold gain value above which systems show a nonlinear behavior was already discussed by Lawandy (2004), who argued that a singularity in the NP local field is expected as the transferred energy compensates the absorptive losses by exciting NP surface plasmon modes. However, the key experiment of this work was performed by measuring the transmission at far field of probe ligthwaves after passing the excited area with different levels of gain, selected by varying the pump rate. Figure 6.10b shows the increase in transmission peaks of the probe signal as a function of the excitation energy, evidencing a clear gain-induced increase in the whole system transparency. The measured transmitted signal is only relative to the probe beam wavelength, selected by pin-hole and notch filters, since stray light checks ruled out any other undesired contribution. A comparative set of measurements (both scattering and transmission) have been performed also on the system G_F, revealing similar threshold values with respect to

NPs-dye solution, even if the total amount of present gain units resulted considerably reduced with respect to the G_A system. We plotted in Fig. 6.10c the enhancement of Rayleigh scattering intensity at the probe wavelength (probe = 532 nm), as a function of the pump rate in the Au@(SiO_2 + $R6G$) sample. A pump energy threshold value of about 1 mJ/pulse represents a clear evidence that the amount of gain molecules, now encapsulated in the shell (10^3 – 10^4 molecules, by considering a few percents of occupied volume ratio), was sufficient to permit an effective nonradiative energy transfer to Au NPs and promote excitation of the surface plasmon modes, by providing an estimated local gain of about 10^4 – 10^5 cm^{-1} [Lawandy (2004)]. This effect can be explained by considering the interdistance dependence of the energy transfer rate [Bhowmick et al. (2006)], due to NP-dye dipolar interactions according to Förster theory (1948). Figure 6.10d shows that the transmission of light with wavelength selected within the resonant band and propagating through the excited volume experienced a considerable enhancement as the gain was increased. A critical behavior of the transmission was observed above a given threshold value of about 0.33 mJ/pulse, revealing that a reduction in optical absorption can be induced only if enough gain is provided. The gain-induced increase in optical transparency of meta-subunits becomes clearer when corroborated by experimental evidences, which provide both direct measurements of the physical quantity (transmission) and effects clearly related to the energy transfer process (fluorescence quenching and Rayleigh scattering enhancement).

To put in evidence the role and the key parameters governing the plasmon–exciton interplay in resonant energy transfer processes, we present the results obtained in terms of loss mitigation effects and strong plasmon–gain coupling in two different G_F systems: as shown in the scheme of Fig. 6.7, we modify the core–shell size ratio (previous case was 2:1; now it is 1:1) and used two different organic dyes as gain media functionalized into the silica shell. In these two new approaches at the nanoscale, metal NPs consist of a gold core (diameter d $\sim$ 12 nm) coated with a silica shell (12 nm thick) containing two photostable organic dyes (Coumarin C500 and DCM by Exciton, chemical formulae in the inset of Fig. 6.11(a) and (b), respectively) chemically encapsulated into the shell. The

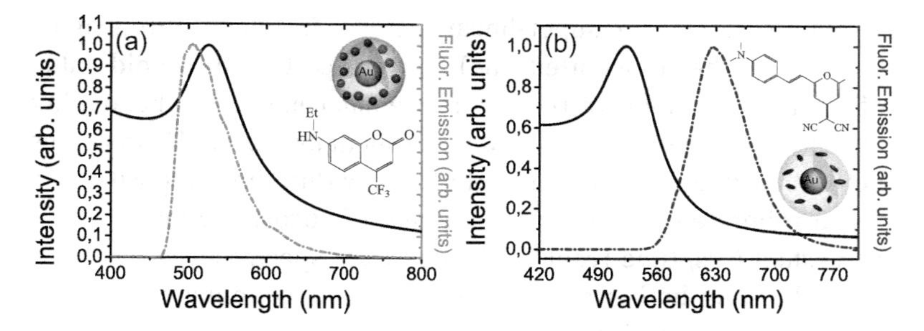

Figure 6.11 (a) Emission (green dash dot) spectrum of C500 dye dissolved in ethanol and plasmon band (black line) of $G_{F\text{-}C500}$ Au NPs. (b) Emission (red dash dot) spectrum of DCM dye dissolved in ethanol and plasmon band (black line) of $G_{F\text{-}DCM}$ Au NPs.

obtained systems are Au@SiO_2/$C500$ ($G_{F\text{-}C500}$) and Au@SiO_2/DCM ($G_{F\text{-}DCM}$). The reciprocal band overlapping between plasmonic NP resonances and emission of gain media are presented in Fig. 6.11. The two systems were optically excited by laser pulses at 355 nm, a wavelength far from the maximum of the NPs extinction band, showing a completely different gain curve overlapping with the plasmon band of Au NPs (black lines in Fig. 6.11). As we can see, the emission curves of the two dyes are considerably different, presenting a maximum at 500 nm (C500) and 630 nm (DCM), a spectral position that results blue shifted with respect to the plasmon band for C500, whereas it is red shifted for the DCM. We studied how the different spectral overlapping affects the coupling strength and the energy transfer processes between gain medium and metal NPs. As observed by Dulkeith et al. (2002), if a donor fluorescent molecule is maintained in the vicinity of a metal NP, a drastic quenching of fluorescence is expected. This effect was predicted by Gersten and Nitzan (1981) by considering a remarkable enhancement of nonradiative resonant energy transfer between donor fluorophore and plasmonic acceptor. In order to perform a comparative analysis of gain-induced optical loss modifications, pump-probe experiments have been carried out on the two systems, along with time-resolved fluorescence spectroscopy. A comparative set of measurements (both scattering and transmission of probe beams at different wavelengths) have been performed on both systems. The super-linear increase in the scattered signal (as well as the transmitted one) above a certain threshold value of gain is a demonstration of the enhancement of the quality factor of surface plasmon resonance mediated by RET processes between active elements and gold cores within the composite NPs. To corroborate evidences of mitigation of absorptive losses in G_F core–shell NPs, time-resolved fluorescence spectroscopy has been carried out on both systems, proving dye-NP coupling and energy transfer processes. The intensity emission decays for the G_{F-C500} system have been fitted as two bi-exponentials. The short-living time constant ($\tau_2 = 0.508$ ns) resulted one order of magnitude lower than the decay of pure dye solution, whereas the long-living decay time ($\tau_1 = 2.78$ ns) resulted in the same range. On the contrary, we had no chance to observe any significant emission in the G_{F-DCM} sample, indicating that DCM dye present into the shell experienced a complete fluorescence quenching [De Luca et al. (2012)]. However, the key experiment was performed by measuring the transmitted intensity at far field of broadband probe light-waves after passing the excited volume with a fixed level of gain, selected by fixing the pump rate. In fact, transmitted probe intensity has been acquired at different wavelengths in presence and absence of the pump beam, to emphasize the changes of absorptive power induced by energy transfer during dye excitation. In Fig. 6.12a, we plotted the difference ($\Delta_T = T_p - T_{wp}$, orange dots) between the transmitted signal in presence (T_p) of pump with respect to that in absence (T_{wp}) of pump beam, as a function of wavelength for G_{F-DCM} system. Same figure shows the plasmon band (black line) and DCM dye emission (green line) in order to identify the overlapping region. In this case, ΔT starts to assume positive values exactly in correspondence of the rise of the DCM emission curve, where spectral overlapping starts to become different from zero, showing a marked distinction between negative and positive values (orange dots). The measured curves related to the absorption bands for G_{F-DCM} system in presence (red dots) and in absence of pump beam (black dots in Fig. 6.12b) clearly show how below 550 nm the whole system experienced higher values for the extinction curve in presence of gain. This is compensated by the modification of the curve above 550 nm, where absorption is reduced by supplying gain to the system as

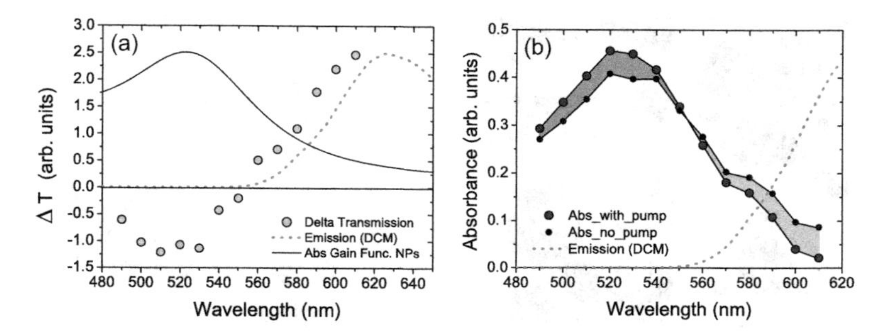

Figure 6.12 Broadband delta of transmission measurements of the G_{F-DCM} system. (a) $\Delta_T = T_p - T_{wp}$, where $T_p = T$ with pump, and $T_{wp} = T$ without pump. (b) represent the retrieved data from (a) utilized to reconstruct the extinction band for the two systems in absence (black dots) and presence (red dots) of gain (pump beam on) calculated as Abs $= \log_{10}(I_T/I_i)$.

evidenced by lower values for extinction curve with respect to those obtained in absence of exciting optical field. These two results obtained in gain-functionalized systems, with the assistance of two different gain media, experimentally emphasize that modifications on the imaginary part of the permittivity $\varepsilon_{im}(\omega)$, which is directly related to the absorption, clearly depend on gain-surface plasmon modes spectral overlapping. A reduction of absorption in a band, that is, between 550 nm and 610 nm, (yellow area in Fig. 6.12b), corresponds to an increase in the same quantity in a complementary wavelength band (490–550 nm, green area in Fig. 6.12b), in good agreement with the results of the theoretical model presented in Section 6.2.2 (Fig. 6.2), accounting for the coupling between gain and plasmonic NPs upon verifying the causal nature of the response of materials via Kramers–Kronig (KK) dispersion relations.

6.3.2 *Mesoscale Approach*

As discussed in the previous section by bringing gain in close proximity to plasmonic nanoresonators, optical losses can be selectively mitigated; the attenuation in this case is indeed limited to a narrow spectral range. Moreover, the plasmon resonance in a bulk material depends on NP spatial distribution and on their reciprocal interactions; in this situation, even the spatial distribution

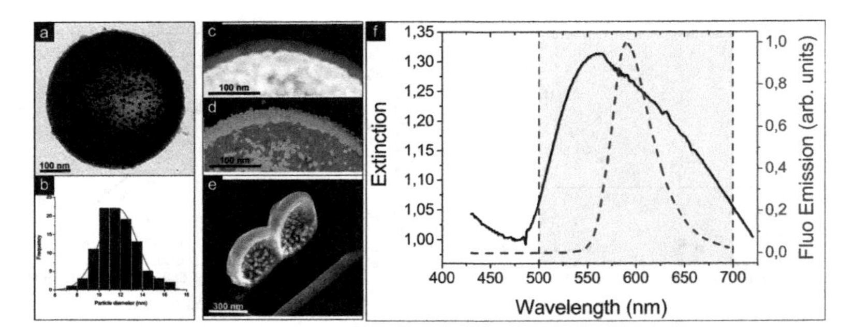

Figure 6.13 (a) TEM image of the typical plasmonic capsules with Au NPs in their inner cavities. (b) Statistical distribution of diameters, with an average of 11.5 ± 1.7 nm. (c) STEM and (d) combined XEDS elemental mapping showing the elemental distribution, Au = red and SiO_2 = green images from the same capsule; (e) FIB cross section image. (f) Mesocapsule plasmonic resonance (black continuous line) and R6G emission spectrum (red dashed line).

of gain elements becomes a key parameter. In this section, we consider a system that is at a scale intermediate between the single plasmonic nanostructure and bulk materials; we exploited this example as a proof of concept to demonstrate that loss compensation can be achieved in more complex systems and that it can be optimized by properly designing their geometry. The system considered is made by sub-micrometric plasmonic structures resembling a reverse bumpy ball configuration in which multiple Au NPs remain engrafted on the inner walls of the porous hollow silica capsules (Fig. 6.13a and e). Their fabrication procedure can be found in [Sanlés-Sobrido et al. (2009, 2012)]. The mesocapsule plasmonic band results from the NPs arrangement, which produces a spectrally wider and red-shifted resonance compared to the one of the single-metal NP (see Fig. 6.13f). It is known that single nanoresonators, which alone would produce a specific plasmon resonance, when in close proximity mix and hybridize creating a completely different response. This effect can be treated just like electron wave functions of simple atomic and molecular orbitals [Prodan et al. (2003); Halas et al. (2011)]. In Section 6.2.3, we theoretically described the hybridization for these structure with a simpler, yet geometrically sound approach [Baudrion et al. (2013)]

by means of a homogenization rule widely used to calculate the collective optical properties of bulk materials.

The gain elements have been easily integrated by soaking the mesostructures in a dye solution. The average dimension of pore diameter (5 nm) allows the dye solution to fill the hollow core, where as it is demonstrated (Section 6.2.3) that the plasmonic field is uniform and intense and all the dye molecules included in this region are able to exchange energy with the plasmon in an efficient way. The dye we have chosen (rhodamine 6G) is characterized by an emission spectrum that suitably overlaps the mesostructure plasmonic resonance (Fig. 6.13f). The concentration ratio between mesocapsules and R6G has been optimized by using time-resolved fluorescence spectroscopy measurements. The reduction of the R6G decay time observed in the solution we used in the experiments we report further in the Section 6.2.3 is around the 41%. The quenching efficiency, calculated as $Q = 1 - (F/F_0)$, where F and F_0 are the fluorescence signal intensity in presence and in absence of mesocapsules, for this system has been proved to be a function of the excitation energy. We ascribe this behavior to the energy transfer processes occurring between gain and plasmon. Figure 6.13a shows the typical transmission electron microscope (TEM) image of the as-prepared plasmonic nanostructures, comprising Au NPs (11 ± 1.7 nm in average) supported on the inner wall of the inorganic hollow silica capsule (Fig. 6.13b). The spatial distribution of the Au NPs growth at the inner wall of the silica mesocapsule was clearly evidenced by the focused ion beam (FIB) cross-sectional analysis of a typical plasmonic capsule (Fig. 6.13e). Scanning transmission electron microscopy (STEM) in 6.13c and X-ray energy dispersive spectroscopy (XEDS) in Fig. 6.13e provide further evidence about the homogeneous nature of the silica shell, and no trace of gold was detected on their outer surface. To demonstrate that the energy transferred is used by the system to reduce metallic losses, an ultra-fast pump-probe experiment for the simultaneous measure of Rayleigh scattering and transmission has been set up. According to Beer-Lambert-Bouguer law, by measuring simultaneously Rayleigh scattering and transmission either in absence or in presence of pump, we are able to understand if the absorptive power of the material is affected by the exciting field. While the scattering

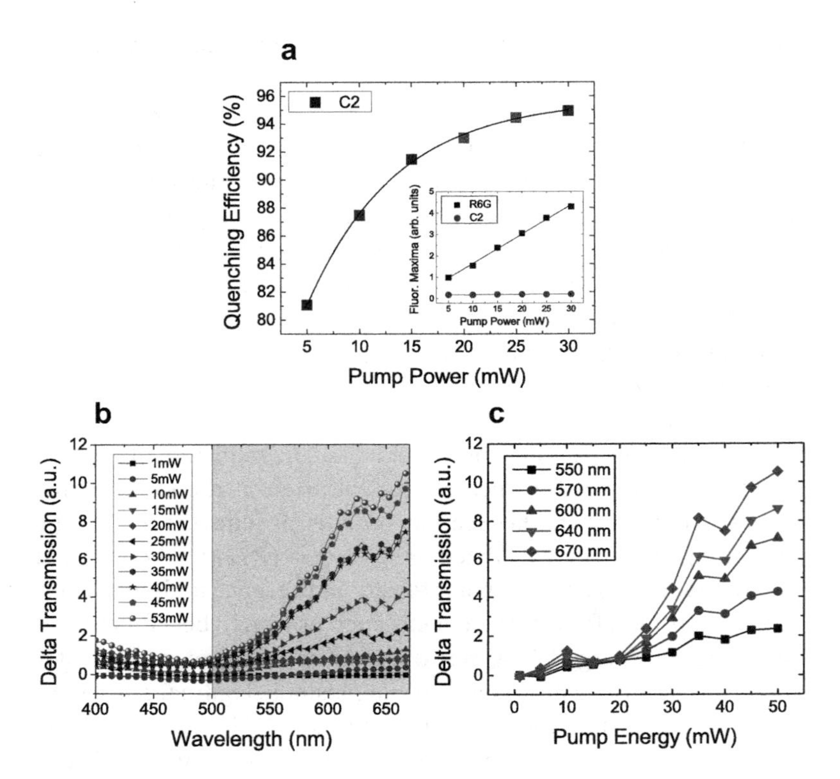

Figure 6.14 Measured percentage change of the light probe that is transmitted (a) by the mesocapsule gain-assisted system as a function of pump energy. (b) Cuts of the previous curves for five different wavelengths of transmitted intensity as a function of pump energy.

intensity of the probe beam seems to be unaffected by excitation energy, a broadband enhancement of the transmitted light has been measured (see Fig. 6.14a).

It shows a typical threshold value for the average pump power (about 20 mW) above which we observe a super-linear increase of the transmission. In fact, the transmitted probe light rises up of an order of magnitude with respect to the absence of the exciting field. This result is in accordance with the model described in Section 6.2.3, in the hypothesis that the gain medium is penetrated within the mesocapsule hollow cores. In fact, the theoretical model shows a complete different behavior, upon considering the chromophores uniformly distributed outside the mesocapsule. We found that the plasmon–gain interplay was still playing a role, but it was much less effective without any evidence of the broadband compensation effect.

6.3.3 *Macroscale Approach*

The approach at the macroscale implies a more complicated theoretical treatment, as well as a redesigning of the experimental setups in order to be able to extract the main physical parameters of real bulk materials. We will present only two distinct examples of plasmonic-exciton interplay and coupling at the macroscale. First, one consists in a flexible host matrix selected to achieve functional nanocomposites based on CdSe@ZnS QDs and Au NPs, simultaneously dispersed in a polymer matrix (polydimethylsil-oxane-PDMS) [De Luca et al. (2013)]. In this case, coherent interactions between QDs and plasmonic Au NPs embedded in PDMS films have demonstrated to lead to a relevant enhancement of the absorption cross section of QDs, remarkably modifying the optical response of the entire system. Optical and time-resolved spectroscopy studies revealed an active plasmon–gain feedback behind the super-absorbing overall effect. Second example is a metallo-dielectric multilayer system that represents a nonmagnetic extremely anisotropic artificial medium, that cannot be found in nature at optical frequencies having an open hyperboloid iso-frequency surface. Being characterized by hyperbolic dispersion, we will refer to it as hyperbolic metamaterial (HMM). In particular, we will present results obtained in a type II HMM, with high anisotropy in the dielectric permittivity tensor components, $\varepsilon_{\parallel}(\varepsilon_x = \varepsilon_y) < 0$ and $\varepsilon_{\perp}(\varepsilon_z) > 0$ [Sreekanth et al. (2013)], that can be used to show the appearance of additional electromagnetic states in the hyperbolic regime, which have wave vectors much larger than those allowed in vacuum (high-k). The existence of high-k modes in the fabricated HMM, through fluorescence lifetime measurements as a function of emission wavelengths, is considered the result of the strong metal-gain interplay. Let's consider the first example, in which the plasmon–exciton interplay in a bulk flexible matrix resulted in an overall super-absorber behavior of gain materials in presence of resonant nano-objects. In particular, a novel synthetic mechanism of the curing process of NPs and QDs incorporation in PDMS has been suggested, highlighting the role of the NP-capping agents, which has been revealed to be fundamental to ensure a good polymer crosslinking. The presence of unsaturated alkyl chain in the ligand molecular structure has been found advantageous to achieve a complete curable nanocomposite under standard curing conditions. The occurrence of small aggregates of QDs and NPs inside the PDMS matrix allowed short-range interactions among the nano-objects, influencing the related optical properties. In Fig. 6.15, the extinction band of pure QDs in PDMS presents the typical behavior of a bare QDs sample, with a relative maximum around 568 nm (solid black line).

The red-shifting of the absorption peak with respect to what has been measured in solution can be accounted to both an electronic reorganization of the quantum dot energy levels, when dispersed

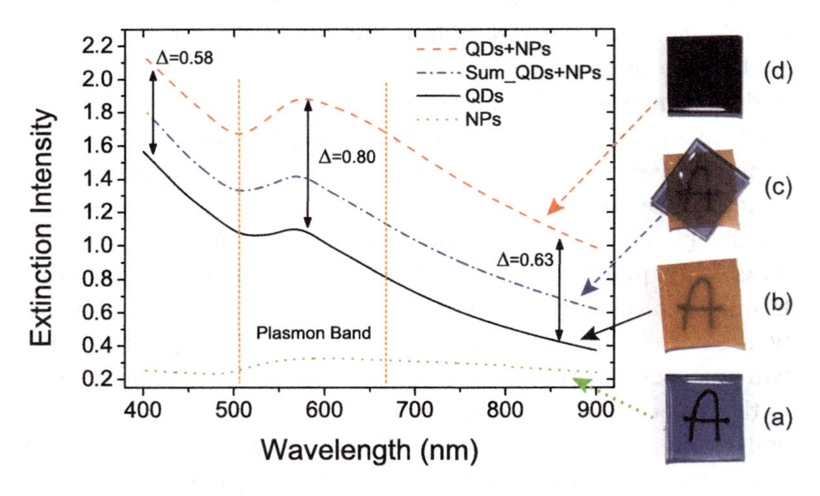

Figure 6.15 Extinction curves for QDs (black solid line), Au NPs (dotted green line) and mixed sample (dashed red line) in PDMS matrix. The mixed sample presents an extinction curve that is much higher than the mathematical summation of the two separated curves (dash-dot blue line). Images (a–d) are referred to pictures of the real plastic matrices over an "A" written on a white paper. (a) Au NPs based nanocomposite, (b) QDs based nanocomposite, (c) this curve corresponds to the overlap between (a) and (b), (d) QDs + Au NPs based nanocomposite.

in a different host environment, and the presence of aggregation states of the QDs inside the PDMS matrix [Shojaei-Zadeh et al. (2011)]. In Fig. 6.15, the green dotted line represents the extinction band of Au NPs in PDMS. The two pictures on the right side show the actual transparency of the two self-standing PDMS-based nanocomposites [incorporating Au NPs (a) and QDs(b)] obtained by acquiring an image of an "A" letter written on a white paper sheet. The mathematical sum (superposition) of the two extinction curves [i.e., sample (a) and sample (b)], represented by the dash-dot blue line [and relative picture (c)] coincided with the spectrum obtained by physically overlapping the two PDMS samples. Instead, the measured extinction spectrum of the PDMS nanocomposite film, simultaneously containing QDs and Au NPs [sample (d)], at the same concentration of the samples (a) and (b), is represented by the dash red line in Fig. 6.15. Strikingly, in the presence of both nano-objects, the extinction curve differs from the bare mathematic sum. In particular, a distinct enhancement of the extinction curve with respect to the addition of the two curves corresponding to the two components is observed. Such an enhancement appears stronger in the overlapping region between the emission band of QDs and Au NPs plasmon band. Indeed the effective overlap between the fluorescence emission of QDs and Au NPs plasmonic band [reported in De Luca et al. (2013)] can originate a strong resonance between the two systems, causing the occurrence of a modified local field able to interact with QDs and modify both their absorption and emission properties. On a quantitative point of view, whereas the mathematical sum presents an increase of about 22% inside the plasmon region (delimited by the orange vertical dash lines), the observed increase became higher than 38% in the case of the mixed sample. Picture (d) clearly shows how it is difficult to distinguish letter "A" across this sample. Experimental evidences of a similar phenomenon observed in the case of organic dyes in the presence of Au NPs were already presented by Lawandy in 2005. A fluorescence quenching of QDs in the presence of Au NPs with respect to the pure QDs has been observed for different excitation energies (steady states). Even the QD quantum yield, defined as the probability of the excited state being deactivated by fluorescence rather than by another nonradiative mechanism, has been observed to decrease

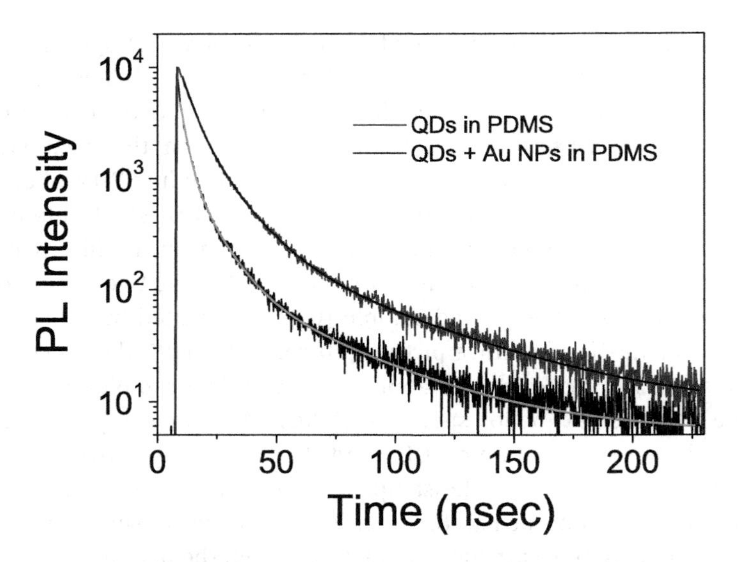

Figure 6.16 Time-resolved fluorescence intensity decays and 3-exponential fitting of QDs in PDMS and of QDs + Au NPs in PDMS.

more than 40 times, moving from 16.5% in the case of bare QDs in PDMS, down to only 0.4% when Au NPs are also present in the nanocomposite [De Luca et al. (2013)]. Time-resolved fluorescence spectroscopy was carried out on both systems [(b) and (d)] to obtain insight on possible QDs-NPs coupling and energy transfer processes. Fluorescence lifetime measurements indicate a faster decay of QDs PL in presence of Au NPs, consistent with a strong resonant coupling between QDs and the NP surface. Figure 6.16 reports TCSPC data obtained by exciting the two samples at 375 nm. Emission intensity was recorded at 575 nm for both QDs and QDs-Au NPs mixed samples. The recorded time-resolved fluorescence intensity was fitted by a multi-exponential function, namely, three exponential decay components were necessary, consistently with relevant literature [Salman et al. (2009); Jones et al. (2009)].

In particular, the fast component is reasonably attributed to the recombination of intrinsic excitons [Tagaya and Nakagawa (2011); Shojaei-Zadeh et al. (2011); Lee et al. (2004)], while the middle/slow ones can be convincingly related to the interplay between exciton states and surface traps. A typical example of the fitting results is shown by the black line in Fig. 6.16. Here the role played by the presence of Au NPs on the de-excitation paths of the QDs can be effectively inferred. Indeed for the QDs in PDMS [sample (b)], an average lifetime value $< \tau >$ of 33.50 ns was calculated, while for the sample containing also NPs [sample (d)], a $< \tau >$ value of 29.3 ns was found. Such an evidence clearly suggests that the PL quenching in presence of Au NPs is related to a faster relaxation dynamics, due to the presence of additional nonradiative de-excitation channels. Therefore, a mechanism of resonant energy transfer from QDs to Au NPs for the QDs in proximity with metal NPs can be thought to take place.

The second system consists of 12 alternating layers of gold (Au) and titanium dioxide (TiO_2) thin films. The proposed HMM was fabricated by sequential deposition of TiO_2 and Au on a glass substrate (Micro slides from Corning) using RF sputtering technique (TiO_2 target from Stanford Materials Corporation) and thermal evaporation of Au pellets (from Kurt J. Lesker Company), with measured thicknesses of 32 nm and 16 nm for TiO_2 and Au, respectively, and a calculated fill fraction of metal is 33% (see schematic in Fig. 6.17a). Since the fabricated HMM belongs to effective medium approximations, we used effective medium theory (see Section 6.2) to calculate the dielectric permittivity of the entire structure.

Then we performed spectroscopic ellipsometry measurements to confirm the dielectric tensor component ($\varepsilon_{\|}$, Fig. 6.17b) in the parallel direction (XY-plane). As experimentally measured, and confirmed by simulated results (see Fig. 6.6, Section 6.2.4), the designed HMM presents a hyperbolic dispersion at optical frequencies, above 530 nm wavelength. To evidence the existence of high-k modes in the fabricated HMM, we performed fluorescence lifetime measurements as a function of emission wavelengths using ultra-fast time-correlated single photon counting (TCSPC) setup (see Fig. 6.10a). Short-living excitonic states of the chromophores placed in close proximity to the metallo-dielectric multilayers and measured in the hyperbolic band would represent a clear signature of the presence of high-k modes. The studied HMM structure

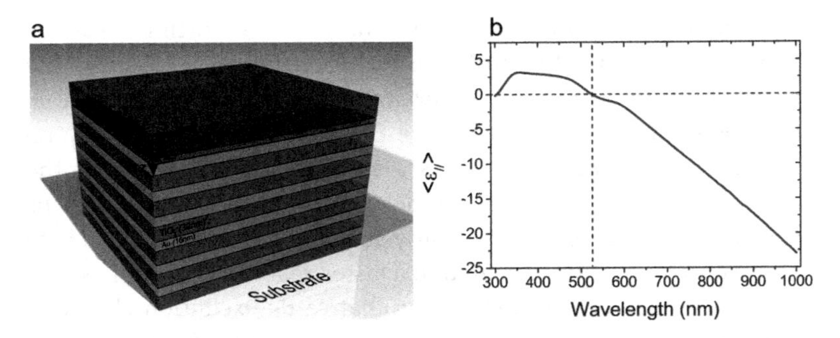

Figure 6.17 (a) Schematic diagram of fabricated hyperbolic metamaterial, which consists of 6 bilayers of Au/TiO_2. (b) Real part of effective permittivity parallel to the plane of HMM ($\varepsilon_x = \varepsilon_y$) measured by spectroscopic ellipsometry measurement.

consists of a DCM dye (0.3% by wt. in ethanol solution) dissolved polymer (PMMA) layer of 100 nm thickness on top of the Au/TiO_2 multilayer, separated by a TiO_2 spacer layer of 10 nm thickness. The maximum emission wavelength of DCM dye dissolved PMMA is observed at 620 nm for an excitation wavelength of 450 nm. The fluorescence lifetime measurements of three samples such as reference sample (DCM on a TiO_2/glass substrate), control sample (DCM on a bilayer of Au/TiO_2) and HMM (DCM on the HMM) were performed and the results are shown in Fig. 6.18.

It is evident from Fig. 6.18a that there is a large variation in spontaneous emission lifetime of DCM onto the HMM compared to reference sample and control sample. This unusual behavior is attributed to the existence of high-k modes as well as nonradiative and SPP modes present in the HMM [Krishnamoorthy et al. (2012)]. Also, dye on reference sample and control sample shows an increase in lifetimes with emission wavelengths; however, lifetime of DCM onto HMM is almost constant in the hyperbolic region of the emission spectra. The coupling of the high-k metamaterial states is responsible for the observed shortest life times of HMM [Krishnamoorthy et al. (2012)]. To further evidence that the high-k modes are responsible for the reduction of lifetime in the hyperbolic region, we calculated the normalized lifetime of dye on HMM with respect to reference sample. As shown in Fig. 6.18b, a considerable reduction

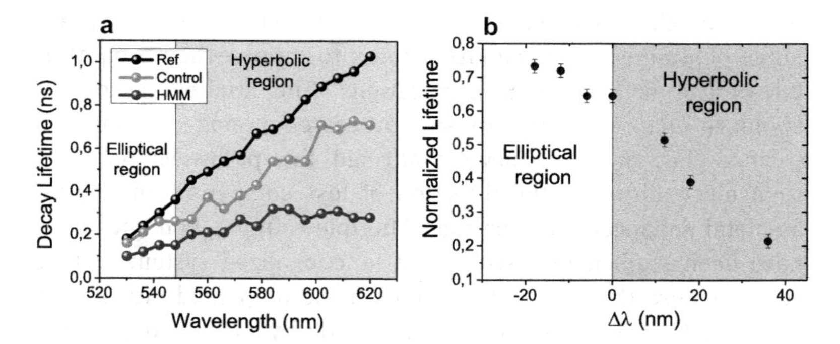

Figure 6.18 Photoluminescence measurements of hyperbolic metamaterial. (a) Experimentally obtained spontaneous emission lifetimes of the DCM dye as a function of emission wavelengths on HMM, control and reference sample and (b) lifetime of dye on HMM normalized with respect to reference sample. The emission wavelengths are normalized with respect to the theoretical transition wavelength (530 nm) with negative and positive regions, respectively, representing the elliptical and hyperbolic dispersion.

in normalized lifetime is observed in the hyperbolic region, that is, when the wavelength crosses the transition wavelength. Note that in Fig. 6.18b, the emission wavelengths are normalized with respect to transition wavelength (530 nm). We also verified the coupling of dye emission into the metamaterial states using steady-state photoluminescence measurements, observing a strong quenching in the emission in the case of HMM [not shown here, see Sreekanth et al. (2013)]. From the above observed properties, we confirmed that the fabricated Au/TiO_2 multilayer is an optical hyperbolic metamaterial.

6.4 Summary

A systematic study of general and specific physical aspects concerning the interplay between plasmon nanostructures and gain elements has been reported in this chapter. It was dedicated to define how the strong coupling of these two classes of materials can bring to advanced optical materials, indeed their proper combination yields to complex systems rich of unpredictable dynamics and fascinating physical mechanisms. The main aim of this research was to provide an effective solution to the optical losses of plasmon-based nanostructures, to improve their properties for optical and photonic applications. The studies performed about resonance energy transfer processes among excitons and exciton–plasmon pairs have confirmed the pathways to apply such mechanisms in the direction of loss compensation as well as metal enhancement purposes. Multiple coupling configurations have been studied in assisted and functionalized systems across scales, where the main interaction was accompanied either by a coherent transfer of excitation energy or by near-field plasmon effects. The former was found to deform the response function of the effective materials in the gain-assisted band, whereas the latter produces a colossal enhancement of the overall extinction cross section. Theory, nanochemistry, and physical experiments have worked synergistically to explore quantum and non-quantum physical phenomena. However, the different sections of the chapter demonstrate that the interaction of the single elements exciton–plasmon, cannot be extended simply to systems at higher scales and symmetries. The presented results confirm that bringing gain (excitons) right at the hearth of metal-based nanostructures (plasmon) represents only the first step to trigger coherent and radiationless transfer of energy to the quasi-static field of the hybrid nanostructure. The significant variations of the imaginary part of the effective dielectric permittivity were found to be associated with the modifications of the quality factor of the plasmon resonances as well as the dispersive relations of the optical material. Most remarkable aspect is the dual valence of the strong coupling exciton–plasmon that in turn enables exciton–exciton coupling via plasmon near-field and plasmon–plasmon because of the enhanced local field. Fascinating physics, not yet fully understood, is revealed by theory and experiments that coherently individuate pathways toward promising applications in nano-optics, photonics and biomedical fields. In particular, gain-assisted vectorial and tensorial metastructures can find applications in advanced imaging for early cancer detection, biosensing, hyperlenses, super-absorbers, super-emitters, and nanolasers.

References

Averitt, R. D., Westcott, S. L., and Halas, N. J. (1999). Linear Optical Properties of Gold Nanoshells, *J. Opt. Soc. Am. B* **16**, pp. 1824–1832.

Avrutsky, I. (2004). Surface Plasmons at Nanoscale Relief Gratings between a Metal and a Dielectric Medium with Optical Gain, *Phys. Rev. B* **70**, p. 155416.

Baudrion, A.-L., Perron, A., Veltri, A., Bouhelier, A., Adam, P.-M., and Bachelot, R. (2013). Reversible Strong Coupling in Silver Nanoparticle Arrays Using Photochromic Molecules, *Nano Lett.* **13**, 1, pp. 282–286, doi:10.1021/nl3040948, URL http://pubs.acs.org/doi/abs/10.1021/nl3040948.

Bergman, D. J., and Stockman, M. I. (2003). Surface Plasmon Amplification by Stimulated Emission of Radiation: Quantum Generation of Coherent Surface Plasmons in Nanosystems, *Phys. Rev. Lett.* **90**, p. 027402.

Bhowmick, S., Saini, S., Shenoy, V. B., and Bagchi, B. (2006). Resonance Energy Transfer from a Fluorescent Dye to a Metal Nanoparticle, *J. Chem. Phys.* **125**, pp. 181102/1–6.

Bohren, F., and Huffman, D. R. (1998). *Absorption and Scattering of Light by Small Particles*, WILEY-VCH Verlag GmbH & Co. KGaA.

Brongersma, M. L., Hartman, J. W., and Atwater, H. A. (2000). Electromagnetic Energy Transfer and Switching in Nanoparticle Chain Arrays Below the Diffraction Limit, *Phys. Rev. B* **62**, pp. R16356–R16359.

Cai, W., Chettiar, U. K., Kildishev, A. V., and Shalaev, V. M. (2008). Optical Cloaking with Metamaterials, *Nature Photon.* **1**, pp. 224–227.

Das, P. C., and Puri, A. (2002). Energy Flow and Fluorescence near a Small Metal Particle, *Phys. Rev. B* **65**, 65, p. 155416.

De Luca, A., Depalo, N., Fanizza, E., Striccoli, M., Curri, M. L., Infusino, M., Rashed, A. R., Deda, M. L., and Strangi, G. (2013). Plasmon-Mediated Super-Absorber Flexible Nanocomposite for Metamaterials, *Nanoscale* **5**, p. 6097.

De Luca, A., Ferrie, M., Ravaine, S., La Deda, M., Infusino, M., Rahimi Rashed, A., Veltri, A., Aradian, A., Scaramuzza, N., and Strangi, G. (2012). Gain Functionalized Core–Shell Nanoparticles: The Way to Selectively Compensate Absorptive Losses, *J. Mater. Chem.* **22**, pp. 8846–8852.

De Luca, A., Grzelczak, M. P., Pastoriza-Santos, I., Liz-Marzán, L. M., Deda, M. L., Striccoli, M., and Strangi, G. (2011). Dispersed and Encapsulated Gain Medium in Plasmonic Nanoparticles: a Multipronged Approach to Mitigate Optical Losses, *ACS Nano* **5**, pp. 5823–5829.

Doussineau, T., Trupp, S., and Mohr, G. J. (2009). Ratiometric pH-Nanosensors based on Rhodamine-Doped Silica Nanoparticles Functionalized with a Naphthalimide Derivative, *Journal of Colloid and Interface Science* **339**, pp. 266–270.

Draine, B. T. (2000). *Light Scattering by Nonspherical Particles: Theory, Measurements, and Applications*, Academic Press, San Diego.

Dulkeith, E., Morteani, A. C., Niedereichholz, T., Klar, T. A., Feldmann, J., Levi, S. A., van Veggel, F. C. J. M., Reinhoudt, D. N., Moller, M., and Gittins, D. I. (2002). Fluorescence Quenching of Dye Molecules near Gold Nanoparticles: Radiative and Nonradiative Effects, *Phys. Rev. Lett.* **89**, p. 203002.

Fang, A., Koschny, T., Wegener, M., and Soukoulis, C. M. (2009). Self-Consistent Calculation of Metamaterials with Gain, *Phys. Rev. B* **79**, p. 241104.

Fernández-López, C., Mateo-Mateo, C., Alvarez-Puebla, R., Pérez-Juste, J., Pastoriza-Santos, I., and Liz-Marzán, L. (2009). Highly Controlled Silica Coating of PEG-Capped Metal Nanoparticles and Preparation of SERS-Encoded Particles, *Langmuir* **25**, pp. 13894–13899.

Förster, T. (1948). Intermolecular Energy Migration and Fluorescence, *Ann. Phys.* **2**, pp. 55–75.

Gersten, J., and Nitzan, A. (1981). Spectroscopic Properties of Molecules Interacting with Small Dielectric Particles, *J. Chem. Phys.* **75**, pp. 1139–1152.

Halas, N. J., Lal, S., Chang, W., Link, S., and Nordlander, P. (2011). Plasmons in Strongly Coupled Metallic Nanostructures, *Chem. Rev.* **111**, pp. 3913–3961.

Imahori, H., and Fukuzumi, S. (2001). Porphyrin Monolayer-Modified Gold Clusters as Photoactive Materials, *Adv. Mater.* **13**, pp. 1197–1199.

Johnson, P., and Christy, R. (1972). Optical Constants of the Noble Metals, *Phys. Rev. B* **6**, pp. 4370–4379.

Jones, M., Lo, S. S., and Scholes, G. D. (2009). Quantitative Modeling of the Role of Surface Traps in CdSe/CdS/ZnS Nanocrystal Photoluminescence Decay Dynamics, *PNAS* **106**, pp. 3011–3016.

Kreibig, U., and Vollmer, M. (1995). *Optical Properties of Metal Clusters* (Springer).

Krishnamoorthy, H. N. S., Jacob, Z., Narimanov, E., Kretzschmar, I., and Menon, V. M. (2012). Topological Transitions in Metamaterials, *Science* **336**, pp. 205–209.

Lakowicz, J. R. (2001). Radiative Decay Engineering: Biophysical and Biomedical Applications, *Anal. Biochem.* **298**, pp. 1–24.

Lawandy, N. M. (2004). Localized Surface Plasmon Singularities in Amplifying Media, *Appl. Phys. Lett.* **85**, pp. 5040–5042.

Lawandy, N. M. (2005). Nano-Particle Plasmonics in Active Media, *Proc. SPIE* **59240**, pp. 59240G-1–13.

Lee, J., Buxton, G., and Balazs, A. (2004). Using Nanoparticles to Create Self-Healing Composites, *J. Chem. Phy.* **121**, p. 5531.

Mock, J. J., Barbic, M., Smith, D. R., Schultz, D. A., and Schultz, S. (2002). Shape Effects in Plasmon Resonance of Individual Colloidal Silver Nanoparticles, *J. Chem. Phys.* **116**, pp. 6755–6759.

Moskovits, M. (1985). Surface-Enhanced Spectroscopy, *Rev. Mod. Phys.* **57**, pp. 783–826.

Nezhad, M. P., Tetz, K., and Fainman, Y. (2004). Gain Assisted Propagation of Surface Plasmon Polaritons on Planar Metallic Waveguides, *Opt. Express* **12**, pp. 4072–4079.

Noginov, M. A., Zhu, G., Bahoura, M., Adegoke, J., Small, C. E., Ritzo, B. A., Drachev, V. P., and Shalaev, V. M. (2006). Enhancement of Surface Plasmons in an Ag Aggregate by Optical Gain in a Dielectric Medium, *Opt. Lett.* **31**, pp. 3022–3024.

Noginov, M. A., Zhu, G., Belgrave, A. M., Bakker, R., Shalaev, V. M., Narimanov, E. E., Stout, S., Herz, E., Suteewong, T., and Wiesner, U. (2009). Demonstration of a Spaser-Based Nanolaser, *Nature* **460**, pp. 1110–1113.

Pendry, J. B. (2000). Negative Refraction Makes a Perfect Lens, *Phys. Rev. Lett.* **85**, pp. 3966–3969.

Powell, R. D., Halsey, C. M. R., Spector, D. L., Kaurin, S. L., McCann, J., and Hainfeld, J. F. (1997). A Covalent Fluorescent-Gold Immunoprobe: Simultaneous Detection of a Pre-mRNA Splicing Factor by Light and Electron Microscopy, *Histochem. Cytochem.* **45**, pp. 947–956.

Prodan, E., Radloff, C., Halas, N. J., and Nordlander, P. (2003). A Hybridization Model for the Plasmon Response of Complex Structures, *Science* **302**, pp. 419–422.

Quinten, M. (1999). Optical Effects Associated with Aggregates of Cluster, *J. Cluster Sci.* **10**, pp. 319–358.

Quinten, M., Leitner, A., Krenn, J. R., and Aussenegg, F. R. (1998). Electromagnetic Energy Transport via Linear Chains of Silver Nanoparticles, *Opt. Lett.* **23**, pp. 1331–1333.

Ritchie, R. H. (1973). Surface Plasmons in Solids, *Surf. Sci.* **34**, pp. 1–19.

Rodríguez-Fernández, J., Pérez-Juste, J., de Abajo, F. J. G., and Liz-Marzán, L. M. (2006). Seeded Growth of Submicron Au Colloids with Quadrupole Plasmon Resonance Modes, *Langmuir* **22**, pp. 7007–7010.

Salman, A. A., Tortschanoff, A., van der Zwan, G., van Mourik, F., and Chergui, M. (2009). A model for the multi-exponential excited-state decay of CdSe nanocrystals, *Chem. Phys.* **357**, pp. 96–101.

Sanlés-Sobrido, M., Exner, W., Rodríguez - Lorenzo, L., Rodríguez - González, B., Correa - Du arte, M. A., Alvarez - Puebla, R. A., and Liz - Marzán, L. (2009). Design of SERS-Encoded, Submicron, Hollow Particles Through Confined Growth of Encapsulated Metal Nanoparticles, *J. Am. Chem. Soc.* **131**, pp. 2699–2705.

Sanlés-Sobrido, M., Pérez - Lorenzo, M., Rodríguez-González, B., Salgueiriño, V., and Correa-Du arte, M. A. (2012). Back Cover: Highly Active Nanoreactors: Nanomaterial Encapsulation Based on Confined Catalysis, *Angew. Chem. Int. Ed.* **51**, pp. 3877–3882.

Schurig, D., Mock, J. J., Justice, B. J., Cummer, S. A., Pendry, J. B., Starr, A. F., and Smith, D. R. (2006). Metamaterial Electromagnetic Cloak at Microwave Frequencies, *Science* **314**, pp. 977–980.

Seidel, J., Grafstrom, S., and Eng, L. (2005). Stimulated Emission of Surface Plasmons at the Interface between a Silver Film and an Optically Pumped Dye Solution, *Phys. Rev. Lett.* **94**, p. 177401.

Shojaei-Zadeh, S., Morris, J. F., Couzis, A., and Maldarelli, C. (2011). Highly Crosslinked Poly(Dimethylsiloxane) Microbeads with Uniformly Dispersed Quantum Dot Nanocrystals, *J. Colloid Interface Sci.* **363**, p. 25.

Sreekanth, K. V., De Luca, A., and Strangi, G. (2013). Experimental demonstration of surface and bulk plasmon polaritons in hypergratings, *Scientific Reports* **3**, p. 3291.

Stockman, M. (2008). Spasers Explained, *Nature Photon.* **2**, pp. 327–329.

Strangi, G., De Luca, A., Ravaine, S., Ferrie, M., and Bartolino, R. (2011). Gain induced optical transparency in metamaterials, *Appl. Phys. Lett.* **98**, p. 251912.

Su, K.-H., Wei, Q.-H., Zhang, X., Mock, J. J., Smith, D. R., and Schultz, S. (2003). Interparticle Coupling Effects on Plasmon Resonances of Nanogold Particles, *Nano Lett.* **3**, pp. 1087–1090.

Sudarkin, A. N., and Demkovich, P. A. (1989). Excitation of Surface Electromagnetic Waves on the Boundary of a Metal with an Amplifying Medium, *Sov. Phys. Tech. Phys.* **34**, pp. 764–766.

Tagaya, M., and Nakagawa, M. (2011). Incorporation of Decanethiol-Passivated Gold Nanoparticles into Cross-Linked Poly(Dimethylsiloxane) Films, *Smart Mater. Res.* Article ID 390273 7 pages.

Tovmachenko, O. G., Graf, C., van den Heuvel, D. J., van Blaaderen, A., and Gerritsen, H. C. (2006). Fluorescence Enhancement by Metal-Core/Silica-Shell Nanoparticles, *Adv. Mat.* **18**, pp. 91–95.

Veltri, A. (2009). Designs for Electromagnetic Cloaking a Three-Dimensional Arbitrary Shaped Star-Domain, *Opt. Express* **17**, pp. 20494–20501.

Veltri, A., and Aradian, A. (2012). Optical Response of Ametallic Nanoparticle Immersed in a Medium with Optical Gain, *Phys. Rev. B* **85**, pp. 1–5.

Veltri, A., Aradian, A., and Chipouline, A. (submitted). Time-Dynamical Model for the Optical Response of a Plasmonic Nanoparticle Immersed in an Active Gain Medium, ***submitted***.

Weitz, D. A., Garoff, S., Gersten, J. I., and Nitzan, A. (1983). The Enhancement of Raman Scattering, Resonance Raman Scattering and Fluorescence from Molecules Adsorbed on a Rough Silver Surface, *J. Chem. Phys.* **78**, pp. 5324–5338.

Zheludev, N., Prosvirnin, S., Papasimakis, N., and Fedotov, V. (2008). Lasing Spaser, *Nature Photon.* **2**, pp. 351–354.

Chapter 7

Localized Surface Plasmons: A Powerful Tool for Sensing

Francesco Todescato
Department of Chemical Sciences, University of Padova, Via Marzolo 1, 35131, Padova, Italy
francesco.todescato@unipd.it

7.1 What Are Plasmons?

Generally speaking, a plasmon is a quantum of a plasma oscillation, namely, a fast oscillation of a free electron cloud (or gas) in a conducting medium. Metals are the most common materials that can sustain plasmons; in fact, they can be seen as a plasma of positive ions (i.e., the nuclei and the core electrons) dressed by free electrons in the conduction band (gas or cloud of free electrons). Two categories of plasmons can be individuated: *volume plasmon* (see Section 7.1.1) and *surface plasmon*.

Even though plasmonic properties of the latter category have been exploited since the age of the Romans (e.g., Lycurgus Cup), the first scientific evidence of a plasmon-related phenomenon have been observed in metallic grating by Wood in 1902 [1]

Active Plasmonic Nanomaterials
Edited by Luciano De Sio

ISBN 978-981-4613-00-2 (Hardcover), 978-981-4613-01-9 (eBook)
www.panstanford.com

and theoretical analysis by Rayleigh in 1907 [2]. In the same years, Gustav Mie published his most famous paper that dwells on the scattering of light by spherical particles, interpreted using Maxwell's electromagnetic theory [3]. Such papers, even though do not directly speak about plasmons, have become the most cited work by the plasmonic community (see Section 7.1.2). Only in 1956, plasmons were introduced by Pines to describe the quantum of an elementary excitation associated with high-frequency collective electronic motion in a plasma [4]. Finally, Ritchie in 1957 defined surface plasmons as plasma modes at the interface between a thin metal film and a dielectric medium corresponding to propagating longitudinal charge density waves [5].

From these pioneering works, surface plasmons have gained ample room in the nanophotonics field, mainly exploring the interactions between an incoming electromagnetic field and the free conduction electrons cloud in metals and its influence on the local field near the metal interface (Section 7.2.2). In fact, when an electromagnetic field with a suitable energy strikes a metallic thin film, the metal free electron gas starts to oscillate with respect to its positive ions backbone. Such collective oscillation is named *surface plasmon resonance* (SPR) and the surface plasmon is the quantum of this charge density oscillation. Such charge displacement generates an electromagnetic wave both inside the metal (i.e., a surface plasmon) and in the outer dielectric medium. So, if retardation effects are not negligible, a surface plasmon cannot exist without being associated with a transverse electromagnetic wave (a photon) [6]. In summary, the generated charge density wave is a mixed mode whose energy is shared between the charge density wave (plasmon) and the dielectric electromagnetic wave (photon), giving rise to so-called *surface plasmon polaritons* (SPPs). SPPs generated in a flat thin metal film can propagate for 10–100 μm along the surface (x- and y-directions) while decaying exponentially in hundreds of nanometers in the metal (z-direction) (Fig. 7.1a).

Such SPPs are usually also called *propagating surface plasmon polaritons* (PSPPs). If SPPs are generated in metallic nanostructures, with smaller dimensions with respect to the wavelength of the incoming radiation, the electron cloud oscillation is now confined around the metals (Fig. 7.1b) and the surface plasmons are called

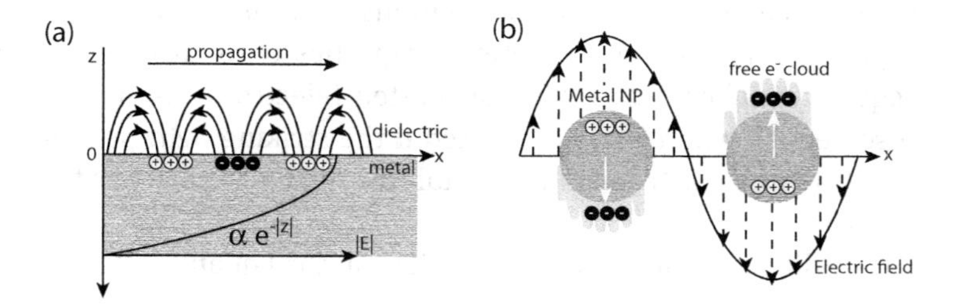

Figure 7.1 (a) Propagation of SPPs along a metal/dielectric interface and corresponding electric field exponential decay in the metal. (b) Electron cloud oscillation in metal nanoparticles under an external electric field.

localized surface plasmons (LSPs, see Section 7.1.2) or *localized surface plasmon polaritons* (LSPPs). An overview of some possible uses of LSPs for sensing is the aim of this chapter.

7.1.1 *Dielectric Functions of Metals: The Drude Model*

In order to understand the optical and surface plasmon properties of a metal, it is crucial to gain knowledge on its dielectric function $\hat{\varepsilon}(\omega)$. In the previous section, the conducting electrons of a metal have been described as cloud able to oscillate with respect to the positive nuclei when an electromagnetic field is applied. Such a system can be described using the Drude dielectric function for a free electron gas, in which the electrons oscillation is damped via collisions that occur with a characteristic collision frequency γ [7]:

$$\varepsilon_{Drude}(\omega) = 1 - \frac{\omega_p^2}{\omega^2 + i\gamma\omega} \tag{7.1}$$

where ω_p is the plasma frequency of the free electron system, defined as:

$$\omega_p = \sqrt{\frac{ne^2}{\varepsilon_0 m}} \tag{7.2}$$

where n is the conduction electron density, e is the elementary charge, ε_0 is the permittivity of the free space, and m is the electron mass. ω_p is the characteristic frequency of electron oscillation (i.e., longitudinal charge density fluctuations, or *volume plasmons*) and plays a crucial role in the optical properties of metals. Light with frequency higher than ω_p is transmitted (electrons are not so fast to screen it), whereas it is reflected if the striking light presents $\omega < \omega_p$ (the free electrons in the metal can screen the electric field of the light).

The dielectric function is a complex physical quantity:

$$\varepsilon(\omega) = \varepsilon_1(\omega) + i\varepsilon_2(\omega). \tag{7.3}$$

The real and imaginary parts of $\varepsilon(\omega)$ are correlated to the refracting and the absorbing properties of metals, respectively. Using Eqs. 7.1 and 7.3, the two components can be written as:

$$\varepsilon_1(\omega) = \mathrm{Re}(\varepsilon(\omega)) = 1 - \frac{\omega_p^2}{\omega^2 + \gamma^2} \tag{7.4}$$

$$\varepsilon_2(\omega) = \mathrm{Im}(\varepsilon(\omega)) = \frac{\omega_p^2 \gamma}{\omega(\omega^2 + \gamma^2)} \tag{7.5}$$

since in the visible part of the electromagnetic spectrum, γ is small with respect to ω and in the great majority of metals ω_p lies in the UV region, most metals show an almost total reflecting behavior. In fact, in such region $\omega < \omega_p$. As a consequence of this, it is clear from Eqs. 7.4 and 7.5 that $\mathrm{Re}[\varepsilon(\omega)] < 0$ and $\mathrm{Im}[\varepsilon(\omega)]$ is small. These conditions enable the excitation of a plasmon resonance, conditions not allowed in classic dielectric media in which $\mathrm{Re}[\varepsilon(\omega)]$ is in the order of a few units.

This simplified explanation does not consider possible inter-band transitions that take place usually at energies in the UV region. Sometimes such transition can strongly alter $\mathrm{Im}[\varepsilon(\omega)]$, increasing $\varepsilon_2(\omega)$, also in the visible part of the spectrum, as for gold, for example. To consider the inter-band presence, a further contribution to $\varepsilon(\omega)$, usually called $\varepsilon_b(\omega)$, is added, which results in a simply additive constant if the inter-band transitions lie in the UV region [6].

7.1.2 *Localized Surface Plasmon: Mie Model*

The classic dispersion relation for an SPP on a metal/dielectric interface is obtained solving Maxwell's equation with suitable boundary conditions and shows that an SPP can be excited only under particular circumstances [8]. Such description is no more valid when the metal structure dimensions are smaller with respect to the wavelength of the excitation light. In this case, a different class of nonpropagating plasmon can be excited: LSPs. The most used and simplest way to understand the optical properties of such plasmon category was proposed by Mie in 1908 [3]. Mie considered the interaction between a small spherical particle immersed in a dielectric medium (with dielectric constant ε_D) and an incident electromagnetic radiation with wavelength bigger with respect to the particle radius. Solving Maxwell's equation for extinction and scattering, with suitable boundary conditions, Mie described the problem as multipolar oscillations. In quasi-static regime (or dipole approximation) [9, 10], the process is dominated by the dipolar term, and the scattering and extinction cross sections are, respectively:

$$\sigma_{sca}(\omega) = 2\left(\frac{\omega}{c}\right)^4 V\left(\frac{\varepsilon_1(\omega) - \varepsilon_D}{\varepsilon_1(\omega) + 2\varepsilon_D}\right) \quad (7.6)$$

$$\sigma_{ext}(\omega) = \frac{9V\varepsilon_D^{3/2}}{c}\frac{\omega\varepsilon_2(\omega)}{[\varepsilon_1(\omega) + 2\varepsilon_D]^2 + \varepsilon_2(\omega)^2} \quad (7.7)$$

where V is the particle volume, ω is the frequency of the exciting wavelength, and c is the speed of light. Considering that the absorption cross section is $\sigma_{abs} = \sigma_{ext} - \sigma_{sca}$, it is possible from Eqs. 7.6 and 7.7 to evince that a resonance condition, which involves a strong absorption band for a metallic sphere, is satisfied for $\varepsilon_1(\omega) = -2\varepsilon_D$ and $\varepsilon_2(\omega) \sim 0$. The particular frequency that fulfills such constrains is called *localized surface plasmon resonance* (LSPR). An example of Au nanosphere extinction spectrum with relative LSPR is given in Fig. 7.2.

In the dipole approximation, the Mie theory also predicts that the incident radiation induces a constant electric dipole moment inside the sphere, proportional to the amplitude of the light electromagnetic field (E_0). As a consequence, the *local* electric field on the sphere surface (i.e., located within a few nanometers from the sphere) becomes equal to [11]:

$$E_{loc} = \frac{3\varepsilon(\omega)}{\varepsilon(\omega) + 2\varepsilon_D}E_0 \quad (7.8)$$

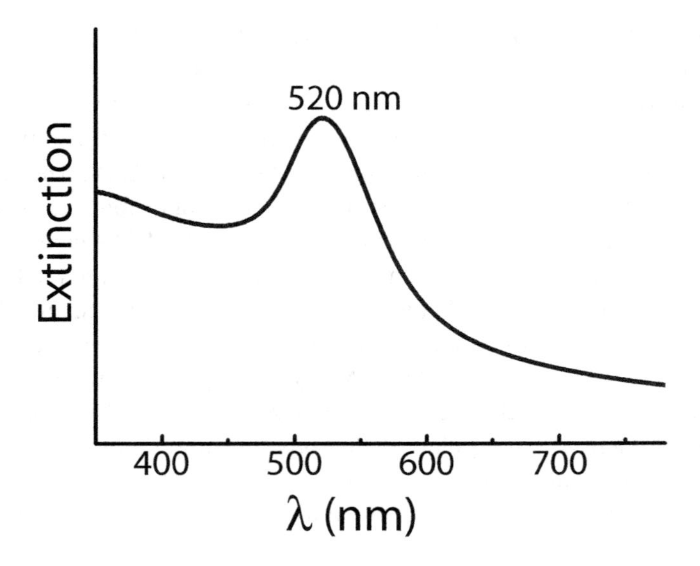

Figure 7.2 Extinction spectrum for a gold spherical nanoparticle with a diameter of about 13 nm showing the typical LSPR at 520 nm.

Under the resonance condition, E_{loc} is strongly enhanced with respect to the far field. This huge increase in the local field around metal nanoparticles (or metal nanostructures, in a more general point of view) is the basis for the sensing applications, which will be described in Sections 7.3 and 7.4.

If the radius of the metal sphere (r) is below a few nanometers, the LSPR band is damped and broadened and in some cases can disappear [10]. This behavior is due to the increase in electron-surface scattering; the mean free path (l) of electrons becomes increasingly smaller. The electrons are scattered at the metal surface in an elastic and completely random way, inducing a fast loss of the plasmonic oscillation coherence. The smaller the metal sphere, the faster the coherence loss and the bigger the band broadening. This phenomenon is called *intrinsic size effect.*

For bulk gold and silver, the two most employed plasmonic metals, l is around 40–50 nm. The Mie theory can be corrected including a dielectric function of the metal dependent on r [10], which justifies the slight LSPR shift varying the particles dimension.

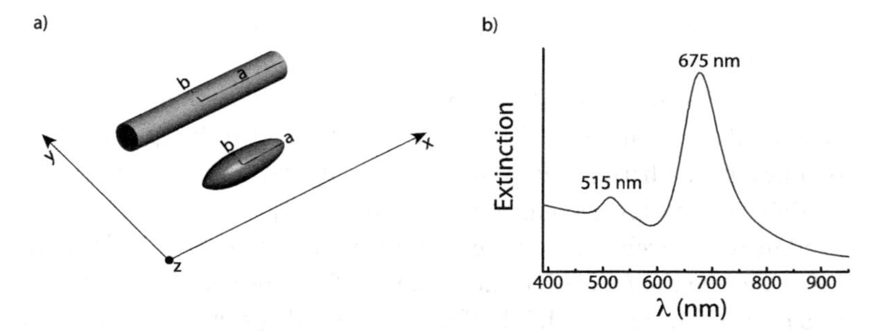

Figure 7.3 (a) Spheroidal metal NP and nanorod with respective axis: a is the principal axis and b is the axis perpendicular to the principal one. (b) Extinction spectrum of Au nanorods with aspect ratio of about 2.5.

The Mie model works well only for spherical nanoparticles (NPs) in solution, or solid state matrix, with low concentration where NPs mutual interactions can be neglected. Increasing the NP concentration, the inter-particle distance becomes smaller (till aggregation) and the LSP mode (and consequently E_{loc}) of each NP is influenced by its neighbors. As a consequence, hybridized modes are formed and the LSPR red-shifts. This phenomenon is well described by the Maxwell–Garnett theory [12, 13].

The second limit of the Mie model resided in the description of elongated or spheroidal NPs (e.g., nanorods, elliptic NPs, etc.). For these nano-objects, the LSPR splits in two different modes: longitudinal and transverse ones. The lower energy longitudinal mode is due to the electron cloud oscillation along the spheroid principal axis (along a-axis, x-direction, Fig. 7.3), while the high energy transverse one is referred to the electrons oscillations perpendicular to the principal axis (along b-axis, y-direction, Fig. 7.3). The aspect ratio R (i.e., a/b) of the nano-object plays a pivotal role in the wavelength position of the LSPR bands. In fact, the longitudinal LSPR band remains constant increasing R, while the transverse LSPR one red-shifts. At the end, the energy separation between the two bands increases with the increase in R. The Gans model well describes the LSPR behavior of these systems [13].

Other very interesting systems that the Mie model is unable to describe are metal structures made of several concentric layers, which can be interesting for sensing applications. Those structures have a dielectric core and a metal outer shell [14].

The LSPR of these and similar systems have been initially described using an extension of the Mie theory (in the quasi-static regime) in which a combination of two (or more) concentric spheres of different materials has been considered [15]. More recently, Prodan has developed a novel mathematical model, the *plasmon hybridization model* [16], to describe LSPR in core–shell system. In this representation, the LSPR of the core–shell system is considered a combination of plasmons supported by a nanosphere and a cavity.

7.1.3 *Plasmon Platforms for Sensing*

In principle, any metal that can fulfill the resonance conditions $\varepsilon_1(\omega) = -2\varepsilon_D$ and $\varepsilon_2(\omega) \sim 0$ at proper wavelength can sustain an LSPR in the visible and can be ideally used as active material in a sensor. Actually, a majority of plasmonic sensor platforms have been developed using gold or silver. Both these materials present, in the visible part of the electromagnetic spectrum, Re($\hat{\varepsilon}(\omega)$) <0 and high in absolute values and a small Im($\hat{\varepsilon}(\omega)$). In particular, Ag presents strong and sharp LSPR in the blue part of the visible spectrum, because of its lower inter-band absorption with respect to Au, and allows to build up highly sensitive devices. Au can sustain less strong LSPR from the red part of the visible spectrum ($>$500 nm) to the near IR and is heavily employed in biosensing because of its chemical stability, resistance to oxidation, and low citotoxicity [17]. Other metals, such as aluminum in the UV or Cu in the red, can sustain LSP but are less studied and employed for sensors.

Based on Au and Ag, a plethora of different plasmonic nanostructures for molecular sensing have been produced [18–20] in solutions via chemical synthesis and subsequent possible deposition or immobilization on supports, or fabricated directly on solid substrates via wet deposition, focused ion beam (FIB) milling, and electron-beam lithography (e-beam). These samples range from spherical NP [21] to nanorods [22], nanoprisms [23], core–shell structures [24], and so on. The production strategies for the majority of the employed nanostructures allow controlling their shapes and thus the position of the LSPR.

Among this jungle of possible plasmonic nanostructure for sensing, it is possible to individuate three main sensor typologies, depending on the physical observable used to probe the system. In these three categories, absorption, scattering, and emission of light are exploited in order to have a plasmonic sensor.

The first category exploits the sensitivity of plasmonic substrates to a refractive index variation. In such configuration, a slight variation of the LSPR, simply observed by UV-absorption spectroscopy, indicates a change in the nanostructures' dielectric environment (and thus refractive index) due to the presence of analytes [25–27]. Such typologies of devices are called LSPR sensors. The second sensor category is based on surface enhanced Raman spectroscopy, in which the scattering of an incoming light by an analyte is greatly increased by the plasmon resonance. Systems based on such phenomenon will be deeply investigated in Section 7.2 and are called SERS sensors. The last category, which will be the topic of Section 7.3, exploits the increase of fluorescence emission of suitable chromophores, bound or able to bind the analytes of interest, in the presence of a plasmonic structure. This event is called *metal enhanced fluorescence* (MEF), and devices based on such phenomenon are called MEF sensors. The last two categories, as will be largely argued in the following, are based on the aforementioned huge increase in the E_{loc} around the metal nanostructure due to the plasmon resonance (see Section 7.1.2).

7.2 Surface Enhanced Raman Spectroscopy for Sensing

7.2.1 *Raman Spectroscopy*

Raman spectroscopy is a powerful technique that allows for an unambiguous identification of chemical and biological molecules. The Raman phenomenon was discovered in 1928 by Sir Chandrasekhra Raman, who used sunlight as light source and a telescope to collect the light on his eyes. Raman observed that part of the light scattered by a transparent material presented a different wavelength with respect to the incoming radiation [28, 29]. In a modern interpretation, this means that a certain molecular vibration

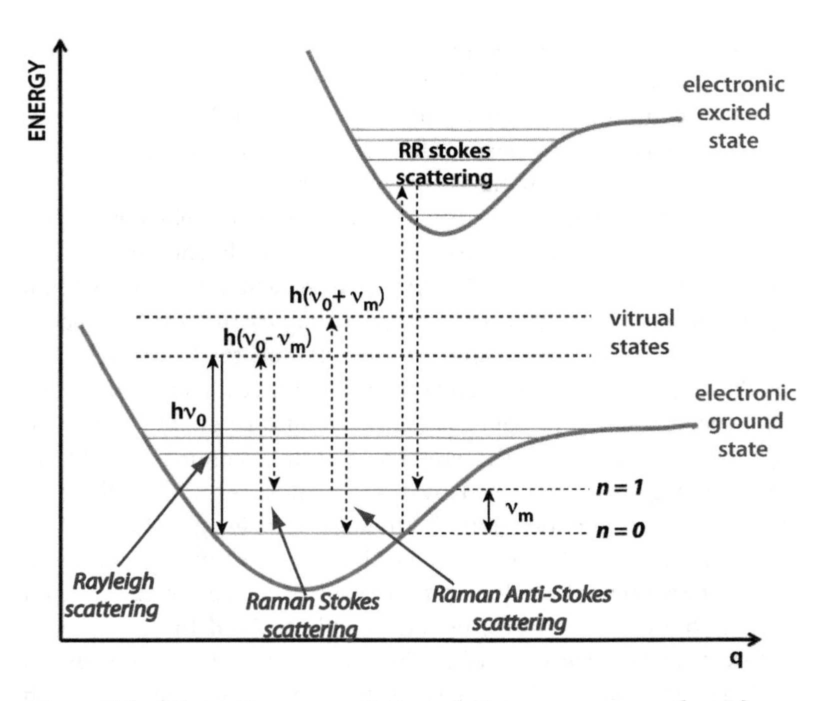

Figure 7.4 Schematic representation of Raman processes through an energy-level diagram of a molecular system: Rayleigh scattering (photon energy $h\nu_0$); Stokes scattering (photon energy $h\nu_0 - h\nu_m$); anti-Stokes scattering (photon energy $h\nu_0 + h\nu_m$). Resonant Raman Stokes scattering is also reported: The intermediate transition state is a proper electronic state of the molecule.

stricken by an intense light (e.g., a laser beam) can scatter the incoming radiation, generating three different signals. The first one, named *Rayleigh scattering,* is strong and with the same frequency of the hitting radiation (ν_0); the other two signals are very weak (about 10^{-6} times the incident light) and have frequencies equal to $\nu_S = \nu_0 - \nu_m$, that is, *Stokes line,* and $\nu_{AS} = \nu_0 + \nu_m$, i.e., *anti-Stokes line,* where ν_m is a typical vibrational frequency of the sampled molecule (Fig. 7.4). These two inelastic scattered radiations are unique "molecular fingerprints" of the sample under analysis and can be successfully employed for sensing applications.

A simplified illustration of Raman scattering is provided in Fig. 7.4. Usually in Raman experiments, the frequency of the incoming light (defined by ν_0) does not allow an electronic transition toward the first-excited state of the probed molecule. Thus, to describe the energetic interaction between the laser beam and the molecule, *excited virtual states* are usually invoked. A virtual state is a not-real intermediate state with a defined energy used to describe a multistep process, that is, Raman scattering. Incident photons can cause transition of a molecule to an excited virtual state from molecule's ground ($n = 0$), or first-excited ($n = 1$) vibrational state of the electronic ground state. The molecule relaxation can give rise to Rayleigh, Stokes, or anti-Stokes signals depending on the final vibrational state involved in the process (Fig. 7.4). Usually, anti-Stokes signal presents a lower intensity with respect to Stokes ones, because at room temperature, only a small fraction of molecules (defined by a Boltzmann population distribution) lies in an excited vibrational state of the electronic ground state [30].

The Raman spectra report the intensity of scattered light as a function of a wavenumber (Fig. 7.5) called *Raman shift* ($\Delta\overline{\upsilon}$), which is the difference between $\tilde{\upsilon}_0$ and $\tilde{\upsilon}_S$ (for Stokes lines) or $\tilde{\upsilon}_{AS}$ (for anti-Stokes lines) (Eq. 7.9).

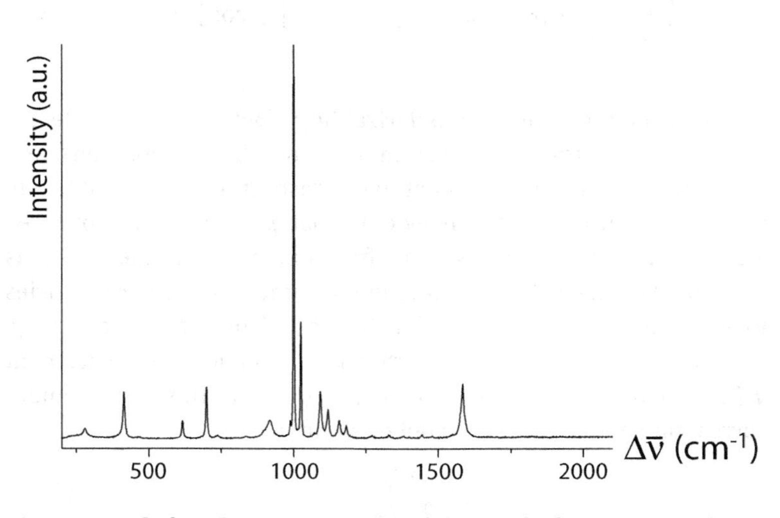

Figure 7.5 Stokes Raman spectrum of Benzenthiol, a common Raman marker. The Raman intensity (y-axis) is plotted as a function of the Raman shift (x-axis).

$$\Delta\bar{\nu} = \tilde{\nu}_0 - \tilde{\nu}_S = \frac{1}{\lambda_0} - \frac{1}{\lambda_S}, \qquad [\text{cm}^{-1}]$$

$$\Delta\bar{\nu} = \tilde{\nu}_0 - \tilde{\nu}_{AS} = \frac{1}{\lambda_0} - \frac{1}{\lambda_{AS}}, \qquad [\text{cm}^{-1}] \qquad (7.9)$$

where λ_0 and λ_S (or λ_{AS}) are the incoming and scattered light wavelengths, respectively. Such representation renders the spectra independent on the excitation frequency and characteristic of the investigated molecular vibrational state.

If ν_0 matches an electronic transition of the molecule, the Raman process is named *Resonant Raman* (RR). In RR spectra, the Raman bands of molecules are strongly enhanced (10^3 to 10^5 times) [31]. The Raman scattering phenomenon is physically explainable using a classical approach, considering a diatomic molecule irradiated by an electromagnetic wave (e.g., a laser beam) with the electric field strength that fluctuates with time: $E(t) = E_0 \cos 2\pi\nu_0 t$, with E_0 and ν_0 the vibrational amplitude and the frequency of the laser, respectively. The induced dielectric dipole moment (p) in the molecule can be expressed as [31]:

$$p = \alpha_0 E_0 \cos 2\pi\nu_0 t + \frac{1}{2}\left(\frac{\partial\alpha}{\partial q}\right)_0 q_o E_0 \{\cos[2\pi(\nu_0 + \nu_m)t] + \cos[2\pi(\nu_0 - \nu_m)t]\} \qquad (7.10)$$

where α is the molecule polarizability (with α_0 the molecule polarizability with the nuclei in the equilibrium position) and q is the nuclear displacement with respect to the equilibrium position. From Eq. 7.10, it is evident that p is the source of three electromagnetic radiations: The first one with frequency ν_0 is ascribable to Rayleigh scattering, and the other two with frequencies $\nu_0 - \nu_\text{m}$ and $\nu_0 + \nu_\text{m}$ are related to Stokes and anti-Stokes scattering.

From Eq. 7.10, it is also clear that a certain nuclear displacement q (i.e., a vibrational mode) is Raman active if to this displacement correspond a variation in the polarizability:

$$\left(\frac{\partial\alpha}{\partial q}\right)_0 \neq 0. \qquad (7.11)$$

Equation 7.11 describes the principal Raman selection rule. Because Raman scattering refers directly to typical vibrational frequencies of the molecule, each Raman spectrum determines in a univocal way a specific molecule. For this reason, the employment of Raman spectroscopy is very interesting in sensing. Unfortunately, the Raman scattering intensity is very low with respect to the incident light; in fact it depends, besides the incident laser intensity and the number of probed molecule (N), on the analyte Raman cross section (σ_R):

$$I_R(\upsilon_s) \propto N I(\upsilon_0)\sigma_R \tag{7.12}$$

where ν_s is the Stokes (or anti-Stokes) line frequency. σ_R is usually very low [6], ranging between 10^{-25} and 10^{-30} cm^{-2}; therefore, the Raman technique presents low sensitivity and needs a huge quantity of analyte.

7.2.2 *Surface Enhanced Raman Spectroscopy*

An intriguing way to overcome the just mentioned Raman drawbacks is the *surface enhanced Raman spectroscopy* (or scattering) technique (SERS technique). The relative increase in importance of Raman spectroscopy for analytical and sensing applications, following the discovery of SERS, is well described by more than 5000 articles, 100 reviews, and several books on this topic from the 1970s [23].

As self-explained by the name, the SERS technique allows for an amplification of a Raman signal, thanks to a surface, and in particular a plasmonic surface. The first evidence of the SERS phenomenon was observed in 1973 by Fischer et al. [32], but they assigned the enhancement of the pyridine signal adsorbed on an Ag electrode to an increase in the surface area due to the metal's high roughness. Only in 1977, Jaenmarie and Van Duyne [33] and Albrecht and Creighton [34] independently pointed out that the enhancement phenomenon (a factor 10^6) cannot be ascribed only to an increase in the electrodes' surface. Finally, in 1978 Moskovits proposed as responsible for such anomalous enhancement an increase in the molecular σ_R due to the presence of surface plasmons [35]. In 1985, Moskovits proposed that two distinct contributions are responsible for the 10^6 increase in the Raman intensity observed [36]: an electromagnetic enhancement mechanism and a chemical one.

The latter contribution considers the possible electronic interaction between the analyte and the metal that can alter the scattering process [23] giving rise to a larger cross section (σ_{SERS}) with respect to the one ascribed to the free molecule (i.e., σ_R). Such contribution enhancement is in the order of 10^2.

The electromagnetic contribution [37] is directly related to the E_{loc} around metal nanostructures (described in Section 7.1.2). Such E_{loc} influences both the incoming and the scattered electromagnetic radiation. An equation similar to Eq. 7.12 can be written for this case [38]:

$$I_{SERS}(\nu_s) \propto N'I(\nu_0)\,|g(\nu_0)|^2 \cdot |g(\nu_S)|^2\,\sigma_{SERS} \tag{7.13}$$

where N' is the number of molecules near the metal surface and $g(\nu_0)$ and $g(\nu_S)$ represent the enhancement of the field felt by the molecules for the incoming and scattered light, respectively. Practically, $g_i \propto E_{loc}/E_i$, where $i = 0$ or s. These two terms account for the higher local field felt by the exciting and scattered light and depend quadratically on their field (i.e., linearly on their intensity).

Since E_{loc} can be very high with respect to the far field, I_{SERS} can be in orders of magnitude higher with respect to I_R. The electromagnetic contribution does not need that the analyte is bonded at the surface; nevertheless, it strongly decays leaving the metal surface ($I_{SERS}\ \alpha\ r^{-10}$ [39], where r is the distance between the metal and the molecule), which implies a close sensing volume.

Different SERS surfaces (or substrates) can exhibit quite different E_{loc} and thus I_{SERS}, keeping constant I_R and the number of molecules involved. To compare different enhancement magnitude for different structures, a figure of merit called *enhancement factor* (EF) is used. Unfortunately, sometimes in literature the term EF is used to describe slightly different things, making difficult the direct comparison of different SERS substrates. In 2007, Le Ru et al. [40] clarified this confusion giving an important contribution to SERS. Among the plethora of different EFs defined by Le Ru, the most useful to compare different SERS substrates is the *SERS substrate enhancement factor* (SSEF, hereafter EF), that is, the ratio between a single-molecule signal in SERS conditions on a single-molecule signal in normal conditions:

$$EF = \frac{I_{\text{SERS}}/N_{\text{surf}}}{I_R/N_{\text{vol}}} \tag{7.14}$$

where N_{surf} is the average number of molecules adsorbed on a metal in the irradiated volume in SERS conditions and N_{vol} is the average number of molecules in classic Raman in the same volume.

Some particular substrates can exhibit incredibly high EF values (ranging from 10^8 to 10^{11}), allowing for Raman spectroscopy of single molecule. In most of these cases, such high EF is reached thanks to a particular metal structural feature present in the substrate: the *hot spot*. A powerful definition of hot spot has been proposed by Kleinman et al. [41]: "a junction or close interaction of two or more plasmonic objects where at least one object has a small radius of curvature on the nanometer scale. This structural motif has the ability to concentrate an incident electromagnetic field and effectively amplify the near field between and around the nanostructures."

The presence of hot spots obviously increases the sensitivity of SERS preserving its selectivity.

7.2.3 *SERS Sensing Applications*

The extremely high sensitivity and selectivity make SERS strongly appealing for the identification of biological and chemical agents. In particular, concerning the biosensing, SERS can employ near IR laser excitation that, even if it implies fewer scattered photons, drastically reduces the fluorescent background present in biological environments. Furthermore, because SERS measurements can be performed in nonresonant conditions, different analytes, or SERS labels, can be probed with the same laser frequency.

Usually, for SERS detection two different strategies have been mainly used: labeled (or *extrinsic* configuration) and label-free (or *intrinsic* configuration) detection [42]. In the latter configuration, the analyte is directly applied on the plasmonic feature (*direct intrinsic detection*) or localized in the field enhancement area by suitable species (i.e., antibodies, aptamers, etc.) bound to the metal (*indirect intrinsic detection*).

The *extrinsic* configuration is based on Raman reporters (molecules with high Raman cross sections in the order of 10^{-18}–10^{-14} cm^2) adsorbed on a plasmonic multifunctional moiety, which is also decorated with capture molecules that work as gathering for the analyte of biological interest.

Following these methodologies, many different analytes have been detected, and exhaustive reviews reported the development of sensors for proteins and antibodies [43, 44], DNA strands [44–46], bacteria and virus [42], in vitro and in vivo analytes (especially glucose) [19, 46, 47], etc. [18, 48, 49].

Several different plasmonic platforms have been employed to develop a huge number of sensor prototypes, but among these different structures, the silver films over nanosphere (or monolayer), Ag-FONs (or Ag-FOMs), proposed by Van Duyne in 1993 [50] (Fig. 7.6) deserve a special mention.

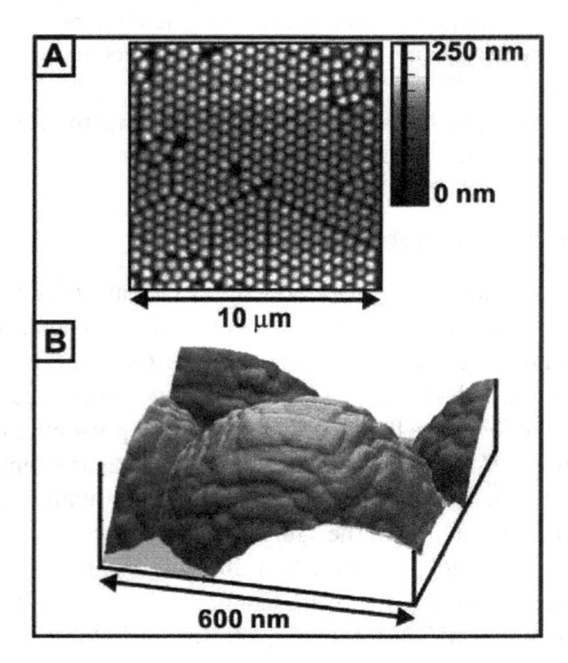

Figure 7.6 Atomic force microscope image of 200 nm Ag over 542 nm diameter polystyrene spheres. (a) Array of spheres (10 μm × 10 μm) and (b) image (600 nm × 600 nm) of one sphere. Reproduced with permission from Ref. [23]. Copyright © 2011, Royal Society of Chemistry.

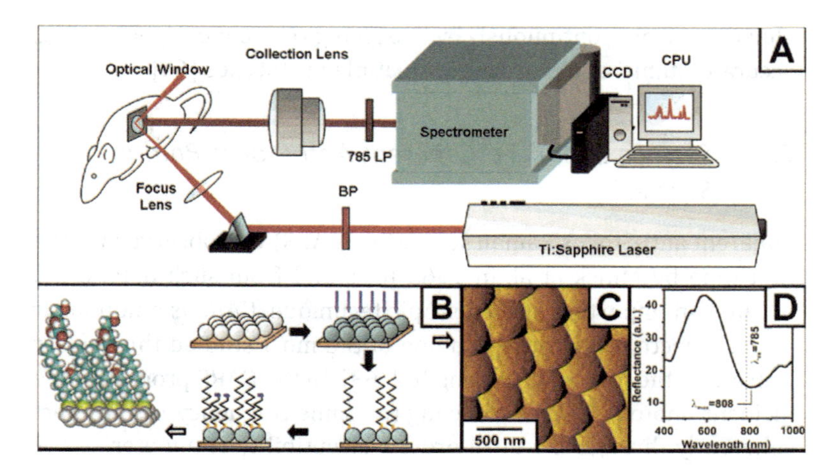

Figure 7.7 (a) Scheme of instrumental apparatus for glucose in vivo sensing. (b) Illustration of the Ag-FON sensor preparation. (c) AFM to study Ag-FON surface morphology. (d) LSPR evaluation by reflectance spectrum. Reprinted with permission from Ref. 58. Copyright © 2006, American Chemical Society.

Such substrates are robust and widely employed for SERS applications (e.g., structural investigation on semiconductor nanocrystals [51]) and especially for biological sensing, thanks to their [23] high enhancement factors (10^5–10^6), high thermal stability (up to $\sim$500 K), high electrochemical stability, long shelf life (>40 days), and, most importantly, their stability in biological environment for almost 10 days [52].

Ag-FONs have been used for probing many chemical and biological analytes, including anthrax spores [53], artists' red dyes [54, 55], and in vitro glucose [52, 56, 57]. Probably, the main goal of Ag-FONs as sensing platform resides in the capability to probe glucose in vivo with similar detection limit of classic sensor, as demonstrated by Stuart in 2006 [58, 59] (Fig. 7.7). The authors claim that with further refinements in the system, the SERS sensor has the potential to replace conventional personal and point-of-care assays.

This step is of crucial importance because, despite continuous innovations in the sensing of blood glucose, a lack of accuracy still remains, and novel and more sensitive glucose sensors are strongly demanded to provide an early diagnosis for diabetes. Ag-FON-based glucose sensors continuously evolve giving rise to more performing, accurate, stable, and more biocompatible prototypes [60].

7.2.4 *Surface Enhanced Coherent Anti-Stokes Raman Scattering*

Coherent anti-Stokes Raman scattering (CARS) was observed for the first time by Minck et al. in 1963 [61], and from such date, CARS became an important spectroscopic technique. CARS is a nonlinear optical spectroscopy based on one of the most studied third-order processes, the four wave mixing [62, 63]. In the CARS process, three incident photons (with two owing the same frequency ω_p) interact coherently through the third-order susceptibility of a material. For example, CARS is observed if two light beams (with frequencies ν_p and ν_s where $\nu_p > \nu_s$) strike a sample with an active Raman mode of frequency $\nu_n = \nu_p - \nu_s$. In such a condition, a new radiation can be generated at the anti-Stokes frequency (Fig. 7.8): $\nu_{as} = \nu_p + \nu_n = 2\nu_p - \nu_s$ [64, 65].

CARS presents much higher sensitivity with respect to classic Raman thanks to its higher order dependence on the incident laser power. Notwithstanding, CARS is still affected by a low absolute sensitivity limit that does not allow for the detection of low concentration analytes and, eventually, to perform single-molecule identification.

To strongly increase the CARS sensitivity, an intriguing possibility is suggested by classic Raman spectroscopy. As in Raman spectroscopy a plasmonic assistance leads to high-sensitivity SERS (Fig. 7.8), the same phenomenon allows moving from CARS to high-sensitivity *surface enhanced coherent anti-Stokes Raman scattering* (SECARS). In SECARS experiments, at least one, or better all, of the frequencies of photons involved in the process is in resonance with the LSPR of the metal nanostructures employed. In the latter case, all the radiations feel the extremely high E_{loc} (due to the plasmonic features) and their fields are enhanced by the g factor (see Section 7.1.2), quadratically dependent on the respective intensities (Fig. 7.8).

Following the first SECARS experimental observation by Liang in 1994 [66], many efforts have been done to increase SECARS

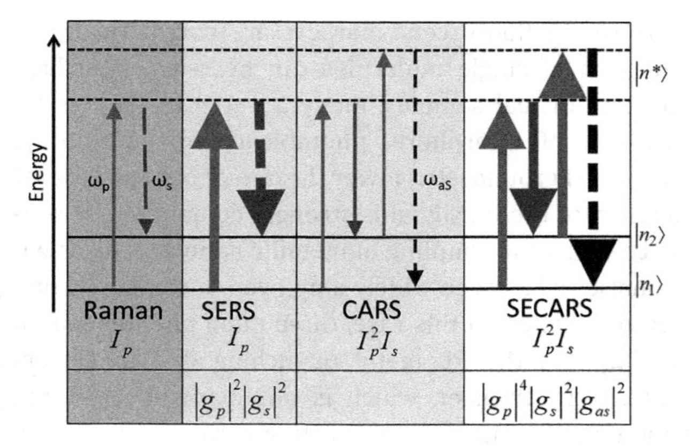

Figure 7.8 Schematic band energy diagram for electronic transitions in various Raman processes (where $\omega_i = 2\pi \nu_i$). Their dependence on the pump (I_p) and Stokes (I_s) intensity is reported and related to respective local electric field enhancement (g_p and g_s). The Dirac notation $|n^*>$denotes virtual states of the molecules, while $|n_1>$and $|n_2>$symbolize its ground and first-excited vibrational states of the ground electronic state. [Reprinted with permission from Ref. 65. Copyright © 2011, American Chemical Society].

efficiency to employ such technique for biological sensing. Ichimura et al. in 2003 [67] observed that CARS intensity of adenosine molecules was increased of 2000 if such an analyte is attached to isolated gold NPs. In 2005, Koo et al. [68] claimed the detection of the first single molecule, by means of SECARS, for deoxyguanosine monophosphate and deoxyadenosine monophosphate. More recently, Machtoub et al. in 2010 [69] and Schlücker et al. in 2011 [70] employed SECARS to probe, in in vitro tissue, biological species of interest directly related to amyotrophic lateral sclerosis and prostate disease, respectively.

7.3 Metal Enhanced Fluorescence for Sensing

7.3.1 *Metal Enhanced Fluorescence Mechanism*

Photo-luminesce (PL) is the most employed technique for sensing in biological and medical applications [71] because of its high sensitivity and in situ possible application. Nowadays, numerous high performing fluorescent markers are commercially available and detection of single molecules can be reached in laboratory. Nevertheless, in real samples, many adverse conditions (e.g., low concentration of fluorphore, photobleaching, auto-fluorescence, interfering background, etc.) lower the detection limit. Therefore, an amplification of the PL signal is strongly demanded for sensing. In the last 10 years, the coupling of metallic nanostructures with fluorescent markers has been widely employed to change chromophore emission properties. In this case, three main phenomena influence the PL [72]: (1) the PL metal quenching due to chromophore to metal energy transfer, which is dominant at short ranges (a few nanometers) and it is proportional to r^{-3}, where r is the metal-chromophore distance; (2) an enhancement of the PL, which becomes most probable for longer ranges; this effect is strongly correlated to an increased rate of excitation, due to a higher transition probability for the chromophore, because of the higher local field around metal nanostructures; (3) the possible increase in the radiative decay rate of the chromophore. These last two effects represent the core of the MEF phenomenon.

Point (3) has been well described by Lackowicz's group [73, 74] using a semiempirical model. In such description, Lackowciz compares the Jablonski diagrams of a chromophore in presence or absence of metal nanostructures (Fig. 7.9). In the latter case, the chromophore quantum yield (Φ_0) and the relative lifetime (τ_0) can be expressed as:

$$\Phi_0 = \frac{\Gamma}{\Gamma + \Gamma_{nr}} \tag{7.15}$$

$$\tau_0 = \frac{1}{\Gamma + \Gamma_{nr}} \tag{7.16}$$

where Γ is the chromophore radiative decay rate and Γ_{nr} is the nonradiative decay rate. Because Γ is almost constant [73], Φ_0 increases if Γ_{nr} decreases and consequently τ_0 increases. If the chromophore gets close to a metal nanostructure, the quantum yield (Φ_{m}) and the lifetime (τ_m) can now be described by:

$$\Phi_m = \frac{(\Gamma + \Gamma_m)}{(\Gamma + \Gamma_{nr} + \Gamma_m)} \tag{7.17}$$

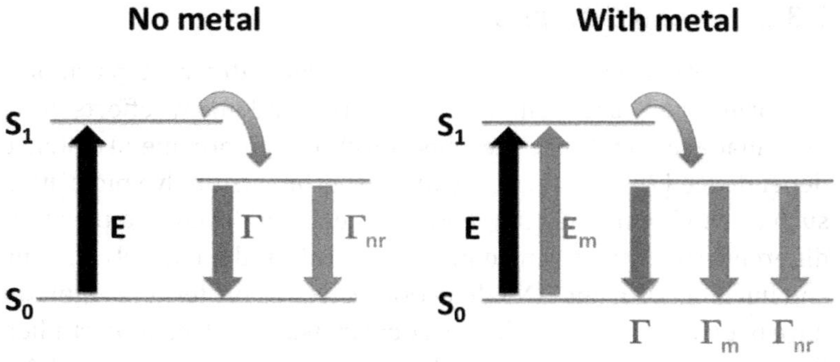

Figure 7.9 Schematic Jablonski diagram for chromophore in free space (left panel) and coupled with a metal nanostructure (right panel). In the latter case, the absorption properties are modified by E_m because of the high local field around metal nanostructures [point (2) in the text]. Γ is the chromophore radiative decay rate and Γ_{nr} is the nonradiative decay rate. Γ_m is an additional radiative rate due to the metal presence.

$$\tau_m = \frac{1}{(\Gamma + \Gamma_{nr} + \Gamma_m)} \tag{7.18}$$

where Γ_m is an additional radiative decay rate due to the metal presence. The higher the Γ_m, the higher the Φ_m and the lower the τ_m.

In conclusion, with high Γ_m values, it is possible to obtain higher chromophore emission (higher Φ_m) and photostability. In fact, the chromophore lies in the excited state for a shorter time (lower τ_m). Unlike the free metal case, now the lifetime decrease is not related to an increase in nonradiative decay rate that leads to a decrease in quantum yield.

From Eqs. 7.17 and 7.18, it is possible to extrapolate that the major relative PL enhancements can be reached for low Φ_0 chromophores [72]. Theoretical [75] and experimental [76] works demonstrate that the higher enhancement (higher Φ_m) for a chromophore is reached when the metal LSPR is slightly red-shifted with respect to the molecule PL maximum wavelength (λ_{em}). If $\lambda_{em} =$ LSPR, radiative and nonradiative decay are equally enhanced, while if $\lambda_{em} >$LSPR, nonradiative rate enhancement prevails with respect to radiative one [77].

7.3.2 *Sensing Examples*

As for SERS sensing (Section 7.2.3), several different plasmonic platforms have been employed for MEF and many efforts have been made to investigate the emission chromophore–metal distance dependence [78–81]. From the first experiments on dye molecules, such as rhodamine 6G [82], MEF has been employed in a plenty of different experiments, including biological (and chemical) sensing and imaging. DNA and RNA detection as well as proteins recognition have been widely explored. In particular, Aslan et al. in 2006 studied fluorescence-marked RNA coupled with Ag islands [83], while DNA coupled with Ag islands, Au NPs and Ag aggregate NPs (to exploit hot spots) have been investigated by Sabanayagam [84], Li [85], and Gill [86], respectively. Also many examples concerning detection of proteins have been reported: studies on bovine serum albumin were performed coupling this protein with Ag or Al nanostructures [87, 88]; analysis on rabbit immunoglobulin G has been done on Ag island [89] as well as on Al nanostructures [88]; streptavidin detection has been performed on Al nanostructures [88] and on Au nanorods [90]; for troponin I recognition, Ag NPs have been employed [91]; human serum albumin has been probed on Ag nanorod [92], etc.

Finally, in 2014 Todescato et al. [93] proposed an easy, sensitive, rapid, and low cost biosensors prototype for ochratoxin A (OTA, a highly diffused food contaminant detrimental to human health) based on MEF. In this work, the aforementioned Ag–FON plasmonic substrates (Section 7.2.3) have been employed to detect OTA in different food commodities, including cereals for infants, milk, and juices. First, OTA detection has been performed in samples spiked with known quantities of OTA marked with Alexa Fluor 647 (AF 647) and then the analysis moves toward samples simulating real OTA detection: OTA has been extracted from food commodities and hence marked with AF 647. For milk, wheat mix, and apple juice matrices, the plasmonic enhancement allowed detection of OTA amounts up to concentrations of one order of magnitude lower than those specified by EU legislation concerning limit of exposure in food for infants and young children (i.e., 0.5 μg/kg) (Fig. 7.10). Concerning grape juice, OTA quantification was not successfully achieved. The

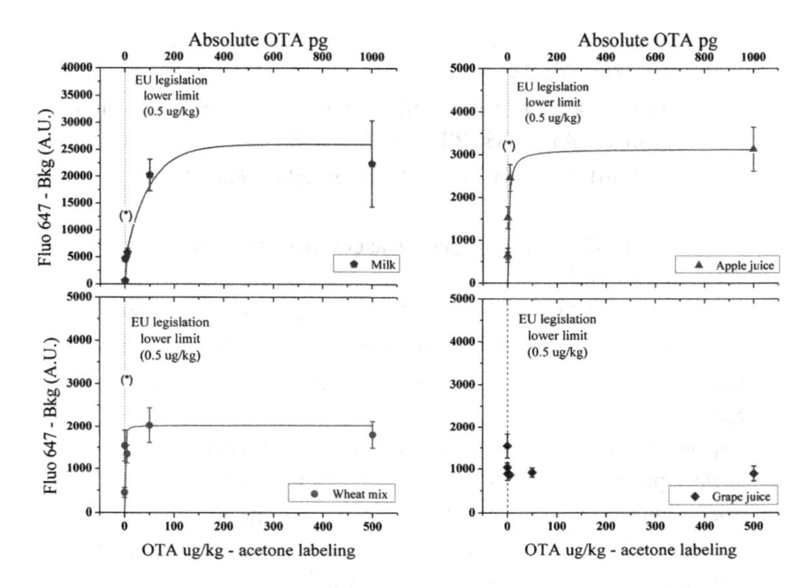

Figure 7.10 Dried milk (blue pentagons), apple juice (green triangles), wheat mix (yellow circles), and grape juice (purple diamonds) samples spiked with known concentrations of AF 647 labeled OTA. The OTA concentration ranges from 0.05 to 500 μg/kg. The lower legislation limit is plotted on the graph as dotted line. Reprinted from Ref. 93, Copyright 2014, with permission from Elsevier.

interference of some components present in the extract gives a higher background and no specific signal in spiked samples.

Acknowledgments

F.T. acknowledges "Assegno di Ricerca Senior" Repertorio 76-2012 of University of Padova. The author is grateful also to Prof. R. Bozio, D. Pedron, Dr. R. Pilot, and W. Veber for the stimulating discussions.

References

1. Wood R.W. (1902) On a remarkable case of uneven distribution of light in a diffraction grating spectrum, *Philos. Mag. Ser.*, **6**, pp. 396–402.

2. Rayleigh L. (1907) On the dynamical theory of gratings, *Proc. R. Soc. A Math. Phys. Eng. Sci.*, **79**, pp. 399–416.
3. Mie G. (1908) Beiträge zur optik trüber medien speziell kolloidaler metalllösungen, *Ann. Phys.*, **330**, pp. 377–445.
4. Pines D. (1956) Collective energy losses in solids, *Rev. Mod. Phys.*, **28**, pp. 184–198.
5. Ritchie R. (1957) Plasma losses by fast electrons in thin films, *Phys. Rev.*, **106**, pp. 874–881.
6. Le Ru E.C. and Etchegoin P.G. (2009) *Principles of Surface-Enhanced Raman Spectroscopy and Related Plasmonic Effects*, 1st ed, Elsevier, UK.
7. Maier S.A. (2007) *Plasmonics: Fundamentals and Applications*, 1st ed, Springer, New York.
8. Zaytas A.V., Smolyaninov I.I., and Maradudin A.A. (2005) Nano-optics of surface plasmons polaritons, *Phys. Rep.*, **408**, pp. 131–314.
9. Alvarez M.M., Khoury J.T., Schaaff T.G., Shafigullin M.N., Vezmar I., and Whetten R.L. (1997) Optical absorption spectra of nanocrystal gold molecules, *J. Phys. Chem. B*, **101**, pp. 3706–3712.
10. Link S. and El-Sayed M.A. (1999) Size and temperature dependence of the plasmon absorption of colloidal gold nanoparticles, *J. Phys. Chem. B*, **103**, pp. 4212–4217.
11. Raether H. (1986) *Surface Plasmons on Smooth and Rough Surfaces and on Gratings*, 1st ed, Springer-Verlag, Berlin.
12. Maxwel-Garnett J.C. (1904) Colours in metal glasses and in metallic films, *Phil. Trans. R. Soc.*, **203**, pp. 385–420.
13. Link S. and El-Sayed M.A. (2000) Shape and size dependence of radiative, non-radiative and photothermal properties of gold nanocrystals, *Int. Rev. Phys. Chem.*, **19**, pp. 409–453.
14. Oldenburg S., Averitt R., Westcott S., and Halas N. (1998) Nanoengineering of optical resonances, *Chem. Phys. Lett.*, **288**, pp. 243–247.
15. Kreibig U. and Vollmer M. (1995) *Optical Properties of Metal Clusters*, 1st ed, Springer-Verlag, Berlin.
16. Prodan E., Radloff C., Halas N.J., and Nordlander P. (2003) A hybridization model for the plasmon response of complex nanostructures, *Science*, **302**, pp. 419–422.
17. Khlebtsov N. and Dykman L. (2011) Biodistribution and toxicity of engineered gold nanoparticles: A review of in vitro and in vivo studies, *Chem. Soc. Rev.*, **40**, pp. 1647–1671.

18. Fan M., Andrade G.F.S., and Brolo A.G. (2011) A review on the fabrication of substrates for surface enhanced Raman spectroscopy and their applications in analytical chemistry, *Anal. Chim. Acta*, **693**, pp. 7–25.
19. Anker J.N., Hall W.P., Lyandres O., Shah N.C., Zhao J., and Van Duyne R.P. (2008) Biosensing with plasmonic nanosensor, *Nat. Mater.*, **7**, pp. 442–453.
20. Baker G.A. and Moore D.S. (2005) Progress in plasmonic engineering of surface-enhanced Raman-scattering substrates toward ultra-trace analysis, *Anal. Bioanal. Chem.*, **382**, pp. 1751–1770.
21. Kimling J., Maier M., Okenve B., Kotaidis V., Ballot H., and Plech A. (2006) Turkevich method for gold nanoparticle synthesis revisited, *J. Phys. Chem. B*, **110**, pp. 15700–15707.
22. Lohse S.E. and Murhpy C.J. (2013) The quest for shape control: A history of gold nanorod synthesis, *Chem. Mater.*, **25**, pp. 1250–1261.
23. Dieringer J.A., McFarland A.D., Shah N.C., Stuart D.A., Whitney A.V., Yonzon C.R., Young M.A., Zhang X., and Van Duyne R.P. (2006) Surface enhanced Raman spectroscopy: New materials, concepts, characterization tools, and applications, *Farad. Discuss.*, **132**, pp. 9–26.
24. Kalele S., Gosavi S.W., Urban J., and Kulkarni S.K. (2006) Nanoshell particles: Synthesis, properties and applications, *Curr. Sci.*, **91**, pp. 1038–1052.
25. Mayer K.M. and Hafner J.H. (2011) Localized surface plasmon resonance sensor, *Chem. Rev.*, **111**, pp. 3828–3857.
26. Sagle L.B., Ruvuna L.K., Ruemmele J.A., and Van Duyne R.P. (2011) Advances in localized surface plasmon resonance spectroscopy biosensing, *Nanomed.*, **6**, pp. 1447–1462.
27. Zhao L., Zhang X., Yonzon C.R., Haes A.J., and R.P. Van Duyne (2006) Localized surface plasmon resonance biosensors, *Future Medicine*, **1**, pp. 219–228.
28. Raman C.V. (1928) A new radiation, *Indian J. Phys.*, **2**, pp. 387–398.
29. Raman C.V. and Krishnan K.S. (1928) A new type of secondary radiation, *Nature*, **121**, pp. 501–502.
30. Long D.A. (2002) *The Raman Effect*, 1st ed, John Wiley and Sons, UK.
31. Ferraro J.R., Nakamoto K., and Brown C.W. (2003) *Introductory Raman Spectroscopy*, 2nd ed, Elsevier.
32. Fleischmann M., Hendra P.J., and McQuillan A.J., (1974) Raman spectra of pyridine adsorbed at a silver electrode, *Chem. Phys. Lett.*, **26**, pp. 163–166.

33. Jeanmarie D.L. and Van Duyne R.P. (1977) Surface Raman spectro-electrochemistry, Part 1: Heterocyclic, aromatic, and aliphatic amines adsorbed on the anodized silver electrode, *J. Electroanal. Chem.*, **84**, pp. 1–20.
34. Albrecht M.G. and Creighton J.A. (1977) Anomalously intense Raman spectra of pyridine at a silver electrode, *J. Am. Chem. Soc.*, **99**, pp. 5215–5217.
35. Moskovits M. (1978) Surface roughness and the enhanced intensity of Raman scattering by molecules adsorbed on metals, *J. Chem. Phys.*, **69**, pp. 4159–4161.
36. Moskovits M. (1985) Surface-enhanced spectroscopy, *Rev. Mod. Phys.*, **57**, pp. 783–826.
37. Schatz G.C. and Van Duyne R.P. (2002) *Handbook of Vibrational Spectroscopy*, 1st ed, Wiley & Sons, Chichester.
38. Kneipp K., Kneipp H., Itzkan I., Dasari R.R., and Feld M.S. (1999) Ultrasenistive chemical analysis by raman spectroscopy, *Chem. Rev.*, **99**, pp. 2957–2975.
39. Stiles P.L., Dieringer J.A., Shah N.C., and R.P. Van Duyne (2008) Surface-enhanced Raman spectroscopy, *Annu. Rev. Anal. Chem.*, **1**, pp. 601–626.
40. Le Ru E.C., Blackie E., Meyer M., and Etchegoin P.G. (2007) Surface enhanced Raman scattering enhancements factors: A comprehensive study, *J. Phys. Chem. C.*, **111**, pp. 13794–13803.
41. Kleinman S.L., Frontiera R.R., Henry A.I., Dieringer J.A., and Van Duyne R.P. (2013) Creating, characterizing, and controlling chemistry with SERS hot spots, *Phys. Chem. Chem. Phys.*, **15**, pp. 21–36.
42. Tripp R.A., Dluhy R.A., and Zhao Y. (2008) Novel nanostructures for SERS biosensing, *Nanotoday*, **3**, pp. 31–37.
43. Han X.X., Zhao B., and Ozaki Y. (2009) Surface-enhanced Raman scattering for protein detection, *Anal. Bioanal. Chem.*, **394**, pp. 1719–1727.
44. Bantz K.C., Meyer A.F., Wittenberg N.J., Im H., Kurtulus O., Lee S.H., Lindquist N.C., Oh S.H., and Haynes C.L. (2011) Recent progress in SERS biosensing, *Phys. Chem. Chem. Phys.*, **13**, pp. 11551–11567.
45. Hering K., Cialla D., Ackermann K., Dörfer T., Möller R., Schneidewind H., Mattheis R., Fritzsche W., Rösch P., and Popp J. (2008) SERS: A versatile tool in chemical and biochemical diagnostics, *Anal. Bioanal. Chem.*, **390**, pp. 113–124.
46. Huh Y.S., Chung A.J., and Erickson D. (2009) Surface enhanced Raman spectroscopy and its application to molecular and cell analysis, *Microfluid Nanofluid*, **6**, pp. 285–297.

47. Kneipp K. (2007) Surface-enhanced Raman scattering, *Phys. Today*, **60**, pp. 40–46.

48. Sharma B., Frontiera R.R., Henry A.I., Ringe E., and Van Duyne R.P. (2012) SERS: Materials, applications, and the future, *Mater. Today*, **15**, pp. 16–25.

49. Cialla D., März A., Böhme R., Theil F., Weber K., Schmitt M., and Popp J. (2012) Surface-enhanced Raman spectroscopy (SERS): Progress and trend, *Anal. Bioanal. Chem.*, **403**, pp. 27–54.

50. Van Duyne R.P., Hulteen J.C., and Treichel D.A. (1993) Atomic force microscopy and surface-enhanced Raman spectroscopy. I. Ag island films and Ag film over polymer nanosphere surfaces supported on glass, *J. Chem. Phys.*, **99**, pp. 2101–2115.

51. Todescato F., Minotto A., Signorini R., Jasieniak J.J., and Bozio R. (2013) Investigation into the heterostructure interface of CdSe-based core-shell quantum dots using surface-enhanced Raman spectroscopy, *ACS Nano*, **7**, pp. 6649–6657.

52. Stuart D.A., Yonzon C.R., Zhang X., Lyandres O., Shah N.C., Glucksberg M.R., Walsh J.T., and Van Duyne R.P. (2005) Glucose sensing using near-infrared surface-enhanced Raman spectroscopy: Gold surfaces, 10-day stability, and improved accuracy, *Anal. Chem.*, **77**, pp. 4013–4019.

53. Zhang X., Young M.A., Lyandres O., and Van Duyne R.P. (2005) Rapid detection of an anthrax biomarker by surface-enhanched Raman spectroscopy, *J. Am. Chem. Soc.*, **127**, pp. 4484–4489.

54. Whitney A.V., Van Duyne R.P., and Casadio F. (2006) An innovative surface-enhanced Raman spectroscopy (SERS) method for the identification of six historical red lakes and dyestuffs, *J. Raman Spectrosc.*, **37**, pp. 993–1002.

55. Whitney A.V., Casadio F., and Van Duyne R.P. (2007) Identification and characterization of artists' red dyes and their mixture by surface-enhanced Raman spectroscopy, *Appl. Spectrosc.*, **61**, pp. 994–999.

56. Shafer-Peltier K.E., Haynes C.L., Glucksberg M.R., and Van Duyne R.P. (2003) Toward a glucose biosensor based in surface-enhanced Raman scattering, *J. Am. Chem. Soc.*, **125**, pp. 588–593.

57. Yonzon C.R., Haynes C.L., Zhang X., Walsh J.T., and Van Duyne R.P. (2004) A glucose biosensor based on surface-enhanced Raman scattering: Improved partition layer, temporal stability, reversibility, and resistance to serum protein interference, *Anal. Chem.*, **76**, pp. 78–85.

58. Stuart D.A., Yuen J.M., Shah N.C., Lyandres O., Yonzon C.R., Glucksberg M.R., Walsh J.T., and Van Duyne R.P. (2006) In vivo glucose measurement

by surface-enhanced Raman spectroscopy, *Anal. Chem.*, **78**, pp. 7211–7215.

59. Lyandres O., Yuen J.M., Shah N.C., Van Duyne R.P., Walsh J.T., and Glucksberg M.R. (2008) Progress toward an in vivo surface-enhanced Raman spectroscopy glucose sensor, *Diabetes Technol. Ther.*, **10**, pp. 257–265.
60. Ma K., Yuen J.M., Shah N.C., Walsh J.T., Glucksberg M.R., and Van Duyne R.P. (2011) In vivo, transcutaneous glucose sensing using surface-enhanced spatially offset Raman spectroscopy: Multiple rats, improved hypoglycemic accuracy, low incident power, and continuous monitoring for greater than 17 days, *Anal. Chem.*, **83**, pp. 9146–9152.
61. Minck R.W., Terhune R.W., and Rado W.G. (1963) Laser-stimulated Raman effect and resonant four-photon interactions in gases H_2, D_2, and CH_4, *Appl. Phys. Lett.*, **3**, pp. 181–184.
62. Shen Y.R. (1984) *The Principle of Nonlinear Optics*, 1st ed, John Wiley & Sons, New York.
63. Levenson M.D. (1982) *Introduction to Nonlinear Laser Spectroscopy*, 1st ed, Academic Press, New York.
64. Druet S. and Taran J.P. (1982) *Chemical and Biochemical Application of Laser*, vol 4, 4th ed (Moore C.B, ed), Academic Press, New York.
65. Steuwe C., Kaminski C.F., Baumberg J.J., and Mahjan S. (2011) Surface enhanced coherent anti-Stokes Raman scattering on nanostructured gold surfaces, *Nano Lett.*, **11**, pp. 5339–5343.
66. Liang E.J., Weippert A., Funk J.M., Materny A., and Kiefer W. (1994) Experimental observation of surface-enhanced coherent anti-Stokes Raman scattering, *Chem. Phys. Lett.*, **227**, pp. 115–120.
67. Ichimura T., Hayazawa N., Hasimoto M., Inouye Y., and Kawata S. (2003) Local enhancement of coherent anti-Stokes Raman scattering by isolated gold nanoparticles, *J. Raman Spectrosc.*, **34**, pp. 651–654.
68. Koo T.W., Chan S., and Berlin A.A. (2005) Single-molecule detection of biomolecules by surface-enhanced coherent anti-Stokes Raman scattering, *Opt. Lett.*, **9**, pp. 1024–1026.
69. Machtoub L.H., Pfeiffer R., Bataveljic D., Andjus P.R., and Jordi H. (2010) Monitoring lipids uptake in neurodegenerative disorders by SECARS microscopy, *Surf. Sci. Nanotech.*, **8**, pp. 362–366.
70. Schlücker S., Salhei M., Bergner G., Schütz M., Ströbel P., Marx A., Petersen I., Dietzek B., and Popp J. (2011) Immuno-surface-enhanced coherent anti-Stokes Raman scattering microscopy: Immunohistochem-

istry with target-specific metallic nanoprobes and nonlinear Raman microscopy, *Anal. Chem.*, **83**, pp. 7081–7085.

71. Goldys E. (2009) *Fluorescence Applications in Biotechnology and Life Science*, 1st ed, Wiley-Blackwell, Hoboken, NY.
72. Geddes C.D. and Lakowicz J.R. (2002) Metal-enhanced fluorescence, *J. Fluoresc.*, **12**, pp. 121–129.
73. Lakowicz J.R. (2006) *Principles of Fluorescence Spectroscopy*, 3rd ed, Springer.
74. Aslan K., Gryczynski I., Malicka J., Matveeva E., Lakowicz J.R., and Geddes C.D. (2005) Metal-enhanced fluorescence: An emerging tool in biotechnology, *Curr. Opin. Biotech.*, **16**, pp. 55–62.
75. Bharadwaj P and Novotny L. (2007) Spectral dependence of single molecule fluorescence enhancement, *Opt. Exp.*, **15**, pp. 14266–14274.
76. Chen Y., Munechika K., and Ginger D.S. (2007) Dependence of fluorescence intensity on the spectral overlap between fluorophores and plasmon resonant single silver nanoparticles, *Nano Lett.*, **7**, pp. 690–696.
77. Petryayeva E. and Krull U.J. (2011) Localized surface plasmon resonance: Nanostructures, bioassays and biosensing—A review, *Anal. Chem. Acta*, **706**, pp. 8–24.
78. Ray K., Badugu R., and Lakowicz J.R. (2006) Distance-dependent metal-enhanced fluorescence from Langmuir–Blodgett monolayers of Alkyl-NBD derivatives on silver island films, *Langmuir*, **22**, pp. 8374–8378.
79. Ray K., Badugu R., and Lakowicz J.R. (2007) Sulforhodamine adsorbed Langmuir–Blodgett layers on silver island films: Effect of probe distance on the metal-enhanced fluorescence, *J. Phys. Chem. C*, **111**, pp. 7091–7097.
80. Ray K., Badugu R., and Lakowicz J.R. (2007) Polyelectrolyte layer-by-layer assembly to control the distance between fluorophores and plasmonic nanostructures, *Chem. Mater.*, **19**, pp. 5902–5909.
81. Kang K.A., Wang J., Jasinski J.B., and Achilefu S. (2011) Fluorescence manipulation by gold nanoparticles: From complete quenching to extensive enhancement, *J. Nanobiotech.*, **9**, pp. 16.
82. Kümmerlen J., Leitner A., Brunner H., Aussenegg F.R., and Wokaun A. (1993) Enhanced dye fluorescence over silver island films: Analysis of the distance dependence, *Molec. Phys.*, **80**, pp. 1031–1046.
83. Aslan K, Huang J., Wilson G.M., and Geddes C.D. (2006) Metal-enhanced fluorescence-based RNA sensing, *J. Am. Chem. Soc.*, **128**, pp. 4206–4207.

84. Sabanayagam C.R. and Lakowicz J.R. (2007) Increasing the sensitivity of DNA microarrays by metal-enhanced fluorescence using surface-bound silver nanoparticles, *Nucleic Acids Res.*, **35**, pp. e13.

85. Li Y.Q., Guan L.Y., Zhangv H.L., Chen J., Lin S., Ma Z.Y., and Zhao Y.D. (2011) Distance-dependent metal-enhanced quantum dots fluorescence analysis in solution by capillary electrophoresis and its application to DNA detection, *Anal. Chem.*, **83**, pp. 34103–4109.

86. Gill R., Tian L., Somerville W.R.C., Le Ru E.C., van Amerongen H., and Subramaniam V. (2012) Silver nanoparticle aggregates as highly efficient plasmonic antennas for fluorescence enhancement, *J. Phys. Chem. C*, **116**, pp. 16687–16693.

87. Aslan K., Holley P., and Geddas C.D. (2006) Microwave-accelerated metal-enhanced fluorescence (MAMEF) with silver colloids in 96-well plates: Application to ultra-fast and sensitive immunoassays, high throughput screening and drug discovery, *J. Imm. Meth.*, **312**, pp. 137–147.

88. Ray K., Szmacinski H., and Lakowicz J.R. (2009) Enhanced fluorescence of proteins and label-free bioassays using aluminum nanostructures, *Anal. Chem.*, **81**, pp. 6049–6054.

89. Zhang J., Matveeva E., Gryczynski I., Leonenko Z., and Lakowicz J.R. (2005) Metal-enhanced fluoroimmunoassay on a silver film by vapor deposition, *J. Phys. Chem. B*, **109**, pp. 7969–7975.

90. Fu Y., Zhang J., and Lakowicz J.R. (2010) Plasmon-enhanced fluorescence from single fluorophores end-linked to gold nanorods, *J. Am. Chem. Soc.*, **132**, pp. 5540–5541.

91. Aslan K. and Grell A.J. (2011) Rapid and sensitive detection of troponin I in human whole blood samples by using silver nanoparticle films and microwave heating, *Clin. Chem.*, **57**, pp. 746–752.

92. Aslan K., Leonenko Z., Lakowicz J.R., and Geddes C.D. (2005) Fast and slow deposition of silver nanorods on planar surfaces: Application to metal-enhanced fluorescence, *J. Phys. Chem. B*, **109**, pp. 3157–3162.

93. Todescato F., Antognoli A., Meneghello A., Cretaio E., Signorini R., and Bozio R. (2014) Sensitive detection of Ochratoxin A in food and drinks using metal–enhanced fluorescence, *Biosens. Bioelectron.*, DOI: 10.1016/j.bios.2014.01.060.

Chapter 8

Self-Organized Plasmonic Nanomaterials Based on Liquid Crystals and Metal Nanoparticles

Toralf Scharf

École polytechnique fédérale de Lausanne EPFL, Optics and Photonics Technology, Laboratory Microcity Rue de la Maladière 71, CH-2000 Neuchâtel, Switzerland
toralf.scharf@epfl.ch

Liquid crystals are known for their self-organization over different length scales. This self-organization capacity can be used to organize nano-entities incorporated into a molecule. The established order is influenced by molecular design, and different factors govern the organization that can lead to complex arrangement. To establish order in such systems, a critical volume is needed. When kept confined, for instance, between substrates in thin films, a homogenous alignment over large surface areas might be achieved. If the nano-entities are particles that carry a plasmon resonance and arranged in an ordered way, particular optical and electrical properties might be expected, leading to active plasmonic materials. To explore the potential of self-organization for active plasmonic materials in application, several conditions concerning the size and arrangement of particles need to be met. We focus here on

Active Plasmonic Nanomaterials
Edited by Luciano De Sio

ISBN 978-981-4613-00-2 (Hardcover), 978-981-4613-01-9 (eBook)
www.panstanford.com

a system where the plasmonic nanoparticle is from a metal and part of the molecule that forms itself a liquid-crystalline phase. The discussion excludes mixtures of nanoparticles in liquid-crystalline hosts and related effects. To obtain a reasonable effective optical effect, the plasmonic nanoparticles should have a certain size. If incorporated into an organic molecule, one needs to assure that the ensemble still shows liquid-crystalline phases, and hence the size of the plasmonic particle should be small enough to allow self-organization. We discuss different molecular design concepts and review first realizations of such materials always with respect to their optical properties. A general introduction will provide the baseline of our research with respect to the optical effects. We will then discuss in more detail specific systems and their characteristics. We close our contribution with an example of a material design scheme that allowed demonstration of a first active plasmonic self-organized material having liquid-crystalline phases.

8.1 Optical Properties of Metallic Plasmonic Entities

The most promising design strategies to create materials with artificial electromagnetic properties are based on using nano-structured matter and metals. That might be layers with metal nanostructures or a combination of thin films. The aim is to create metamaterials with properties that are not different from that of the constituents. Metamaterials are defined as composite structures that influence the response of the matter to external electromagnetic fields dominantly as a function of the unit cell geometry. If the metamaterial consists of periodically arranged unit elements much smaller than the wavelength of the electromagnetic radiation, effective material properties for the metamaterial can be determined [1]. Unit elements can be designed that influence the effective permittivity [2], the effective permeability [3], or both material parameters simultaneously [4]. In this way, values of permittivity and permeability that are not observed in naturally occurring matter can be obtained, which potentially allows to obtain materials with even negative effective refractive index [5]. To obtain metamaterials with effective material parameters that strongly deviate from the spatially averaged material properties of the unit cell constituents, one has to make use of resonance effects within the unit cell, which are not associated with the periodic arrangement [6, 7]. Typical examples are plasmonic resonances or Mie type resonances in spheres made of dielectric materials having a large refractive index capable of influencing the effective permittivity or permeability depending on the resonant excitation of an electric or magnetic mode [8]. Common design approaches for metamaterials based on metals are split-ring resonators [6, 9], pairs of coupled metallic wires [3, 10, 11], or coupled perforated thin metal plates [12]. The first two aforementioned metamaterials influence both effective material parameters, although their resonances are usually spectrally well separated and a negative refractive index is hence hardly achievable. The last example provides a negative permittivity over an extended spectral domain according to a Drude-type resonance as the structure behaves essentially like a diluted metal. Negative refraction was also demonstrated for structures that influence the effective permeability according to a Lorentz-type resonance.

8.1.1 *Plasmon Resonance of Nanoparticles and Assemblies*

Plasmon resonances in metallic nanoparticles are determined by the electronic properties of the electrons inside a confided volume. It is a movement of the electron ensemble against the lattice of atoms in the metal. The effect is localized due to the confinement to the nanoparticle and leads to nonpropagating excitations of the conduction electrons of the metallic nanoparticles when coupled to the electromagnetic field. One calls this localized excitation a localized surface plasmon in contrary to surface plasmons in extended geometries where propagating dispersive electromagnetic waves couple to the electron plasma of a conductor at a dielectric interface. In localized surface plasmon, modes arise naturally from the scattering problem of a small, sub-wavelength conductive nanoparticle in an oscillating electromagnetic field. The electrons in resonance feel a restoring force again given by the particularity of the confinement in nanoparticles by the curved surface of the particle. The resonance will lead to field amplification both inside and in the near-field zone outside the particle.

The localized surface plasmon resonances can be excited by direct light illumination. For nanoparticles made from silver of gold, the resonance frequency falls in the visible light spectral region and can be observed by the naked eye. If active optical materials for the visible region are aimed on, silver and gold nanoparticles seem the best qualified candidates also because chemistry is available to incorporate such particles into macromolecules.

8.1.2 *Properties of Localized Plasmonic Resonance of Small Nanoparticles*

The concept of self-organization of plasmonic nanoparticles in liquid crystal host is based on the assumption that the nanoparticles are not too large. Large particles would be more difficult to implement into the molecular matrix. That is the reason why we revisit shortly the peculiarities of the optics of small particles of size below 10 nm in this section.

Optical properties of nanoparticles can be theoretically treated by Mie theory. The dependence of different parameters is usually expressed in a scattering cross section of particles of a given complex refractive index in a surrounding material. For small particles with dimensions smaller than the wavelength, the electrostatic limit can be applied to find approximate solutions [13]. One finds two important quantities to be considered: the scattering cross section C_{sca} and the absorption cross section C_{abs}. Both are given below using the incident intensity I_{inc} and scattered intensity I_{sca} as measures:

$$C_{\text{sca}} = \frac{I_{\text{sca}}}{I_{\text{inc}}} = \frac{k^4}{6\pi} \left|\alpha_{\text{particle}}\right|^2$$

$$C_{\text{abs}} = \frac{I_{\text{abs}}}{I_{\text{inc}}} = k\text{Im}\left\{\alpha_{\text{particle}}\right\}$$

The wavenumber k depends on the wavelength λ and the refractive index n of the surrounding material. It reads

$$k = \frac{2\pi n}{\lambda}$$

The missing important parameter α_{particle} is the dipolar polarizability of the particle, which needs to be found. If we assume a spherical particle with radius a, the classical description of the polarizability of a particle in a surrounding material can be found when the complex dielectric constants of both components are known. One finds

$$\alpha_{\text{sphere}} = 4\pi\varepsilon_0 a^3 \frac{\varepsilon_p - \varepsilon_m}{\varepsilon_p + 2\varepsilon_m}$$

with the dielectric permittivity of the particle material ε_p and ε_m of the surrounding material. The radius of the particle is a. Often the frequency-dependent dielectric function of the metallic sphere can be modeled with the Drude model that describes a resonant behavior of an electron plasma in the metal with a characteristic plasma frequency ω_p. A good approximation of such a resonant behavior leads to a dielectric permittivity as

$$\varepsilon_p = \varepsilon_{\text{metal}} = 1 - \frac{\omega_p^2}{\omega(\omega - i\Gamma)}$$

The damping is introduced by a factor Γ, and ω is the frequency of the light in free space. The damping constant Γ can be related to the mean free path l of the electrons and the speed of the electrons, which is called the Fermi velocity v_F, by the relation

$$\Gamma = \frac{v_F}{l}$$

The Drude model provides a link between the electronic properties and the optical properties via the plasma resonance frequency ω_p given as

$$\omega_p = \left(\frac{n_e e^2}{\varepsilon_0 m_e}\right)^2$$

introducing n_e as the electron density, e as the electronic charge, ε_0 as the dielectric constant of vacuum, and m_e as the mass of the electron.

Until know we have discussed bulk metals. If the electrons are heavily confined, that is, for small particles, the dielectric constant changes and corrections need to be done. This becomes evident in the case when the size of the particle gets smaller than the mean free path of the free electrons; the electrons' movement is restricted

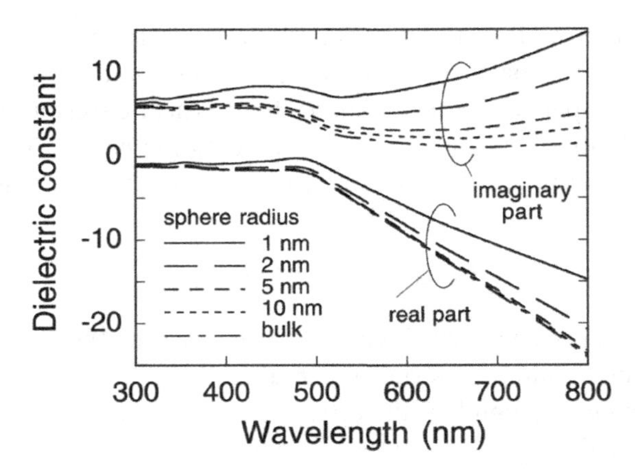

Figure 8.1 Dielectric constant of gold nanoparticles (here called sphere) of different radius. Reprinted from Ref. [14] with kind permission from Springer Science and Business Media.

within the particle. If this effect is included, the damping constant for particles, Γ_{particle}, can be modeled as

$$\Gamma_{\text{particle}} = \Gamma + \frac{Av_F}{a}$$

where Γ is the damping constant for the bulk, a is the radius of the particle, and A is a constant describing the effect and has a value that depends on the particle shape. The parameter A can safely assumed to be close to unity [14].

With this model, dielectric constants can be calculated for nanoparticles based on material parameters found for bulk metals. Figure 8.1 gives typical values as calculated by Okamoto [14]. One recognizes imaginary parts of the dielectric constants to be positive with values between 0 and 10 in the visible spectral region. The real part of the dielectric constant is negative. It is important to note that the imaginary part which describes the absorption is increasing if the particle size decreases. Small particles are, therefore, not favorable to design materials with low loss.

For instance, assume for a moment we keep the solid material content of the metal in a composite material constant and we calculate the absorption change for a fixed thickness when the imaginary part of the dielectric constant is doubled. Figure 8.1 shows what happens at around 600 nm when the radius of the nanoparticle changes from 5 nm to 1 nm. Because the host material, in our case an organic substance like a liquid crystal, has virtually no absorption, the effect is translated directly to the whole composite. Absorption for bulk materials has an exponential dependence of the imaginary part of refractive index. The refractive index is the square of the dielectric constant, and by doubling the imaginary part of the dielectric constant and having only little change in the real part, we decrease the light transmission through bulk material by a factor $e^2 = 7.39$, almost one order of magnitude. This is a high increase in loss that heats the sample up, and it is important to find material design schemes that allow incorporation of large particles with diameters larger than 3 nm.

Coming back to the single particle case, one can now determine basic relations for the scattering and absorption of single particle. Having the dielectric constant at hand, one can calculate the scattering cross section as

$$C_{\text{sca}} = \frac{I_{\text{sca}}}{I_{\text{inc}}} = \frac{k^4}{6\pi}\left|\alpha_{\text{particle}}\right|^2 = \frac{8}{3}\pi a^6 k^4 \left|\frac{\varepsilon_p - \varepsilon_m}{\varepsilon_p + 2\varepsilon_m}\right|^2$$

The wavenumber k is inversely proportional to the wavelength. One can easily see with the following relation that the scattering efficiency scales with respect to the particle size:

$$C_{\text{sca}} \sim \frac{a^6}{\lambda^4}$$

Here again we find that small particles have much less interaction as bigger ones. If absorption is considered, one finds

$$C_{\text{abs}} = \frac{I_{\text{abs}}}{I_{\text{inc}}} = k\text{Im}\left\{\alpha_{\text{particle}}\right\} = 4\pi k a^3 \text{Im}\left\{\frac{\varepsilon_p - \varepsilon_m}{\varepsilon_p + 2\varepsilon_m}\right\}$$

The absorption coefficient C_{abs} is proportional to the third power of the particle radius a and inversely proportional to the wavelength, hence much less dependent on the wavelength as the scattering cross section C_{sca}. One reads

$$C_{\text{abs}} \sim \frac{a^3}{\lambda}$$

In composite materials, nanoparticles are embedded in a matrix made from a different material. Detailed simulations need to consider this, and different models exist to calculate the electromagnetic properties of composite materials. A basic introduction with respect to optical metamaterials is given by Cai [15].

8.1.3 *Density, Arrangements, and Optical Efficiency*

To create materials with nonconventional material properties, it will not be sufficient to incorporate nanoparticles into a host material at low density. The nonconventional effect comes from coupling between closely packed nanoparticles leading to interaction of their electromagnetic resonances when excited, but they should not touch. The coupling of resonances becomes remarkable when the distance (center to center) of coupling entities is smaller than the double of the diameter of the entity itself [16]. In such a case, spectral features that are different from the original signature of a single nanoparticle can be found. Touching needs to be avoided to keep the electromagnetic properties of the nanoparticles. In the case of small nanoparticles, for example a 3 nm nanoparticle, its next neighbor should only be 6 nm apart with a 3 nm surface-to-surface gap between them to see such effects. If one considers a volume packaging of nanoparticles to fulfill such a condition, one can calculate the load in volume percent, c_V, and mass percent, c_M, for a model substance. This is of interest if one needs to judge the conditions of real use of a designed material and is usually neglected when surfaces are considered or effectiveness is of no importance. To give an example, a heavily loaded metal nanoparticle ink might have 60% mass concentration of silver nanoparticles (for instance Harima Nanopaste NPS-J [17]), which represents a volume concentration of only $c_V = 11\%$. To better understand this aspect, we calculate the distance of nanoparticles in a hexagonal packing. The closest hexagonal packing density of mono-dispersed spheres gives a volume-filling fraction of 0.74 [$= \pi/(3\sqrt{2})$], and the center-to-center distance between touching particles is equal to the diameter. The volume of a sphere scales with the third power of its radius, that is, r^3. The maximum filling fraction is given for the touching spheres. If we introduce a distance a between the spheres of radius r, which is always larger than the diameter $d = 2r$ of the spheres, we can establish a relation for the volume-filling fraction as

$$c_V = \frac{\pi}{3\sqrt{2}}\left(\frac{d}{a}\right)^3$$

For the touching spheres, we find 0.74, and if we double the distance between the spheres, that is, $a = 2d$, we find a volume-filling fraction of less than 10%:

$$c_V = \frac{\pi}{3\sqrt{2}}\left(\frac{d}{2d}\right)^3 = \frac{\pi}{3\sqrt{2}}\left(\frac{1}{2}\right)^3 = 9.25\%$$

In the case of 3 nm diameters, the distance of nanoparticles is 6 nm (center to center), which is too large to observe strongly coupled plasmon resonance effects [16].

Usually chemists measure the solid content in mass percent, c_M. The electromagnetic interaction is managed by the distance between entities and linked to the volume-filling fraction c_V. The conversion between these entities can be done when the density of materials is known. Silver has a density of $\rho_{\mathrm{AG}} = 10.52$ (g/ml) and a conventional solvent for silver nanoparticles like Tetradecane has a density of ρ_{Solvent} =0.762 g/ml. For a certain mass percentage, one can calculate the volume percentage with

$$c_V = \frac{c_M/\rho_{AG}}{\frac{c_M}{\rho_{\mathrm{AG}}} + \frac{1-c_M}{\rho_{\mathrm{Solvent}}}}$$

If we refer to our example of a very dense nanoparticle ink with $c_M = 60\%$, we find

$$c_V = \frac{0.60/10.52}{\frac{0.60}{10.52} + \frac{0.40}{0.762}} = 0.098 = 9.8\%$$

A material with a mass concentration of 60% metal can realize suspension with extreme difficulty, and it is still at the limit of what is needed to bring small particles close together to allow for electromagnetic coupling. For best performance, the density should reach values where the mean distance is smaller than 1.5 times the diameter d (<4.5 nm for our 3 nm nanoparticle example) to create strong electromagnetic coupling; that represents more than 20% in volume load.

The condition could be somewhat relaxed if one allows anisotropy to the structure, so those electromagnetic couplings exist only in one direction/orientation of the structure and not in the others. The distance in one direction needs to be short and could be larger in the others, leading to a chain-like arrangement. In this case, the filling factor can drop without losing the nonconventional optical properties, but the resulting material would show strong polarization dependence in the optical response. Just remember that to keep the special optical properties of nanoparticles alive, they should not touch.

In summary, one sees here that the concept of dispersing nanoparticles in solvents to realize a highly efficient material based on nano-plasmonic composites is difficult to realize because of the high density of nanoparticles needed. Synthetizing and preparing such highly loaded composites is very challenging. If done there is still the problem of optical absorption, which is probably very high. One possible solution to the density problem is the chemical synthesis of a nanoparticle containing macromolecules that can self-organize at close distances. This concept is discussed in the next chapter.

8.2 Liquid Crystal Self-Organization and Nanoparticles

The most promising way of realizing a structure of nontouching nano-entities that getting very close with metamaterial properties is a bottom-up approach that relies on organic chemistry of mesogens and macromolecules, plasmon resonance effects, and physical chemistry principles of self-organization. To design organic-inorganic composite materials with tailored properties, several parameters dominating the system's plasmonic properties have to be controlled: particle size (including size distribution) and shape; particle distance (period); and organizations in one, two, or three dimensions (statistical distributions, wires, plates, and crystals). Mesogenic rod-like, disk-like, and taper-shaped (dendrimer) molecules form liquid-crystalline phases even with particles included. At the same time, the presence of the particle may induce a higher degree of order than exhibited by the native mesogen. The distance between the entities should be anything between 0.3 nm and 10 nm, but touching particles need to be avoided. Therefore, macromolecules such as dendrimers and mesogens of controlled shapes seem well suited to organize small metal particles whose size can range from several nanometers to tens of nanometers. A potential scheme is shown in Fig. 8.2.

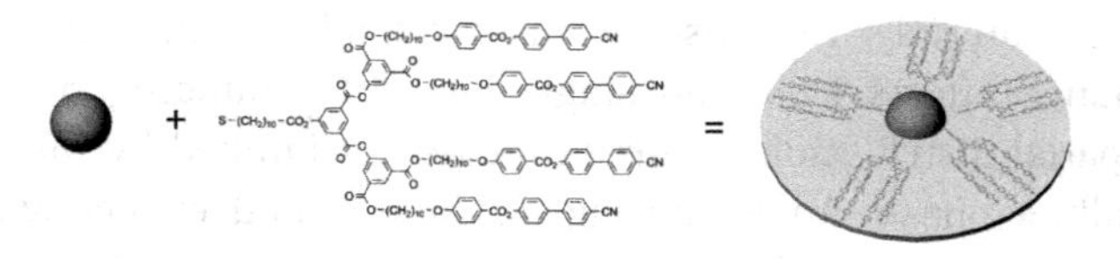

Figure 8.2 One hypothetical building principle of disk-shaped macromolecules containing inorganic particles. In this example, dendrimer-like macromolecules are attached to the particle and might form disk-like entities. The particle is in the center without being completely wrapped in all directions.

Macromolecular and low-molar liquid crystal ligands can be attached to the particles by chemical methods and arrangements such as disks or rods can be envisaged where the particle sits in the center and is surrounded by radially or tangentially arranged organic molecules. In the example of dendritic molecules, the shapes obtained will strongly depend on the ligand structure, the number of dendrons attached to the nanoparticles, and the volume filled around the particle. Thus, for example, principles of close packing and radial distribution of density, introduced for spherical [18] and rod-like [19] supramolecular dendrimers, should be applied to guide the design of specific types of one-dimensional and three-dimensional ordered arrays of nanoparticles. Disk-like arrangements are shown as examples in Fig. 8.3.

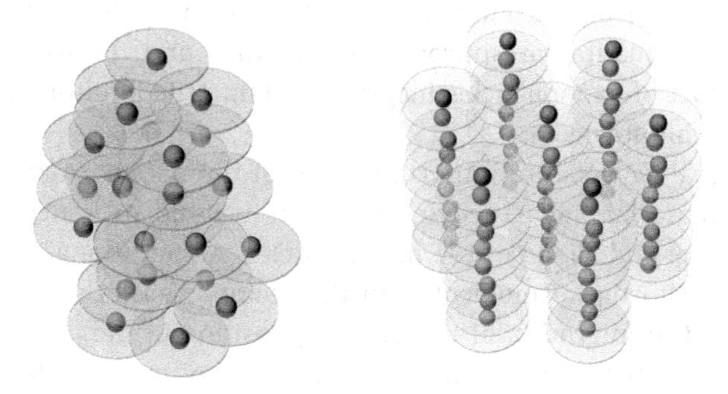

Figure 8.3 Liquid-crystalline phases of disc-like materials. A nematic discotic phase is shown on the left. A columnar hexagonal phase is illustrated on the right. The yellow (darker) dots represent the incorporated nanoparticles.

The material's response to electric fields is characterized by resonance effects from the single molecular building block and the spatial distribution in three dimensions implied by the liquid-crystalline phase. Analyzing Fig. 8.3 one would expect wire-like arrangement of particles in the case of columnar phases.

The challenge is to design macromolecules and mesogens incorporating metal nanoparticles of a certain size to give effective electromagnetic response and to assure that these macromolecules still show distinct mesophases with two- and three-dimensional order. The distance between the particles should be small. If we consider the arrangement in Fig. 8.3, the nematic discotic phase sketched to the left will have relatively large distances between nanoparticles, while a columnar arrangement like the one on the right in Fig. 8.3 would allow close distance at least in one direction. To understand better the different factors influencing the order, it is instructive to discuss the size of different components.

8.2.1 *Sizes of Components*

Liquid crystals are known for their force of self-organization. Several models can be applied to understand the resulting structures. One of the simpler models is based on the steric interaction of shape anisotropy of rigid molecular cores. For instance, rigid rods could be used to model the phase behavior of nematic phases as done in the Huggins Flory theory [20]. We will not explore this path in a qualitative approach any further and discuss instead the impact of nanoparticles incorporated into the liquid-crystalline phase qualitatively.

Several aspects are important: load of the nanoparticles in the entire material (measured in weight or volume percent) as well as the size and shape of the nanoparticles. An additional difficulty is the anchoring of molecules on the nanoparticles that might be subjected to migration of the bonding position of the molecule on the nanoparticle. To design a material made of macromolecules of a single type, mesogens can be attached to the surface via a linker unit. Figure 8.4 shows the basic components in a schematic manner and tries at the same time to introduce the relative length scale of the different components.

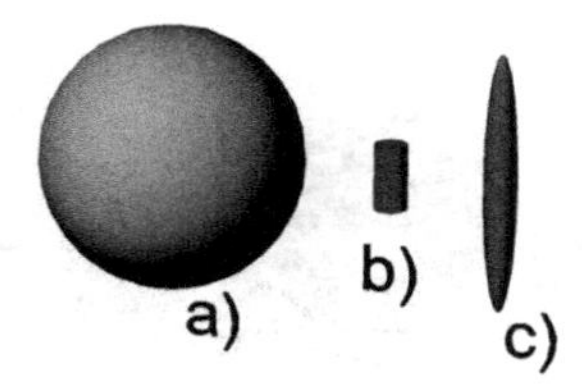

Figure 8.4 (a) Nanoparticle with a typical size of 3 nm in diameter, (b) typical linker molecule presented as a rod, and (c) mesogen molecule core here shown as rotational ellipsoid.

In a typical scenario, the nanoparticle has a diameter of 3 nm, the mesogen core is 3 nm, and the spacer or linker molecule is around 1 nm.

The spacer or linker is a special molecule introduced to get more design freedom to attach the molecule responsible for self-organization, which liquid crystal specialists call a mesogen [21].

If a linker is used to attach the mesogen to the nanoparticles surface, the mesogen might be held at different positions, at the side or at the end. If the rotational freedom is mainly at the junction between the linker and the mesogen, a multitude of configurations is possible. Figure 8.5 illustrates this concept.

The linker molecule concept has the advantage of giving more freedom to the chemistry of fixing the organic components onto the metal surface without redesigning the mesogen. In general, the use of larger/longer molecules for the linker and mesogenic part leads to high viscosity, because the molecular interaction becomes

Figure 8.5 Extreme position to connect the mesogen molecule (ellipsoid) with a linker molecule (cylinder). The freedom to rotate and tilt in this model is linked to the point of contact. On the left, a top-end contact is illustrated, and to the right, a side connection is made in the middle of the molecule.

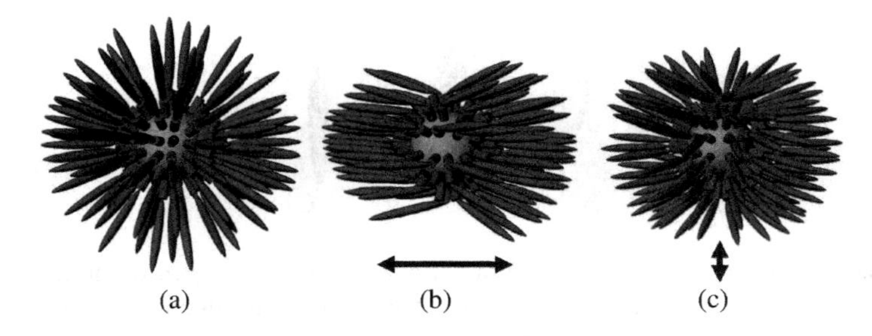

Figure 8.6 Arrangement of molecules in a top-end configuration between the linker and the mesogen in three configurations. In (a), the nanoparticle is wrapped and at each position, a mesogen is normally aligned with respect to the spheres surface. Configurations where molecules align along a certain direction due to different forces are shown in (b) and (c). The flash indicates the axis of rotational symmetry.

much more complex and the freedom to move for large molecules is hindered by several mechanisms. As a result, one looks for small molecules that just provide enough flexibility to organize the system.

In the next step, the linker mesogen molecule is attached to the nanoparticle. The particularity here is the curvature introduced by the geometry of the nanoparticle, which has to be seen in relation to the steric interaction of the mesogens that, in the simplest case, leads to parallel orientation. As one can imagine, the interplay between these two effects can lead to a number of configurations and phases that are very difficult to predict [22]. To illustrate the problem, the sketch in Fig. 8.6 shows the situation where only a few molecules are assembled for only one configuration—the top-end bonding of linker and mesogen.

The lazy man model in Fig. 8.6 gives an impression of how liquid crystals might form if nanoparticles are wrapped with mesogenic molecules. In matter with high density, the macromolecule will be deformed and mesogens of one unit will interact with the mesogens of the neighboring unit. Out of this, alignment of mesogens and order may result. With respect to a nematic alignment, the configuration in Fig. 8.6b looks very promising, but one should not forget that we are speaking of a molecule here and the forces leading to orientations are manifold and difficult to interpret. The orientation of mesogen units illustrated in Fig. 8.6 seems to have a defect as described by the classical continuum theory of liquid crystals. The continuum theory based on a minimum order volume cannot be applied to the molecular regime for reorientation at the molecular level. However, the macromolecules designed here develop characteristics leading to similar topological structures. It is, therefore, interesting to compare the alignment of liquid crystal around defects and its characteristic lengths scales with our problem of molecular design.

8.2.2 *Alignment, Texture, and Characteristic Scales*

Liquid crystal order is established only for a minimum number of molecules and hence for a certain volume [23]. Classical theories describing the mechanical and optical properties of liquid crystals are based on this fact and attribute to each ordered domain or minimum volume a so-called director that describes local properties such as direction of anisotropy. In the case of a nanoparticle put into such an environment, different scenarios exist: The nanoparticles are small compared to the minimum order volume, similar in size or much bigger. In each case, a different theoretical description is needed. Macroscopically, the liquid crystal behavior can be described with elastic theory and liquid crystal confinement, including anchoring at substrate surface or interfaces. The resulting alignment of molecular directors causes a texture with macroscopic feature visible under the polarized microscope [24]. Liquid crystal textures are spatial distributions of the director and show characteristic appearances for different phases of liquid crystals. Liquid crystal textures show defects that are often points or lines. Due to long distance elastic forces in liquid crystals, the defects deform the liquid crystal director field over large distances, even visible by the naked eye [26]. It is assumed that such points and lines have a non-ordered or isotropic core, small in size. The defects have a core with a diameter that can be linked to the basic properties and order of liquid crystals: the minimum order volume. The minimum order volume could serve as a reference parameter to describe the interaction of nanoparticles with liquid crystal hosts.

To evaluate the minimum order volume, one can consider the situation of a point defect and the core size of such a defect [27, 28]. Outside the core, macroscopic elastic theories apply [29–31].

The core of the defect forms a singularity where it is assumed that no order exists. It is a result of the fact that mechanical deformation energy outside the core can reach only a certain value before the ordered state breaks and an isotropic volume is formed. An interesting concept to describe the phenomenon uses the different coherence lengths (nematic coherence length ξ, mechanical coherence length ζ due to surface interaction, and electrical coherence length). The longer the coherence length for a certain parameter, the weaker the energy that is attributed to it. For instance, the mechanical coherence length ζ in a liquid crystal confinement is given by a finite surface anchoring energy W and the elastic properties of the system given by the elastic constant K. In a constant approximation, one finds

$$\zeta = \frac{K}{W}$$

with typical values for ζ ranging from nanometers to micrometers [typical values for $K = 10^{-11}$ N (J/m), $W = 10^{-4}$ to 10^{-6} J/m^2, and $\zeta = 10^{-7}$ to 10^{-5} m]. The nematic coherence length is given by interaction parameters A and the elastic constant K as

$$\xi = \sqrt{\frac{K}{A}}$$

If the characteristic scale of distortions is much larger than the nematic coherence length ξ, the orientation is close to its equilibrium. If the distortion is strong and the characteristic length is smaller than the nematic coherence length, it will dominate. This leads, for instance, to the breaking of anchoring by external electric or magnetic fields in liquid crystal physics. But it will also lead to a change in order if the characteristic mechanical length ζ gets smaller than the nematic coherence length ξ. The concept can be used to deduce transitions between states, although one should keep in mind that quantitative explanation would need complex molecular simulation. If we apply the concept to the incorporation of nanoparticles in liquid crystal matrices, we need to compare the nematic coherence length with the nanoparticle diameter. The unknown parameter here is A, which is strongly temperature dependent at phase transition and assumed to be in the order of $A = 10^6$ J/m^3 far away from the transition temperature [32]. We use the elastic constant value $K = 10^{-11}$ N (J/m) (also strongly temperature dependent at phase transition) and find

$$\xi = \sqrt{\frac{10^{-11}\mathrm{Jm}^3}{10^6\mathrm{Jm}}} = \sqrt{10^{-17}\,\mathrm{m}^2} = 3.16 \times 10^{-9}\,\mathrm{m} = 3.16\,\mathrm{nm}$$

The value found here suggests that different regimes exist depending on the particle size: Large particles with diameters much larger than 3 nm could be treated with elastic theory, while small particles below 3 nm need to be considered on a molecular simulation level. When compared with the typical size regime for optically efficient metal nanoparticles, that is, the particles need to have sizes above 3 nm, we see that it falls exactly in the size regime of the coherence length, making the analysis almost impossible with simple models of unified theory. Large particles with diameter much larger than 3 nm will disturb the macroscopic director field that maintains its order over 3 nm distances. We observe a situation where liquid crystal molecules wrap the nanoparticle, and the exact alignment on the nanoparticle will determine the interaction and structure formation. Small nanoparticles with diameter below the coherence length might be incorporated into the liquid crystal without visible distortion of the director field.

While the first case of large particles leads to several interesting scientific discussions [33–36], the last situation is more interesting from the viewpoint of optical material design. Because only for small particles, a homogenous material can be expected that shows no particular granulation at the length scale of the wavelength of visible light, which is around 500 nm. In the following, we will, therefore, concentrate on the size regime of small particles with diameters of metallic nanoparticles below 10 nm and discuss the concept of incorporating such particles into liquid crystal matrices.

8.3 Material Design Parameters and Concepts

Material design is governed by several parameters and based on the experience of structure formation known from liquid crystals [37–40]: the number of molecules attached to the nanoparticle, molecular interaction of mesogens, stiffness and flexibility of the molecules, the connection between the linker and the mesogen, and of course the statistics describing how many molecules are linked to the nanoparticle and under what condition. In real-world situation, the linker molecules will not be homogenously distributed on the surface and density variation might occur. It might happen that all molecules are found at the poles of a nanoparticle, which makes it much easier to form liquid crystal phases by steric interaction. In a more complex situation, different linkers and mesogens might be used and the design freedoms become virtually infinite.

8.3.1 *Design of Building Blocks*

To achieve an entire material with nonconventional electromagnetic properties, it seems favorable to incorporate the nanoparticle by chemical-binding mesogenic molecules to the particle and find a molecular design so that the whole matter will self-organize. To get flexibility, one produces the nanoparticles, linker molecules, and mesogenic molecules separately. Basically, three components are, therefore, used, which can be modified and show different properties and opens a large parameter space to play with. Table 8.1 summarizes the main parameters and the impact on the formation of matter.

Until now, only very little is known about the influence of these parameters on self-organization and the ensuing spatial assembly of gold nanoparticles wrapped with liquid crystals. Liquid crystal multipodes with inorganic cores [41, 42] and nematogen-coated gold nanoparticles [43] are first designs that point to realistic strategies. The key is careful characterization of materials to

Table 8.1 Parameters to consider for the design of self-organizing nanocomposites with nonconventional electromagnetic properties

Component	Role	Parameters
Plasmonic nanoparticle	Active optical component	Size, chemical functionalization, volume (mass) filling fraction
Linker	Binds mesogen onto the metallic nanoparticle	Number of linkers bound on nanoparticle, migration at the surface, length-size, stiffness
Mesogen	Stimulates self-organization	Length-size, form, stiffness, phase behavior, position of attachment, number of mesogens attached

understand their structure and properties. Molecular simulations are necessary to define routes for material design and might lead to a more detailed understanding of the influence of different parameters involved. Besides the typical molecular parameters, the type of ordering, lattice symmetry and size as well as macroscopic alignment need to be controlled to achieve useful materials. It has already become clear that incorporation of functional groups is necessary for the bonding between the particles and the liquid crystal mesogens, as well as for post alignment fixation, e.g., by in situ polymerization. The functional groups and linker molecules are also needed to control stiffness of molecules, volume-filling fractions, and other parameters. To be more specific, we discuss the case when a linker molecule (red) binds to a mesogen (green) and to a plasmonic nanoparticle (yellow) as shown in Fig. 8.6. Figure 8.7 takes up the situation and develops the model further. A model that describes organization can be based on shells of different characteristics as proposed by Ungar et al. [44]. As sketched in Fig. 8.6a,b, the volume occupied by different functional molecules could be symbolized by volumes that have a certain mechanical characteristics. In Fig. 8.6b, the mesogen-determined volume, called the mesogen shell (or corona), is in green color and the linker forms thin shells on the nanoparticles' surface, here in red. For instance, the shell with the linker could be considered to be inflexible and solid without any possibility to move the position of the linker molecules. In contrast, the mesogen shell might be assumed to have a certain capability to deform. Such a shell deformation leads to interesting new packing variants and symmetries to minimize the packing volume with respect to the deformation of shells [18, 19].

The picture could still be based on pure mechanical approaches to deliver results. For instance, the mesogen shell mechanics of deformation is mainly given by the density of mesogen molecules in this shell; the more it is filled, the stiffer it becomes and the less the packing might be open for highly oriented phases. Of course this is linked to a number of anchor points on the nanoparticle surfaces.

The chemistry will lead to a high yield of linker mesogen connections, and almost all linkers that are at disposition will be used as binding site. If one controls the number of binding sites, one gets access to the properties of the mesogen shell. If only a few

Figure 8.7 Schematics of order formation process in different molecular arrangements. The different rows describe different cases. In (a), the molecule shown has mesogen units and forms a building unit under the influence of external forces and might have different shapes as visualized in (b). A loose packing as in (c) will lead to less ordered structures that can gain more order when conditions are changed. Closed packing of spheres, layers, and chains are only a few possible arrangements that might be found.

binding sites installed on the nanoparticle are available, the position of sites and the number on each particle are difficult to control and migration of binding sites is likely, an unfavorable situation. A concept to prevent this is to use not only linker molecules but also spacer molecules on the nanoparticle. Figure 8.8 gives an illustration of this concept. The spacer molecules would allow stabilizing configurations and open a possibility to safely change parameters such as mesogen content with losing control and for one and the same chemical environment. This is important because of molecular forces that play a crucial role in phase sequences of liquid crystal substances.

If we come back to our goal of designed molecules able to deliver precise parameters of nanoparticle structures, we need to

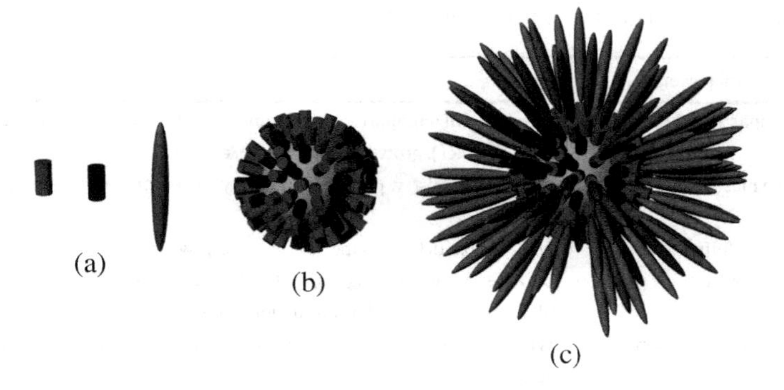

Figure 8.8 Besides the already known component linker molecules (red) and mesogen (green), we introduce a spacer of the same lengths as the linker. Using a mix of spacer and linker, the binding sites for mesogens are reduced in number in (b) and there will be less mesogens on the final molecule. The mesogen shell gets "diluted" and has different characteristics with respect to their packing characteristics in ordered phases.

remember first what was of advantage. The question how one can manage density of nanoparticles can be attacked via the length molecules and the spacer approach. A second problem that arises is the short distance between nanoparticles that should not outreach the nanoparticles' size to assure electromagnetic coupling. It seems that chains of molecules are favored here, but organization at this close distance level is managed by intermolecular forces between mesogens, and predictions are almost impossible. Here the physical chemistry of self-organization of liquid crystal mesogens [39, 45], an art in itself, needs to be applied. To provide an overview of different possibilities, we present in Table 8.2 parameters and possible mechanisms to influence them.

At the final stage, quantification of different parameters would be needed, such as filling fraction, softness of shells, etc. The early attempts to establish such a model are visible in literature [44], and time will show how powerful such models can be to predict the design of phases.

To conclude, we would like to discuss in the next chapter specific examples of materials that were realized and discuss the outcome with respect to the ideas discussed so far.

Table 8.2 Structural parameters to consider and means to influence them

Structural parameter	Mean
Nanoparticle size	Chemistry of nanoparticle, mechanism of fabrication (from inorganic salts or other), growth rate and time, environment
Type mesogen	Organic chemistry, physical chemistry of self-organization of liquid crystals
Volume-filling fraction of nanoparticles	Size of particles, shape of particles, wrapping agents around particles, type of mesogens, molecular weight of molecules, number of molecules, molecular interaction and forces
Nanoparticle distance	Wrapping chemistry, type of mesogen, number of mesogens, type of structure
Structure and order	Size of particle, type of mesogen, molecular forces between mesogens, number of mesogens

8.3.2 *Liquid-Crystalline Materials Incorporating Metal Nanoparticles*

Before discussing specific mesogen-coated nanoparticles, one should notice that preparation of metallic nanoparticles is presently a very dynamic field of research and a variety of methods exist for the preparation of gold nanoparticles and belts in particular [46–49]. This interest is associated with the size-dependent properties of matter at the nanoscale and the potential applications emerging thereof, in particular in biosensing, optics, catalysis, and nanotechnology [50, 51]. Such particles can be viewed as inorganic–organic composite materials containing a metal core and a passivating or functional organic shell. After preparation of the shell of nanoparticles, their chemical properties can be engineered via different strategies. Thiolate (SR) ligands on the nanoparticle shell can be exchanged at least partly against other ligands by simple addition of the corresponding thiol HSR' to a solution of the Au:SR nanoparticles [52]. Recently, a number of techniques have been reported of how edge and vertex sides can be functionalized and of the selective attachment of thiols at the poles of gold nanoparticles [53]. The thiol shell can be functionalized further (e.g., amide and ester couplings [54]) depending on the substituent.

Interest in adding colloid and nanoparticles to liquid crystals has largely been based on their ability to significantly alter the dielectric behavior of liquid crystals, which could potentially result in faster switching devices [55–58]. The striking ability of nematic liquid crystals to induce one- and two-dimensional order in dispersed particles on the colloidal scale has been the subject of a number of experimental and theoretical studies [59, 60]; chains, braids, and rafts of micron-sized colloidal spheres were assembled by laser tweezers and held together by dipolar and quadrupolar defects in the director field [61]. Assembling metallic nanoparticles is of great interest for electronic [62], optical [63], and photonics [64] applications.

Metal nanoparticles derivatized with alkane thiols, acids, or amines could only be made to form bulk lattices with long-range order if they were highly monodisperse [65, 66]. The lattices were those expected from packing of spheres, that is, body-centered cubic, hexagonal close packed, or face-centered cubic. Derivatization of gold nanoparticles with bulkier monodendron thiols also produced a body-centered cubic lattice [67]. Using bimodal particle size distribution, interesting planar structures have been achieved on surfaces [68]. Self-assembly of nanoparticles of two different materials has yielded a considerable diversity of binary nanoparticle superlattices [69], the structure being determined primarily by close-packing criteria, but also modified by dipolar or magnetic forces.

Attempts to order nanoparticles using liquid crystals are relatively recent. It leads to increased activity, and several excellent review articles provide overview of the activities inspired mainly by chemistry [70–72]. First attempts were made more than 10 years ago when a nematic phase and the formation of disordered chains of gold nanoparticles coated with cyanobiphenyl mesogens end-attached via a thioalkyl spacer was observed [73, 74]. Bent core [75] and discotic [76] mesogens have also been attached to gold nanoparticles. In both cases, the functionalized particles were studied neat as well as dispersed as dopants in a host liquid crystal. Only one pure bent core derivative formed a metastable liquid crystal phase of unknown nature [75]. The discotic-capped nanoparticles were thought to cluster at the grain boundaries of the columnar phase. Other reports in literature support these results [77–81].

In recent years, advances in the preparation of different kinds of mesogenic nanoparticles have been made. All have used nanoparticles smaller than 3 nm. The main subject was the chemistry and structural investigation of the materials, and optics was not considered. Three different ideas were followed:

- Combining nanoparticles with mesogen attachment along the chain, [72–74, 82–86, 94, 95]
- Use side-chain mesogen attachment and nanoparticles [41–44, 82, 87, 88]
- Introduce dendrimers of low order to wrap the nanoparticle [79, 89–92]

All of these approaches led to liquid-crystalline phases. Different types of mesophases are observed. A review of the material properties and liquid-crystalline phase behavior is given in [70] and need not be reassessed here. More experimental details with respect to optical properties are provided in Table 8.3, which indicates key parameters such as nanoparticle size and content.

In Table 8.3 one discovers that only two publications were able to proof particular optical properties based on metal nanoparticle plasmons.

Very recently, slightly larger gold nanoparticles (larger than 3 nm in diameter) covered by a mixture of thioalkane and a rod-like mesogen attached laterally via a thioalkyl spacer were successfully studied optically. Such materials preserved the ability of the mesogens to form a liquid crystal phase under the condition to have a recognizable optical plasmon signature [93]. The phase was thought to be nematic, but more detailed investigations have shown that the structure of the liquid crystal phases is more complex, with the nanoparticles forming anisotropic periodic lattices with pronounced short-range order [93]. Long-range order seems to be only established if external alignment forces such as shearing are applied. Nevertheless for the first time, a measurable electromagnetic coupling between neighboring nanoparticles could be observed as a dichroism of the sample and a shift of the resonant wavelength. The result is very encouraging and gives hope that more such materials will be synthesized and a design of structures

Table 8.3 Overview of different materials as could be found in the literature today

Type	Nanoparticle size [nm]	Transition °C	Nanoparticle content (% cm)	Optical plasmon	Optical image of texture	Ref.
Long chain attachment	2.4	<150	61.7	no	YES	72
	3.0	<130	66	no	no	73
	2.7	<130	—	no	no	74
	1.8	<150	60.2	no	no	82
	1.0	<150	—	no	no	83
	—	—	—	no	no	84
	2	<170	—	no	no	85
	2.2	<100	—	no	YES	86
	1.5	<130	—	no	YES	94
	4.4	<170	—	YES	no	95
Side-chain attachment	1.6	<80	—	no	YES	41
	1.7–2.0	<150	—	no	YES	42
	1.89	<130	58.6	no	YES	43
	1.8–2.3	<90	53	no	YES	44
	9.9	<120	—	no	YES	87
	3.4	<80	70.8	YES	YES	93
Dendrimers	5.9–6.8	<80	69.3	no	no	79
	1.7	<200	-	no	YES	89
	2.1	<180	—	no	no	90
	1.6	<220	—	no	no	92

becomes possible where specific electromagnetic features could be brought up.

A second example was realized with long chain attachment and even larger particles [95]. It was possible to create macroscopically ordered, anisotropic structures made from different metal nanoparticles. Silver and gold were used having mesogenic ligands grafted onto their surface. The silver and gold compound showed liquid-crystalline behavior and the chemistry was based on stilbene-comprising molecular architecture.

The successful material had a mercapto-functionalized 15-carbon-length alkyl chain that provides flexibility and allows covalent bonding of the ligand to the nanoparticle surface. A terminal oleylalkoxy chain was used to facilitate lamellar structure formation. Such type of molecules showed higher ordered layered phases (SmC and SmF) with layer spacing in the order of 5 nm. The optical properties were studied by UV-VIS spectroscopy, and a small shift of the absorption peak was found in aligned samples. The authors interpret this as an effect mainly due to the optical birefringence of the material.

In summary, one realizes that the concept of liquid-crystalline plasmonic nanoparticles leads to first active plasmonic materials and seems to be the right way to realize plasmonic paints. The basic structural parameters such as nanoparticle size and distance are within reach and self-assembly is still maintained.

8.4 Conclusion and Outlook

Active plasmonic materials based on metallic nanoparticles and very active field of research [96] and especially mesogen-coated particles attract a lot of attention. For the moment chemists dominate the field and synthetize materials that will very soon be available for the design of artificial index materials. The difficulties are manifold, and a rational design is still in its early stages [44, 72]. Based on the technology to use liquid crystal materials in real-world applications, the transfer of ideas [45, 97, 98] into devices is a rather short path and we can expect spectacular materials within the near future leading to extraordinary possibilities of refractive index engineering.

References

1. D. R. Smith, S. Schultz, P. Markoš, and C. M. Soukoulis. (2002). Determination of effective permittivity and permeability of metamaterials from reflection and transmission coefficients, *Phys. Rev. B*, **65,** pp. 195104.
2. W. Rotman. (1962). Plasma simulation by artificial dielectrics and parallel-plate media IRE, *Trans. Antennas Propag.* **10**, 82.
3. V. M. Shalaev, W. Cai, U. K. Chettiar, H.-K. Yuan, A. K. Sarychev, V. P. Drachev, and A. V. Kildishev. (2005). Negative index of refraction in optical metamaterials, *Opt. Lett.*, **30**, pp. 3356–3358.

4. D. R. Smith, W. J. Padilla, D. C. Vier, S. C. Nemat-Nasser, and S. Schultz. (2000). Composite medium with simultaneously negative permeability and permittivity, *Phys. Rev. Lett.* **84**, pp. 4184–4187 DOI: 10.1103/PhysRevLett.84.4184.
5. V. G. Veselago. (1968). The electrodynamics of substances with simultaneously negative values of ε and μ, *Soviet Physics Uspekhi-Ussr,* **10,** pp. 509.
6. J. B. Pendry, A. J. Holden, D. J. Robbins, and W. J. Stewart. (1999). Magnetism from conductors and enhanced nonlinear phenomena, *IEEE Tran. Microwave Theory Tech.,* **47**(11), pp. 2075–2084, DOI: 10.1109/22.798002.
7. C. Rockstuhl, F. Lederer, C. Etrich, T. Pertsch, and T. Scharf. (2007). Design of an artificial three-dimensional composite metamaterial with magnetic resonances in the visible range of the electromagnetic spectrum, *Phys. Rev. Lett.,* **99,** pp. 017401, DOI: 10.1103/PhysRevLett.99.017401.
8. S. O'Brien and J. B. Pendry. (2002). Photonic band-gap effects and magnetic activity in dielectric composites, *J. Phys. Cond. Matter,* **14**, pp. 4035, DOI: 10.1088/0953-8984/14/15/317.
9. C. Rockstuhl, T. Zentgraf, H. Guo, N. Liu, C. Etrich, I. Loa, K. Syassen, J. Kuhl, F. Lederer, and H. Giessen. (2006). Resonances of split-ring resonator metamaterials in the near infrared, *Appl. Phys. B: Lasers Opt.,* **84**, pp. 219–227.
10. F. Garwe, C. Rockstuhl, C. Etrich, U. Hübner, U. Bauerschäfer, F. Setzpfandt, M. Augustin, T. Pertsch, A. Tünnermann, and F. Lederer. (2006). Evaluation of gold nanowire pairs as a potential negative index material, *Appl. Phys. B: Lasers Opt.,* **84,** pp. 139–148.
11. J. Yao, Z. Liu, Y. Liu, Y. Wang, C. Sun, G. Bartal, A. M. Stacy, and X. Zhang. (2008). Optical negative refraction in bulk metamaterials of nanowires, *Science,* **321**, pp. 930.
12. S. Zhang, W. Fan, N. C. Panoiu, K. J. Malloy, R. M. Osgood, and S. R. J. Brueck. (2005). Experimental demonstration of near-infrared negative-index metamaterials, *Phys. Rev. Lett.,* **95,** pp. 137404.
13. C. F. Bohren and D. R. Huffman. (1998). *Absorption and Scattering of Light by Small Particles,* Wiley, New York.
14. S. Kawata (ed.). (2001). *Near-Field Optics and Surface Plasmon Polaritons: Topics in Applied Physics,* Springer-Verlag, Berlin, Heidelberg, New York, ISBN: 3540415025 (alk. paper).
15. W. Cia and V. Shalaev. (2010). *Optical Metamaterials,* Springer, New York, ISBN: 978-1-4419-1150-6.

16. C. Rockstuhl, M. G. Salt, and H. P. Herzig. (2004). Analyzing the scattering properties of coupled metallic nanoparticles, *J. Opt. Soc. Am. A*, **21**, pp. 1761–1768.
17. http://www.harima.co.jp/, Nanopaste series.
18. G. Ungar, Y. Liu, X. Zeng, V. Percec, and W.-D. Cho. (2003). Giant supramolecular liquid crystal lattice, *Science*, **299**(5610), pp. 1208–1211, DOI: 10.1126/science.1078849.
19. X. Zeng, G. Ungar, and M. Impéror-Clerc. (2005). A triple-network tricontinuous cubic liquid crystal, *Nature Mater.*, **4**, pp. 562–567, DOI: 10.1038/nmat1413.
20. P. J. Flory. (1953). *Principles of Polymers Chemistry*, Cornell University Press, Ithaca, NY.
21. M. Barón. (2001). Definitions for basic terms relating to low-molar-mass and polymer liquid crystals, *Pure Appl. Chem.*, **73**(5), pp. 845–895.
22. S. Orlandi and C. Zannoni. (2013). Phase organization of mesogen-decorated spherical nanoparticles, *Mol. Cryst. Liq. Cryst.*, **573**, pp. 1–9, DOI: 10.1080/15421406.2012.763213.
23. M. Kleman and O. D. Lavrentovich. (2003). *Soft Matter Physics: An Introduction*, Springer Verlag, New York, pp. 638, ISBN: 9780387952673.
24. T. Scharf. (2006). *Polarized Light in Liquid Crystal and Polymers*, Wiley Hoboken, ISBN: 978-0-471-74064-3.
25. I. Dierking. (2003). *Textures of Liquid Crystals*, Wiley-VCH, Verlag Weinheim, ISBN: 9783527307258.
26. O. D. Lavrentovich. (2003). Defects in liquid crystals: Surface and interfacial anchoring effects, in *Patterns of Symmetry Breaking* (H. Arodz, J. Dziarmaga, and W. H. Zurek, eds.), NATO Science Series, II. Mathematics, Physics, and Chemistry, vol. 127, Kluwer Academic Publishers, Dordrecht, the Netherlands, pp. 161–195.
27. M. Kleman and O. D. Lavrentovich. (2006). Topological point defects in nematic liquid crystals, *Philo. Mag.*, **86**(25–26), pp. 4117–4137, DOI: 10.1080/14786430600593016.
28. S. Zhang, E. M. Terentjev, and A. M. Donald. (2005). Nature of disclination cores in liquid crystals, *Liq. Cryst.*, **32**(1), pp. 69–75.
29. S. Chandrasekhar. (1993). *Liquid crystals*, 2nd ed, Cambridge University Press, Cambridge, ISBN: 978-0521427418.
30. P. G. de Gennes. (1995). *The Physics of Liquid Crystals*, 2nd ed, Oxford University Press, USA, ISBN: 978-0198517856.

31. I. W. Stewart. (2004). *The Static and Dynamic Continuum Theory of Liquid Crystals: A Mathematical Introduction*, Taylor & Francis, USA, ISBN: 978-0748408962.

32. S. Singh. (2002). *Liquid Crystals: Fundamentals*, World Scientific Publishing, Singapore, ISBN: 981-02-4250-6.

33. C. P. Lapointe, T. G. Mason, and I. I. Smalyukh. (2009). Shape-controlled colloidal interactions in nematic liquid crystals, *Science*, **1083**, pp. 326, DOI: 10.1126/science.1176587.

34. Q. Liu, Y. Cui, D. Gardner, X. Li, S. He, and I. I. Smalyukh. (2010). Self-alignment of plasmonic gold nanorods in reconfigurable anisotropic fluids for tunable bulk metamaterial applications, *Nano Lett.*, **10**(4), pp. 1347–1353, DOI: 10.1021/nl9042104.

35. B. Senyuk, J. S. Evans, P. J. Ackerman, T. Lee, P. Manna, L. Vigderman, E. R. Zubarev, J. van de Lagemaat, and I. I. Smalyukh. (2012). Shape-dependent oriented trapping and scaffolding of plasmonic nanoparticles by topological defects for self-assembly of colloidal dimers in liquid crystals, *Nano Lett.*, **12**(2), pp. 955–963, DOI: 10.1021/nl204030t.

36. C. Blanc, D. Coursault, and E. Lacaze. (2013). Ordering nano- and microparticles assemblies with liquid crystals, *Liq. Cryst. Rev.*, **1**(2), pp. 83–109, DOI: 10.1080/21680396.2013.818515.

37. I. M. Saez and J. W. Goodby. (2005). Supermolecular liquid crystals, *J. Mater. Chem.*, **15**, pp. 26–40, DOI: 10.1039/b413416h.

38. J. W. Goodby, I. M. Saez, S. J. Cowling, V. Görtz, M. Draper, A. W. Hall, S. Sia, G. Cosquer, S.-E. Lee, and E. P. Raynes. (2008). Transmission and amplification of information and properties in nanostructured liquid crystals. *Angew. Chem. Int. Ed.*, **47**, pp. 2754–2787, DOI: 10.1002/anie.200701111.

39. C. Tschierske. (2013). Development of structural complexity by liquid-crystal self-assembly, *Angew. Chem. Int. Ed.*, **52**, pp. 8828–8878, DOI: 10.1002/anie.201300872.

40. J. W. Goodby, I. M. Saez, S. J. Cowling, J. S. Gasowska, R. A. MacDonald, S. Sia, P. Watson, K. J. Toyne, M. Hird, R. A. Lewis, S.-E. Lee, and V. Vaschenko. (2009). Molecular complexity and the control of self-organising processes, *Liq. Cryst.*, **36**(6–7), pp. 567–605, DOI: 10.1080/02678290903146060.

41. L. Cseh and G. H. Mehl. (2006). The design and investigation of room temperature thermotropic nematic gold nanoparticles, *J. Am. Chem. Soc.*, **128**(41), pp. 13376–13377, DOI: 10.1021/ja066099c.

42. L. Cseh and G. H. Mehl. (2007). Structure–property relationships in nematic gold nanoparticles, *J. Mater. Chem.*, **17**, pp. 311–315, DOI: 10.1039/B614046G.
43. X. Zeng, F. Liu, A. G. Fowler, G. Ungar, L. Cseh, G. H. Mehl, and J. E. Macdonald. (2009). 3D ordered gold strings by coating nanoparticles with mesogens, *Adv. Mater.*, **21**, pp. 1746–1750, DOI: 10.1002/adma.200803403.
44. X. Mang, X. Zeng, B. Tang, F. Liu, G. Ungar, R. Zhang, L. Cseh, and G. H. Mehl. (2012). Control of anisotropic self-assembly of gold nanoparticles coated with mesogens, *J. Mater. Chem.*, **22**, pp. 11101, DOI: 10.1039/c2jm16794h.
45. A. Guerrero-Martínez, M. Grzelczak, and L. M. Liz-Marzán. (2012). Molecular thinking for nanoplasmonic design, *ACS Nano*, **6**(5), pp. 3655–3662, DOI: 10.1021/nn301390s.
46. J. Turkevich, P. C. Stevenson, and J. Hillier. (1951). A study of the nucleation and growth processes in the synthesis of colloidal gold, *Discuss. Faraday Soc.*, **11**, pp. 55–75, DOI: 10.1039/DF9511100055.
47. M. Brust, M. Walker, D. Bethell, D. J. Schiffrin, and R. Whyman. (1994). Synthesis of thiol-derivatised gold nanoparticles in a two-phase liquid–liquid system, *J. Chem. Soc. Chem. Commun.*, **801**, DOI: 10.1039/C39940000801.
48. J. Zhuang, H. Wu, Y. Yang, and, Y. C. Cao. (2007). Supercrystalline colloidal particles from artificial atoms, *J. Am. Chem. Soc.*, **129**(46), pp. 14166–14167, DOI: 10.1021/ja076494i.
49. J. Zhang, J. Du, B. Han, Z. Liu, T. Jiang, and Z. Zhang. (2006). Sonochemical formation of single-crystalline gold nanobelts, *Angew. Chem. Int. Ed.*, **45**, pp. 1116–1119, DOI: 10.1002/anie.200503762.
50. D. A. Giljohann, D. S. Seferos, W. L. Daniel, M. D. Massich, P. C. Patel, and C. A. Mirkin. (2010). Gold nanoparticles for biology and medicine, *Angew. Chem. Int. Ed.*, **49**, pp. 3280–3294, DOI: 10.1002/anie.200904359.
51. N. J. Halas, S. Lal, W.-S. Chang, S. Link, and P. Nordlander. (2011). Plasmons in strongly coupled metallic nanostructurers, *Chem. Rev.*, **111**(6), pp. 3913–3961, DOI: 10.1021/cr200061k.
52. A. C. Templeton, D. E. Cliffel, and R. W. Murray. (1999). Redox and fluorophore functionalization of water-soluble, tiopronin-protected gold clusters, *J. Am. Chem. Soc.*, **121**(30), pp. 7081–7089, DOI: 10.1021/ja990513+.
53. G. A. DeVries, M. Brunnbauer, Y. Hu, A. M. Jackson, B. Long, B. T. Neltner, O. Uzun, B. H. Wunsch, and F. Stellacci. (2007). Divalent metal nanoparticles, *Science*, **358**, pp. 315, DOI: 10.1126/science.1133162.

54. A. C. Templeton, M. J. Hostetler, E. K. Warmoth, S. Chen, C. M. Hartshorn, V. M. Krishnamurthy, M. D. E. Forbes, and R. W. Murray. (1998). Gateway reactions to diverse, polyfunctional monolayer-protected gold clusters, *J. Am. Chem. Soc.*, **120**(19), pp. 4845–4849, DOI: 10.1021/ja980177h.

55. D. F. Gardner, J. S. Evans, and I. I. Smalyukh. (2011). Towards reconfigurable optical metamaterials: Colloidal nanoparticle self-assembly and self-alignment in liquid crystals, *Mol. Cryst. Liq. Cryst.*, **545**(1), 3/[1227]–21/[1245].

56. H. Yoshikawa, K. Maeda, Y. Shiraishi, J. Xu, Y. Shiraiki, N. Toshima, and S. Kobayashi. (2002). Frequency modulation response of a tunable birefringent mode nematic liquid crystal electrooptic device fabricated by doping nanoparticles of Pd covered with liquid-crystal molecules, *Jpn. J. Appl. Phys., Part 2*, **41**, pp. L1315–L1317.

57. Y. Reznikov, O. Buchnev, O. Tereshchenko, V. Reshetnyak, A. Glushchenko, and J. West. (2003). Ferroelectric nematic suspension, *Appl. Phys. Lett.*, **82**, pp. 1917–1919.

58. F. H. Li, O. Buchnev, C. I. Cheon, A. Glushchenko, V. Reshetnyak, Y. Reznikov, T. J. Sluckin, and J. L. West. (2006). Orientational coupling amplification in ferroelectric nematic colloids, *Phys. Rev. Lett.*, **97**, pp. 147801.

59. P. Poulin, H. Stark, T. C. Lubensky, and D. A. Weltz. (1997). Novel colloidal interactions in anisotropic fluids, *Science*, **275**, pp. 1770–1773.

60. I. Smalyukh, O. D. Lavrentovich, A. N. Kuzmin, A. V. Kachinski, and P. N. Prasad. (2005). Elasticity-mediated self-organization and colloidal interactions of solid spheres with tangential anchoring in a nematic liquid crystal, *Phys. Rev. Lett.*, **95**, pp. 157801.

61. I. Musevic, M. Skarabot, U. Tkalec, M. Ravnik, S. Zumer. (2006). Two-dimensional nematic colloidal crystals self-assembled by topological defects, *Science*, **313**, pp. 954–958.

62. S. W. Boettcher, N. C. Strandwitz, M. Schierhorn, N. Lock, M. C. Lonergan, and G. D. Stucky. (2007). Tunable electronic interfaces between bulk semiconductors and ligand-stabilized nanoparticle assemblies, *Nat. Mater.*, **6**, pp. 592–596.

63. S. Eustis and M. A. El-Sayed. (2006). Why gold nanoparticles are more precious than pretty gold: Noble metal surface plasmon resonance and its enhancement of the radiative and nonradiative properties of nanocrystals of different shapes, *Chem. Soc. Rev.*, **35**, pp. 209–217.

64. P. N. Prasad. (2004). *Nanophotonics*, John Wiley & Sons, Inc., Hoboken, New Jersey.

65. R L. Whetten, J. T. Khoury, M. M. Alvarez, S. Murthy, I. Vezmar, Z. L. Wang, P. W. Stephens, C. L. Cleveland, W. D. Luedtke, and U. Landman. (1996). Nanocrystal gold molecules, *Adv. Mater.*, **8**, pp. 428–433.

66. S. Sun, C. B. Murray, D. Weller, L. Folks, and A. Moser. (2000). Monodisperse FePt nanoparticles and ferromagnetic FePt nanocrystal superlattices, *Science*, **287**, pp. 1989–1992.

67. B. Donnio, P. García-Vázquez, J.-L. Gallani, D. Guillon, and E. Terazzi, and D. Ferromagnetic. (2007). Gold nanoparticles self-organized in a thermotropic cubic phase, *Adv. Mater.*, **19**, pp. 3534–3539.

68. C. J. Kiely, M. Brust, J. Fink, D. Bethell, and D. J. Schiffrin. (1998). Spontaneous ordering of bimodal ensembles of nanoscopic gold clusters, *Nature*, **396**, pp. 444–446.

69. E. V. Shevchenko, D. V. Talapin, N. A. Kotov, S. O'Brien, and C. B. Murray. (2006). Structural diversity in binary nanoparticle superlattices, *Nature*, **439**, pp. 55–59.

70. G. L. Nealon, R. Greget, C. Dominguez, Z. T. Nagy, D. Guillon, J.-L. Gallani, and B. Donnio. (2012). Liquid-crystalline nanoparticles: Hybrid design and mesophase structures, *Beilstein J. Org. Chem.*, **8**, pp. 349–370, DOI: 10.3762/bjoc.8.39.

71. S. Saliba, C. Mingotaud, M. L. Kahn, and J.-D. Marty. (2013). Liquid crystalline thermotropic and lyotropic nanohybrids, *Nanoscale*, **5**, pp. 6641–6661, DOI: 10.1039/C3NR01175E.

72. M. Draper, I. M. Saez, S. J. Cowling, P. Gai, B. Heinrich, B. Donnio, D. Guillon, and J. W. Goodby. (2011). Self-assembly and shape morphology of liquid crystalline gold metamaterials, *Adv. Funct. Mater.*, **21**, pp. 1260–1278, DOI: 10.1002/adfm.201001606.

73. N. Kanayama, O. Tsutsumi, A. Kanazawa, and T. Ikeda. (2001). Distinct thermodynamic behaviour of a mesomorphic gold nanoparticle covered with a liquid-crystalline compound, *Chem. Commun.*, **24**, pp. 2640–2641.

74. I. In, Y.-W. Jun, Y. J. Kim, and S. Y. Kim. (2004). Spontaneous one dimensional arrangement of spherical Au nanoparticles with liquid crystal ligands, *Chem. Commun.*, **2005**, pp. 800–801, DOI: 10.1039/b413510e.

75. V. M. Marx, H. Girgis, P. A. Heineyb, and T. Hegmann. (2008). Bent-core liquid crystal (LC) decorated gold nanoclusters: synthesis, self-assembly, and effects in mixtures with bent-core LC hosts, *J. Mater. Chem.*, **18**, pp. 2983–2994.

76. S. Kumar, S. K. Pal, P. S. Kumar, and V. Lakshminarayanan. (2007). Novel conducting nanocomposites: Synthesis of triphenylene-covered gold

nanoparticles and their insertion into a columnar matrix, *Soft Matter*, **3**, pp. 896–900.

77. R. H. Terril, T. A. Postlethwaite, C. H. Chen, C. D. Poon, A. Tzerzis, A. D. Hutchinson, M. R. Clark, G. Wignall, J. D. Londono, R. Superfine, M. Falvo, C. S. Johnson, E. T. Samulski, and R. W. Murray. (1995). Monolayers in three dimensions: NMR, SAXS, thermal, and electron hopping studies of alkanethiol stabilized gold clusters, *J. Am. Chem. Soc.*, **117**, pp. 12537–12548.
78. A. Badia, L. Demers, L. Guccia, F. Morin, and R. B. Lennox. (1997). Structure and dynamics in alkanethiolate monolayers self-assembled on gold nanoparticles: A DSC, FT-IR, and deuterium NMR study, *J. Am. Chem. Soc.*, **119**, pp. 2682–2692.
79. K. Kanie, M. Matsubara, X. Zeng, F. Liu, G. Ungar, H. Nakamura, and A. Muramatsu. (2011). Simple cubic packing of gold nanoparticles through rational design of their dendrimeric corona, *J. Am. Chem. Soc.*, **134**(2), pp. 808–811, DOI: 10.1021/ja2095816.
80. N. R. Jana, L. A. Gearheart, S. O. Obare, C. J. Johnson, K. J. S. Mann, and C. J. Murphy. (2002). Liquid crystalline assemblies of ordered gold nanorods, *J. Mater. Chem.*, **12**, pp. 2909–2912.
81. R. T. M. Jakobs, J. van Herrikhuyzen, J. C. Gielen, P. C. M. Christianen, S. C. J. Meskers, and A. P. H. J. Schenning. (2008). Self-assembly of amphiphilic gold nanoparticles decorated with a mixed shell of oligo(p-phenylene vinylene)s and ethyleneoxide ligands, *J. Mater. Chem.*, **18**, pp. 3438–3441.
82. M. Wojcik, W. Lewandowski, J. Matraszek, J. Mieczkowski, J. Borysiuk, D. Pociecha, and E. Gorecka. (2009). Liquid-crystalline phases made of gold nanoparticles, *Angew. Chem. Int. Ed.*, **48**, pp. 5167–5169, DOI: 10.1002/anie.200901206.
83. M. Wojcik, M. Kolpaczynska, D. Pociecha, J. Mieczkowski, and E. Gorecka. (2010). Multidimensional structures made by gold nanoparticles with shape-adaptive grafting layers, *Soft Matter*, **6**, pp. 5397–5400, DOI: 10.1039/C0SM00539H.
84. M. M. Wojcik, M. Gora, J. Mieczkowski, J. Romiszewski, E. Gorecka, and D. Pociecha. (2011). Temperature-controlled liquid crystalline polymorphism of gold nanoparticles, *Soft Matter*, **7**, pp. 10561, DOI: 10.1039/c1sm06436c.
85. T.-Y. Ye, X.-F. Chen, K. Qian, Z. Shen, L. Qi, and X.-H. Fan. (2012). Controlling the packing of gold nanoparticles with grafted liquid crystals, *J. Nanopart. Res.*, **14**, 1055, DOI: 10.1007/s11051-012-1055-6.

86. W. Lewandowski, K. Jatczak, D. Pociecha, and J. Mieczkowski. (2013). Control of gold nanoparticle superlattice properties via mesogenic ligand architecture, *Langmuir*, **29**(10), pp. 3404–3410, DOI: 10.1021/la3043236.

87. C. H. Yu, C. P. J. Schubert, C. Welch, B. J. Tang, M.-G. Tamba, and G. H. Mehl. (2012). Design, synthesis, and characterization of mesogenic amine-capped nematic gold nanoparticles with surface-enhanced plasmonic resonances, *J. Am. Chem. Soc.*, **134**(11), pp. 5076–5079, DOI: 10.1021/ja300492d.

88. R. Breckon, S. Chakraborty, C. Zhang, N. Diorio, J. T. Gleeson, S. Sprunt, R. J. Twieg, and A. Jákli. (2013). Nanostructures of nematic materials of laterally branched molecules, *Chem. Phys. Chem.*, **15**, 1457–1462, DOI: 10.1002/cphc.201300578.

89. S. Frein, J. Boudon, M. Vonlanthen, T. Scharf, J. Barberá, G. Süss-Fink, T. Bürgi, and R. Deschenaux. (2008). Liquid-crystalline thiol- and disulfide-based dendrimers for the functionalization of gold nanoparticles, *Preliminary Commun. HCA*, **91**, pp. 2321–2337, DOI: 10.1002/hlca.200890253.

90. B. Donnio, P. García-Vázquez, J.-L Gallani, D Guillon, and E. Terazzi. (2007). Dendronized ferromagnetic gold nanoparticles self-organized in a thermotropic cubic phase, *Adv. Mater.*, **19**, pp. 3534–3539, DOI: 10.1002/adma.200701252.

91. B. Donnio, A. Derory, E. Terazzi, M. Drillon, D. Guillo, and J.-L. Gallani. (2010). Very slow high-temperature relaxation of the remnant magnetic moment in 2 nm mesomorphic gold nanoparticles, *Soft Matter*, **6**, pp. 965–970, DOI: 10.1039/B918602F.

92. S. Mischler, S. Guerra, and R. Deschenaux. (2012). Design of liquid-crystalline gold nanoparticles by click chemistry, *Chem. Commun.*, **48**, pp. 2183–2185, DOI: 10.1039/c2cc17375a.

93. J. Dintinger, B.-J. Tang, X. Zeng, F. Liu, T. Kienzler, G. H. Mehl, G. Ungar, C. Rockstuhl, and T. Scharf. (2013). A self-organized anisotropic liquid-crystal plasmonic metamaterial, *Adv. Mater.*, **25**, 1999–2004, DOI: 10.1002/adma.201203965.

94. J. M. Wolska, D. Pociecha, J. Mieczkowski, and E. Gorecka. (2013). Gold nanoparticles with flexible mesogenic grafting layers, *Soft Matter*, **9**, pp. 3005–3008, DOI: 10.1039/C3SM27882D.

95. W. Lewandowski, D. Constantin, K. Walicka, D. Pociecha, J. Mieczkowskia, and E. Górecka. (2013). Smectic mesophases of functionalized silver

and gold nanoparticles with anisotropic plasmonic properties, *Chem. Commun.*, **49**, pp. 7845–7847, DOI: 10.1039/C3CC43166E.

96. S. Mühlig, A. Cunningham, J. Dintinger, T. Scharf, T. Bürgi, F. Lederer, and C. Rockstuhl. (2013). Self-assembled plasmonic metamaterials, *Nanophotonics*, **2**(3), pp. 211–240, DOI: 10.1515/nanoph-2012-0036.
97. O. D. Lavrentovich. (2011). Liquid crystals, photonic crystals, metamaterials, and transformation optics, *PNAS*, **108**(13), pp. 5143–5144, DOI/10.1073/pnas.1102130108.
98. J. Xiang and O. D. Lavrentovich. (2012). Liquid crystal structures for transformation optics, *Mol. Cryst. Liq. Cryst.*, 559(1), pp. 106–114, DOI: 10.1080/15421406.2012.658692.

Chapter 9

Tunable Plasmonics Based on Liquid Crystals

Yan Jun Liu,[a] Guangyuan Si,[b] Yanhui Zhao,[c] and Eunice Sok Ping Leong[a]

[a] *Institute of Materials Research and Engineering, 3 Research Link, Singapore 117602, Singapore*
[b] *College of Information Science and Engineering, Northeastern University, Shenyang 110004, China*
[c] *Department of Engineering Science and Mechanics, The Pennsylvania State University, University Park, PA 16802, USA*
liuy@imre.a-star.edu.sg

9.1 Introduction

Plasmonics [1–13], the study of the interaction between electromagnetic field and free electrons in a metal, has drawn increasing attention recently due to its huge potential for solving many eminent issues encountered by our world. Up to now, exciting plasmonic applications, for instance, super-resolution imaging [14–19], optical cloaking [20–23], and energy harvesting [24–29], have been reported. Many other potential applications are under

Active Plasmonic Nanomaterials
Edited by Luciano De Sio

ISBN 978-981-4613-00-2 (Hardcover), 978-981-4613-01-9 (eBook)
www.panstanford.com

development. All these developments are attributed to the advanced nanofabrication techniques. Top-down nanofabrication techniques such as electron-beam lithography [30–33] and focus ion beam milling [34–40] allow the accurate fabrication of structures at nanoscale with desirable trade-off on the high equipment expenses, as well as the considerably long time during sample preparation. Bottom-up techniques like self-assembly [41–43] can easily achieve regular patterns at a large scale at a rather low cost but are not preferred for the cases that require accurate positioning and alignment with nanometer precision. Despite their strengths and weaknesses, both top-down and bottom-up techniques make their unique contributions to plasmonics by providing various nanostructures for plasmonic applications for different purposes. However, most plasmonic devices fabricated through top-down and bottom-up techniques are passive, which greatly limits their application because of the additional investments required in fabricating another similar device with little change in the sample design. Thus, the new research field of active plasmonics emerges, which deals with reconfigurable function after the devices are fabricated, with the help of active mediums responsive to certain stimulus. The first original active plasmonic device is gallium-coated plasmonic waveguides proposed by Krasavin and Zheludev [44]. Since then, many other active mediums have been used to build active plasmonic devices, including liquid crystals [45–60], molecular machines [61, 62], elastic polymers [63–65], and chemical oxidation/reduction [66–68]. Among all the mentioned active mediums, liquid crystal stands out from all the rest because of its large birefringence on refractive index, low threshold on transition among different states, and versatile driven methods to cause the transitions.

Liquid crystals are materials that represent a phase of matter whose properties lie between those of a conventional liquid and a solid crystal. They are a class of materials particularly attractive for liquid crystal displays and optoelectronic applications due to their high sensitivity to the external stimulus and have been extensively studied [69, 70]. There are three main kinds of liquid crystals: nematic liquid crystals, cholesteric liquid crystals, and smectic liquid crystals. In a liquid crystal, the molecules have no positional order but tend to align along the same direction. Due to thermal random motion, friction, and collision between molecules, not all molecules align along a certain direction and their directions vary around the average direction randomly. This average direction is referred to as the orientation of the liquid crystal, which stands for the average direction of most molecules. This parameter, that is, the director, is an important factor that denotes the liquid crystal's properties. Under proper treatment, a slab of liquid crystal can be obtained with a uniform alignment of the director. Such a sample exhibits uniaxial optical symmetry with two principal refractive indices n_o and n_e. The ordinary refractive index n_o is for light with electric field polarization perpendicular to the director, and the extraordinary refractive index n_e is for light with electric field polarization parallel to the director. The birefringence (or optical anisotropy) is defined as $\Delta n = n_e - n_o$. Therefore, the refractive index of a liquid crystal can be changed between n_o and n_e by controlling the orientation of the directors. As a truly unique gift from nature, liquid crystals possess the smallest elastic constants and the largest birefringence among all known materials. In addition, their large birefringence spans the entire visible-infrared spectrum and beyond, which was first reported by Wu in 1986 [71]. By virtue of their organic nature, they can be chemically synthesized and processed on a very large scale; they are also compatible with almost all technologically important optoelectronic materials. The alignment of liquid crystals can be easily controlled by many means, such as electricity, light, and acoustic waves, thus making them an excellent candidate for the development of active nanophotonics.

By integrating liquid crystals with plasmonic nanostructures, active plasmonic materials and devices with enhanced performance have been demonstrated. In this chapter, we summarize the recent research progress and achievements in liquid-crystal-based plasmonics. We hope the contents covered in this chapter can serve as a tutorial introduction to readers with little background in either plasmonics or liquid crystals. We also hope experienced researchers can further expand their knowledge of the given topic and be inspired to take this field to new horizons.

9.2 Different Tuning Schemes

The birefringence of liquid crystal is demonstrated as a difference in refractive index, making it an ideal active medium for active plasmonic devices for different applications, such as plasmonic switches [51–55], active plasmonic color filters [30, 35, 39], and plasmonic waveguides [72]. While the concept of applying liquid crystal to plasmonic nanostructure is straightforward, there are various tuning schemes of the liquid crystals to achieve a noticeable refractive index change. In the following, representative liquid-crystal-based active plasmonic devices will be categorized and discussed according to the tuning schemes.

9.2.1 *Electric Tuning*

Electric fields have very strong impacts on the rod shape of liquid crystal molecules. When the liquid crystal molecules are subject to an electric field, one end of a molecule has positive charges, while the other end is negatively charged, forming an electric dipole. As a result, the director of the liquid crystal molecules will be reoriented along the direction of an external electric field. Therefore, electric field is the most commonly used method to drive the liquid crystal devices. Utilizing this characteristic of the liquid crystal, efforts have been made to combine liquid crystals with periodically nanostructured metal films to develop electronically controlled transmission, reflection, and absorption of the plasmonic structures, concerning their applications on switches, filters, and modulators. One example is given by Dickson et al. [56], who have demonstrated precise control over surface plasmon dispersion and transmission of gold nanohole arrays using liquid crystals. A schematic of the experimental setup is shown in Fig. 9.1a. A liquid crystal layer with thickness controlled by spacers is sandwiched between a conductive indium-tin-oxide (ITO) glass substrate and a gold film with nanohole arrays milled by using a focused ion beam. White light impinging from the ITO side of the sample will pass through the liquid crystal layer and then reach the nanostructures to excite different surface plasmon polariton modes. Upon applying voltages, electric fields are built up across the ITO substrate and the gold film, and the liquid crystal molecules tend to realign along the electric field direction. As a result, effective refractive index at the interface of gold/liquid crystal is changed and subsequently results in a variation on the surface plasmon dispersion relation, thus changing the excitation conditions of certain plasmon modes. This change is phenomenally reflected as a transmission modulation from the spectrum and provides us an active control of surface plasmon modes. De Sio et al. also reported that a carrier accumulation layer in the proximity of the ITO substrate can be formed under an external electric field, hence modifying the effective refractive index around the metallic nanostructures and affecting their plasmonic resonances subsequently [58]. To achieve an additional degree of control freedom, dual-frequency liquid crystal (DFLC) is, therefore, much preferred in designing an active plasmonic device, because it responds to the applied voltages with different frequencies. DFLC can change its sign of dielectric anisotropy from positive to negative, or negative to positive, upon the frequency change of the applied electric fields [73, 74]. Complementary gold nanodisk and nanohole arrays have been demonstrated using electron-beam lithography followed by lift-off process [49, 51]. By overlaying a DFLC layer on these nanodisk and nanohole structures, reversible tuning of plasmonic resonances and transmission have been demonstrated. Figure 9.1b,c shows the typical gold nanodisk arrays and corresponding plasmonic resonances with the DFLC overlayer. In this hybrid system, homeotropic alignment of DFLC was set at the initial state by using a self-assembly alignment layer of hexadecyl trimethyl ammonium bromide (HTAB). When the frequency of the applied electric field is lower than the crossover frequency of DFLC, all the liquid crystal molecules will keep the homeotropic alignment, that is, perpendicular to the substrate, which is independent of applied voltages. Once the applied frequency exceeds the crossover frequency, the liquid crystal molecules tend to prefer the homogenous alignment, that is, parallel to the substrate, which is dependent on applied voltages. As a result, the switching/tuning effect with various extinctions/transmissions can be easily obtained by varying applied voltages.

Although the electrically tunable plasmonic devices give clear resonance shift or intensity modulation, there is still much room

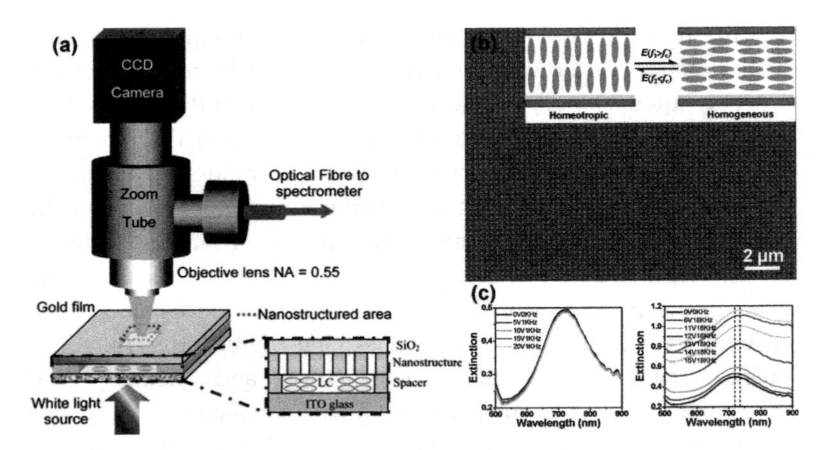

Figure 9.1 (a) Schematic and experimental setup of a plasmonic switch consisting of nanostructures and liquid crystals. Electric field is generated by applying voltages across the gold film and bottom ITO glass. (b) SEM image of gold nanodisk arrays. The insets show the working mechanism of DFLC. (c) The plasmonic resonance change under different voltage and frequency combinations. Figure (a) reprinted with permission from Ref. [56]. Copyright 2008, American Chemical Society, and figures (b) and (c) reprinted with permission from Ref. 51. Copyright 2010, AIP Publishing LLC.

for further improvement. Khatua and coworkers have demonstrated a new approach to modulate the polarized scattering intensities of individual gold nanorods by 100% using liquid crystals with applied voltages [59]. A critical step in their method is to employ a planar electrode geometry that allows for in-plane switching of the liquid crystal director. With the applied external voltage as low as 4 V, the intensity of the longitudinal surface plasmon resonance in the gold nanorod can be reversibly modulated with 100% efficiency (complete on/off switching), as shown in Fig. 9.2. This efficient intensity modulation is realized by an electric-field-induced phase transition from a homogenous to a twisted nematic phase of the liquid crystal, which causes an orthogonal rotation of the scattered light polarization independent of the nanorod orientation. This strategy could be readily translated to other plasmonic architectures with more complex designs and various plasmonic elements for the electrical manipulation of light in structures with nanoscale dimensions.

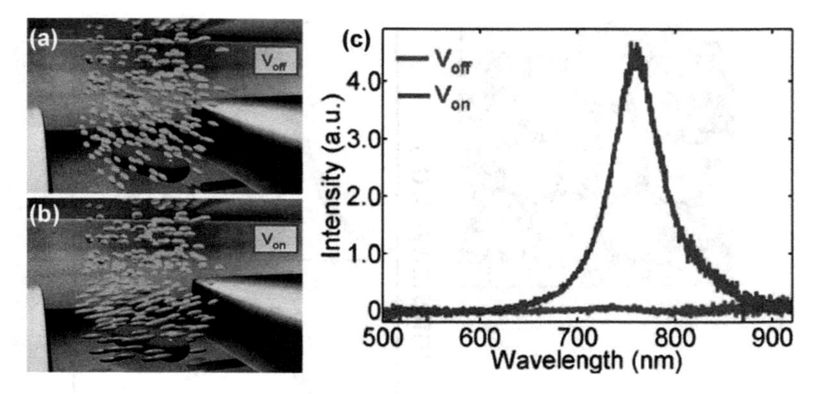

Figure 9.2 (a) The homogenous nematic phase in the V_{off} state; (b) the twisted nematic phase in the V_{on} state; and (c) the measured scattering spectra in the V_{on} and V_{off} states. Reprinted with permission from Ref. [59]. Copyright 2011 American Chemical Society.

Recently, a plasmonic Fano switch has also been demonstrated with a more specific design [75]. The device consists of a specifically designed cluster of gold nanoparticles fabricated using electron-beam lithography. The cluster comprises a large hemi-circular disc surrounded by seven smaller nanodisks as shown in Fig. 9.3a. Interactions between LSPRs of the individual nanoparticles within a cluster lead to a so-called Fano resonance, which is a result of the near-field coupling between collective "bright" and "dark" plasmon modes of the cluster. By breaking the symmetry of the nanoparticle cluster through the hemi-circular center disc, the Fano resonance is polarization-dependent and can only be observed for one polarization of the incident light. As a result, no Fano resonance appears in the light spectrum, for incident light polarized at 90° to this direction. The nanoparticle clusters are incorporated into liquid crystals in which the molecules at the device interface can be rotated in plane by 90° when an ac voltage of about 6 V is applied. The field creates a twist in the overall alignment direction of the crystals, which leads to a phase transition from "homogenous nematic" (voltage off) to "twisted nematic" (voltage on). Due to the birefringence of the liquid crystal, the voltage-induced phase transition causes an orthogonal rotation of the scattered light from the plasmonic clusters as it travels through the device. This results in

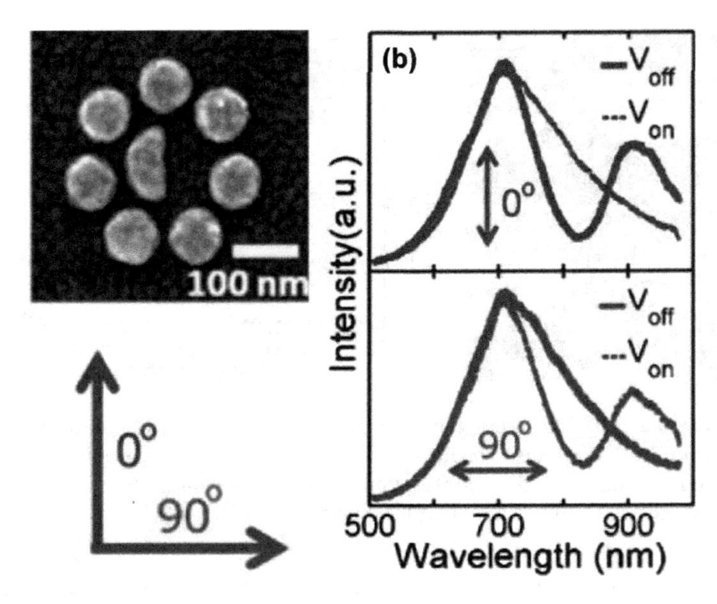

Figure 9.3 (a) SEM image of the octamer structure (top), which shows Fano-like and non-Fano-like spectra with polarization at 0° and 90° (bottom); (b) the scattering spectra of the octamer structure in the V_{on} and V_{off} states measured at a detection: the homogenous nematic phase in the V_{off} state (top) and the twisted nematic phase in the V_{on} state (bottom). Reprinted with permission from Ref. [75]. Copyright 2012 American Chemical Society.

switching between the optical response with and without the Fano resonance, as shown in Fig. 9.3b.

9.2.2 *Optical Tuning*

All-optical tuning method has been widely applied in liquid-crystal-based optical elements such as spatial light modulator, filter, reflector, etc.; and it has many advantages such as noncontact tuning, low power consumption, and friendly integration, making it an exciting concept to be applied on light-driven liquid-crystal-based devices. As compared to the electrical tuning method, the light-driven method requires: (1) no conductive ITO substrates, (2) low power consumptions, and (3) large operation windows covering UV to mid-IR. Azobenzene and its derivatives are a widely used guest in a liquid crystal host. They have a trans-cis reversible isomerization dynamic behavior upon exposure to a UV or visible light beam. The isomerization will disrupt the local order of the surrounding liquid crystal molecules in the mixture, resulting in realignment of those liquid crystal molecules to exhibit a refractive index change. Such an index change has been utilized for dynamic control of many photonic devices, such as switchable gratings [76–80] and photonic crystals [81, 82]. It is also straightforward for dynamic control of surface plasmons [53, 83, 84]. One example of a light-driven plasmonic switch has been demonstrated by Hsiao et al. [53] with a light-responsive liquid mixture consisting of nematic liquid crystals and an azobenzene derivative. 4-butyl-4′-methyl-azobenzene (BMAB) is used to induce the switching effect. Figure 9.4a shows the spectral changes of BMAB under UV exposure that reflects its trans-cis isomerization process. Figure 9.4b shows the extinction spectra of a photoresponsive liquid crystal/gold nanodisk array (see the inset of Fig. 9.4b) cell at normal incidence of a probing beam before (solid curve) and after (dashed curve) the light pump ($\lambda = 420$ nm, $I =$ 20 mW) at the incident angle of 45°. One can observe a 30 nm blue-shift of the extinction peak. Figure 9.4c shows the spectral changes of extinction in another azo dye (methyl red) doped grating integrated with the same gold nanodisk array [84]. The switching effect before and after the light pump is shown in Fig. 9.4d with both switching "ON" and "OFF" time less than 4 s, which is the typical response time for the azobenzene and its derivatives under the continuous light pump. However, under the pulsed laser pump, the response time of azobenzene-doped liquid crystals could reach nanoseconds since the photoisomerization of azobenzene from trans-state to cis-state could undergo the pathway of either $\pi - \pi^*$ rotation or $n - \pi^*$ inversion with different response speeds [85].

Based on the same working mechanism, Liu et al. have also demonstrated light-driven plasmonic color filters with high optical transmission and narrow bandwidth by overlaying a layer of photoresponsive liquid crystals on gold annular aperture arrays (AAAs). The schematic of the light-driven plasmonic color filters and experimental setup is shown in Fig. 9.5a. The enlarged Part I in Fig. 9.5a shows the fabricated square pattern of gold AAAs using focused ion beam lithography. The inner and outer radii of each individual aperture are labeled as r_{in} and r_{out}, respectively. The magnified Part II in Fig. 9.5a shows the working mechanism of the optical driving

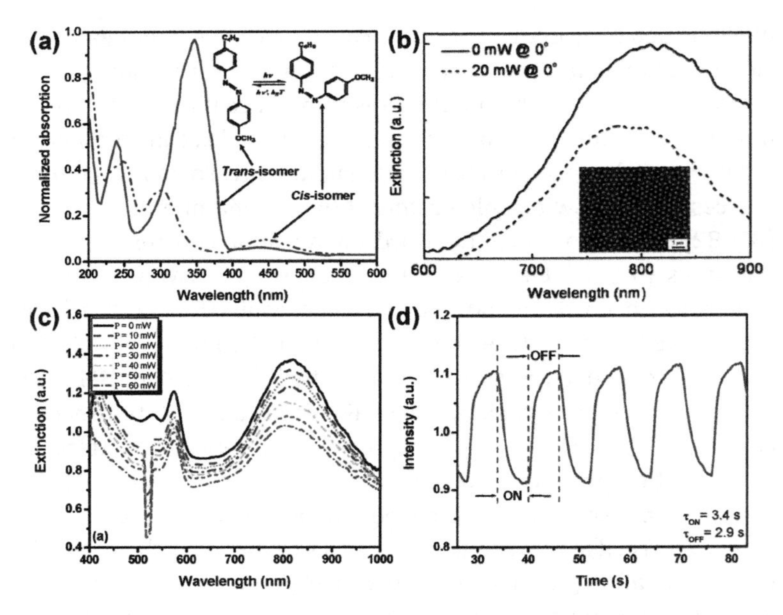

Figure 9.4 (a) The spectral changes of BMAB under UV exposure that reflects its trans-cis isomerization process. (b) Extinction spectra (normal incidence) of a photoresponsive liquid crystal/gold nanodisk array cell before (solid curve) and after (dashed curve) the application of a 20 mW pump light ($\lambda = 420$ nm) at an incident angle of 45°; blue-shift of 30 nm in the extinction peak can be observed. The alignment of liquid crystal molecules is achieved mechanically using photoresponsive azobenzene. (c) Spectral changes of extinction in another azo dye (methyl red) doped grating integrated with the same gold nanodisk array. (d) Switching effect of azo-dyedoped holographic gratings upon laser excitation. Figure (a) reprinted from Ref. [39], with permission from John Wiley & Sons, Inc., Copyright 2012; figure (b) reprinted from Ref. [53], with permission from John Wiley & Sons, Inc., Copyright 2008; figure (c) and (d) reprinted from Ref. [84]. Copyright 2011, American Chemical Society.

process: a reversible nematic–isotropic phase transition induced by the *trans-cis* photoisomerization of the photochromic liquid crystals. The gold AAAs with various aperture sizes and periods have been reported to generate different colors in the visible range. The addition of photoresponsive liquid crystals will then make the transmission of color filters optically tunable. The photosensitive liquid crystal mixture has similar composition as other mixtures discussed above.

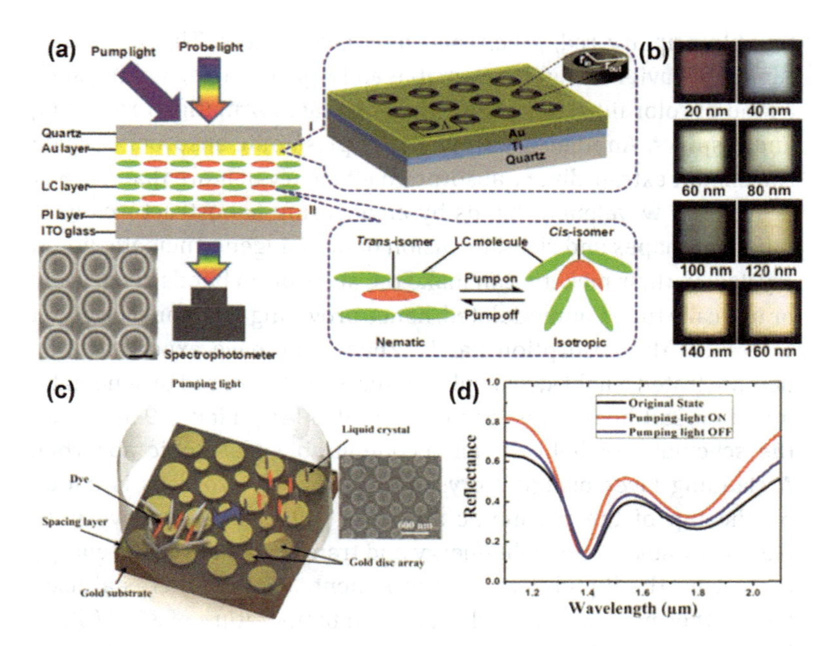

Figure 9.5 (a) Light-driven plasmonic color filter tuned by liquid crystals. The nanostructure used is a new type AAA that can generate different colors by changing the aperture size and period of AAA. One example of AAA is given in the inset. (b) Different colors generated using plasmonic color filters. (c) One representative case of reconfigurable plasmonic absorber using liquid crystal. The small and big nanodisk arrays in the inset are designed to produce two absorption maximums to improve the performance of the plasmonic absorbers. (d) Light-sensitive liquid crystal mixture is used to tune the absorption dips in real time. Experimental results confirm a tuning range around 25 nm in the near-infrared range. Figures (a) and (b) reprinted from Ref. [39], with permission from John Wiley & Sons, Inc., Copyright 2012, and figures (c) and (d) reprinted with permission from Ref. [46]. Copyright 2012, AIP Publishing LLC.

With its large birefringence $\Delta n = 0.225$ (at the wavelength of 589 nm, $n_o = 1.521$, $n_e = 1.746$), it offers major intensity modulation of the transmitted color through different aperture arrays, as shown in Fig. 9.5b, where the intensity of each individual color generated from plasmonic structures can be further tuned, as confirmed by both simulations and experiments in the reported work [39]. Therefore, any color could be achieved through the composition of three tunable primary red, green, and blue color filters. This all-optical tuning behavior is highly reversible and reproducible, making such a kind of color filter promising in all-optical information processing and displays. Another useful device is plasmonic absorbers, which can exhibit extraordinary absorption efficiency of more than 90% at designated wavelength bands by engineering nanostructures with different shapes and sizes. Inclusion of birefringence nematic liquid crystals in their makeup can make the absorption bands electrically or optically tuned or modulated, hence providing additional freedom to control the absorption bands. Zhao et al. have experimentally demonstrated a light-driven plasmonic absorber based on a nematic liquid crystal host doped with an azo dye [46]. Figure 9.5c shows the schematic of light-driven reconfigurable plasmonic absorber. A cladding layer of liquid crystals mixed with azo dyes is added on the top of the plasmonic absorber. The use of azo dye is to effectively absorb the light energy and transfer the absorbed energy to heat up the liquid crystal environment. The liquid crystal used here is thermo-sensitive with transition temperature of 35°C (5CB). Liquid crystal will transform from a nematic to isotropic state after heating up, along with a gradual refractive index change from 1.61 to 1.56 with the increase in temperature. The refractive index change modifies the frequency selected surfaces as well as the resonance frequency of the resonant cells formed by top gold nanostructures, and a bottom gold layer. A noticeable shift in the absorption dips of around 25 nm has been confirmed in Fig. 9.5d. Note that the photosensitive liquid crystals with an azo group in their mesogenic structure exhibit a higher solubility compared to non-mesogenic azo dyes [86–88]. Furthermore, the orientation order of liquid crystals still remains high in the presence of mesogenic azo dopants, and their photoisomerization effect on the host liquid crystal is stronger than that of non-mesogenic dopants.

9.2.3 *Magnetic Tuning*

Nanosystems that combine magnetic and plasmonic functionalities have become a hot topic for active plasmonics research in recent years. Magnetic-field-induced modifications of the optical properties of materials were first observed by Faraday [89] and Kerr [90,

91]. They detected a change in the polarization state of the transmitted (Faraday effect) or reflected light (Kerr effect) when a magnetic field was applied to a glass or to a ferromagnetic metal, respectively. Since then, magneto-optics has been playing an important technological role in different areas, especially in information storage. The marriage of plasmonics and magnetism is promising to enable the development of active plasmonics. Different kinds of magneto-plasmonic materials have been proposed so far, with the plasmonic material being a transition metal (Au, Ag, etc.) and the ferromagnetic component [92–98]. As a matter of fact, all materials exhibit magneto-optical activity, but the intensity of this response depends mainly on their magnetic nature. It is well known that liquid crystals have pronounced magneto-optical properties. An external magnetic field can induce realignment of the liquid crystal molecules. Liu and his coworkers have demonstrated the bulk self-alignment of dispersed gold nanorods imposed by the intrinsic cylindrical micelle self-assembly in nematic and columnar hexagonal lyotropic liquid crystals [99]. With the help of an external magnetic field, alignment and realignment of the liquid crystal matrix with the ensuing long-range orientational order of well-dispersed plasmonic nanorods have been achieved. Figure 9.6a,b shows the schematic illumination of the gold nanorods in the liquid crystal matrix. The unidirectional alignment of nanorods with high three-dimensional order parameter values approaching 0.9 has been achieved in bulk samples of thickness ranging from several micrometers to millimeters, over large (approximately square inch) sample areas. This unidirectional alignment of nanorods gives rise to the strong polarization sensitivity of surface plasmon resonance effects, as shown in Fig. 9.6c,d. From Fig. 9.6c,d, the aligned gold nanorods in the liquid crystal matrix exhibit extinction spectra varying with the angle between the polarizer and gold nanorods aligned along director **n**. This results in a switchable polarization-sensitive plasmon resonance exhibiting stark differences from that of the same nanorods in isotropic fluids.

9.2.4 *Acoustic Tuning*

Acoustic tuning is another promising way to control the plasmonics, which is based on the modulation of refractive index profile of

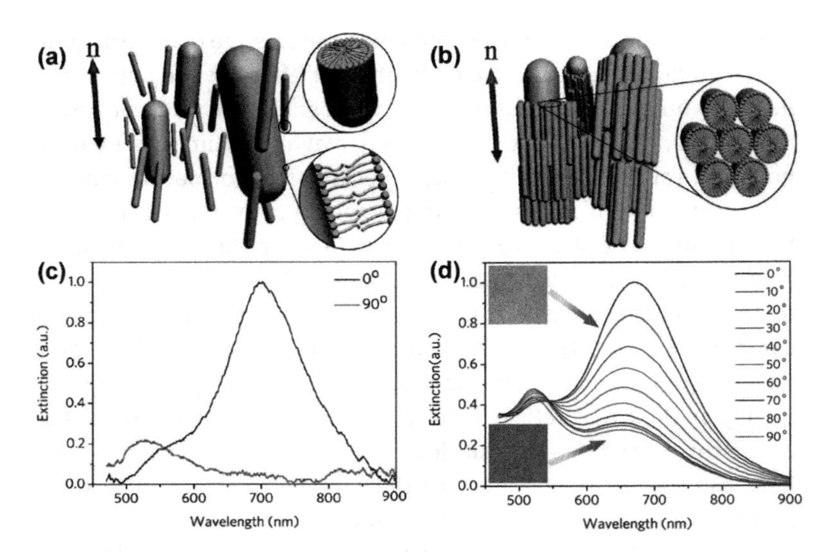

Figure 9.6 Schematic illustration of god nanorods alignment in a nematic liquid crystal (a) and a columnar hexagonal lyotropic liquid crystal (b). Experimental extinction spectra of god nanorods in nematic (c) and lyotropic (d) liquid crystals realigned by a magnetic field. Reprinted with permission from Ref. [99]. Copyright 2010 American Chemical Society.

the propagation medium. Examples are acousto-optical devices in their bulk configuration, such as acousto-optical tunable filtering of laser sources or pulse shaping in ultrashort lasers [100]. In integrated optics, most of the reported solutions rely on the use of surface acoustic waves (SAWs) that can be easily generated atop a piezoelectric surface. Gérard et al. have numerically studied the acoustic effect in periodically nanostructured metallic films that exhibit extraordinary optical transmission (EOT) [101]. Their numerical results showed that low frequency acoustic waves can significantly tune the resonance frequency of the EOT structure and modulate the transmitted intensity. In their studies, the position of the transmission maximum follows a linear law, and an elasto-optic variation of the refractive index of $\Delta n = 0.0015$ is necessary to shift the resonance by 1 nm. Figure 9.7a shows the maximum displacement of the resonance, that is, the spectrum when $\Delta n = \pm \Delta n_0$. In order to see the modulation effect, Fig. 9.7b shows the transmission efficiency versus time at two given wavelengths, taking

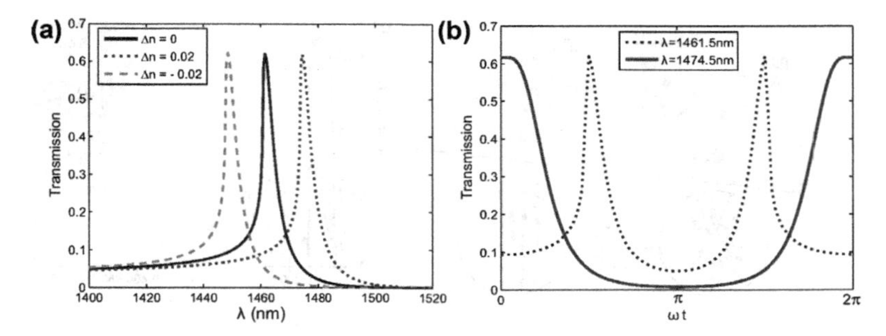

Figure 9.7 Modulation and tunability of the EOT by a low frequency SAW. (a) Transmission efficiency through the EOT structure for different values of the refractive index of $LiNbO_3$: no variation (solid line), $\Delta n = 0.02$ (dotted line), and $\Delta n = -0.02$ (dashed line). (b) Transmission efficiency versus time for a maximum modulation $\Delta n = 0.02$ at $\lambda = 1461.5$ nm (dotted line) and $\lambda = 1447.5$ nm (solid line). Reprinted figures with permission from Ref. [101]. Copyright 2007 by the American Physical Society.

account of the temporal evolution of the elasto-optic modulation $\Delta n = \Delta n_0 \cos(\omega t)$. It appears that for $\Delta n_0 = 0.02$, a significant modulation of the transmission is observed. For $\lambda = 1461.5$ nm, the transmission efficiency is modulated from about 61% at maximum to less than 10%. The rejection power is significantly improved if the wavelength is set to $\lambda = 1474.5$ nm. In this case, the transmission drops to about 0.8% at $\omega t = \pi$.

Although there is no experimental demonstration of acoustically tunable plasmonic devices so far, acoustic waves appear to be an interesting route to realize controllable plasmonic devices. A big challenge for acoustic tuning on a piezoelectric surface is that the refractive index change of the piezoelectric materials is quite small, which gives rise to a very limited resonance shift. To achieve a large resonance shift, researchers have to explore other mechanism to obtain a large refractive index change. A possible way is to use SAW to drive liquid crystals. It is well known that liquid crystal alignment can be conveniently controlled by an acoustic wave. The realignment of liquid crystals caused by the acoustic wave can produce a refractive index change an order of magnitude larger than that in a piezoelectric material. Although liquid crystal driven by SAWs has not been utilized to control plasmon resonances, we put it here as an

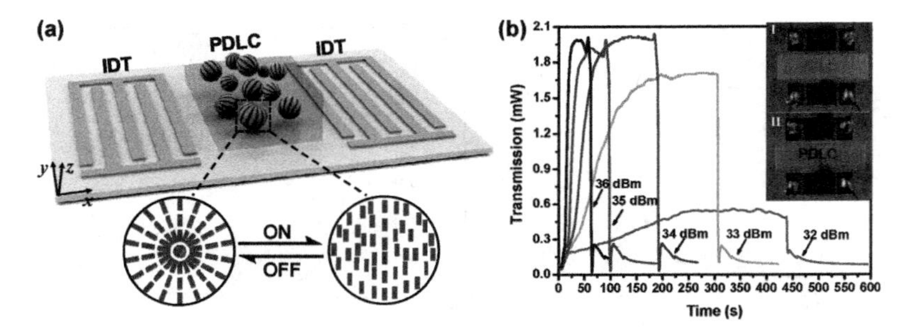

Figure 9.8 (a) Light shutter based on acoustic modulation of PDLC. The shutter effect is caused by realignment of liquid crystal molecules. (b) Experimental demonstration of acoustic-driven light shutter. Letters of "PDLC" beneath the PDLC film can be clearly observed after SAW applied. Transmission changes before and after application of SAW are shown. Response time can be estimated. Figure reprinted from Ref. [50], with permission from John Wiley & Sons, Inc., Copyright 2011.

insight into prospective integration of liquid crystal with micro, and nanotechnologies. Liu et al. [50] have demonstrated a SAW-driven light shutter based on polymer-dispersed liquid crystal (PDLC). A schematic of the designed light shutter is shown in Fig. 9.8a. It consists of a PDLC film layer and two inter-digital transducers (IDTs) on a piezoelectric substrate. The PDLC film consists of liquid crystal droplets randomly distributed in a polymer matrix. Before applying SAW, the PDLC film exhibits strong scattering due to the refractive index mismatch between the polymer matrix and liquid crystal droplets, thus demonstrating a nontransparent state. When a radio frequency signal is applied to IDT, SAW gets excited and then propagates along the surface. When the propagating SAW encounters the PDLCs, a longitudinal wave is induced and leaks into the PDLCs. The longitudinal wave will cause the liquid crystal molecules to realign to eliminate the refractive index mismatch, that is, their ordinary refractive index matches that of the polymer. As a result, the PDLC film becomes completely transparent. The acoustic-wave-induced streaming inside the liquid-crystal-rich regions as well as the temperature change caused by the attenuation of this longitudinal wave are believed to be key factors to reorientate the alignment of liquid crystal molecules. Figure 9.8b shows the experimental results marking the "ON" and "OFF" states of the SAW-driven shutter. The word "PDLC" can be clearly observed by turning "ON" the shutter, and the film is opaque when the shutter is "OFF." We believe that such a SAW-driven liquid crystal mechanism will be applicable to plasmonic devices as well. Various tunable plasmonic devices based on such a mechanism will be demonstrated in near future.

9.2.5 *Thermal Tuning*

Temperature also plays an important role in affecting the liquid crystal refractive indices. As the temperature increases, the extraordinary refractive index n_e behaves differently from the ordinary refractive index n_o. The derivative of n_e (i.e., $\partial n_e/\partial T$) is always negative. However, $\partial n_o/\partial T$ changes from negative to positive as the temperature exceeds the crossover temperature [102, 103]. For many liquid crystals, the temperature-dependent refractive indices can be accurately controlled, hence providing another effective means to develop active plasmonics. Figure 9.9 gives a typical temperature-dependent behavior of the liquid crystal refractive index changes [104]. We can clearly observe that the optical anisotropy decreases as the temperature increases and finally disappears (i.e., $n_e = n_o$) at the isotropic phase regardless of wavelength. Based on this mechanism, the Altug group has demonstrated the thermal tuning of SPPs using liquid crystals [105]. The experimental setup is illustrated in Fig. 9.10a,b. By varying the temperature within the nematic phase from 15°C to 33°C, they demonstrated a refractive index change as large as $\approx$0.0317 (Fig. 9.10c,d), thus enabling a tuning of plasmonic wavelength as large as $\approx$19 nm. The ability to control the order of liquid crystal molecules from nematic to isotropic phase provides an efficient way of spectral tuning. At the phase transition temperature, more than 12 nm shift has been achieved via changing the temperature by only $\approx$1°C corresponding to a refractive index change of $\approx$0.02 (Fig. 9.10c,d). It has been known that plasmonic nanoparticles can induce a photo-thermal heating effect, which depends on the geometries of the nanoparticles strongly and has been extensively investigated [106–108]. It is expected that with the involvement of liquid crystals, such a photo-thermal heating effect will also affect the effective refractive index of the liquid crystals and hence

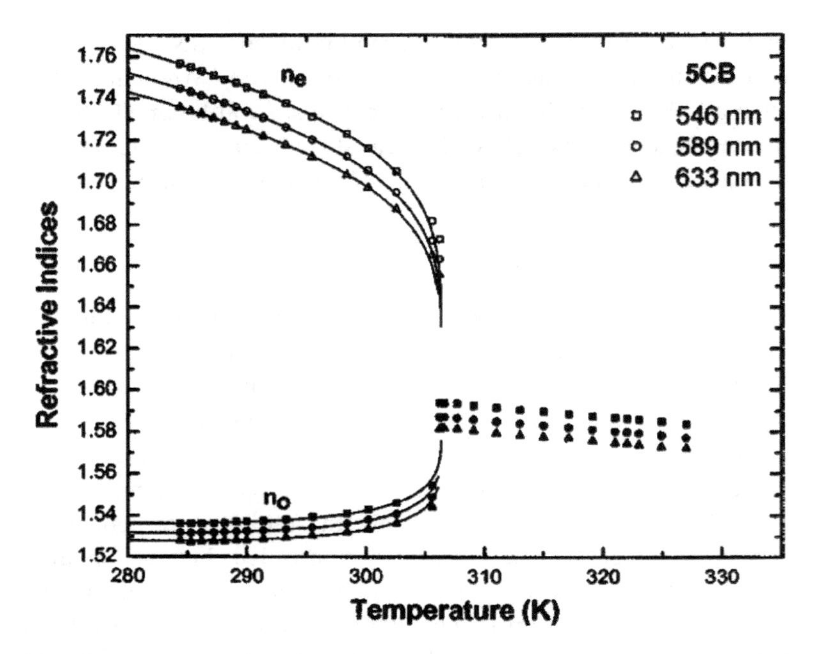

Figure 9.9 Temperature-dependent refractive indices of 5CB at $\lambda = 546$, 589, and 633 nm. Squares, circles, and triangles are experimental data for refractive indices measured at $\lambda = 546$, 589, and 633 nm, respectively. Reprinted with permission from Ref. [104]. Copyright [2004], AIP Publishing LLC.

tune the plasmonic resonances as well. De Sio and his coworkers have developed an innovative method for simultaneously achieving optical control of both the selective reflection of a cholesteric liquid crystal and the plasmonic resonance of gold nanorods [109]. By surface modification of the gold nanorods, when the gold nanorods are dispersed in a cholesteric liquid crystal, the self-organization process of cholesteric liquid crystals is able to effectively induce confinement and topological organization of gold nanorods along the sample structural defects. Figure 9.11a shows the schematic of all-optical setup for sample characterization. Under an optical pump, a local photo-thermal effect is induced by the presence of a near-infrared selective plasmonic resonance. The local heating will then cause a variation of the surrounding medium refractive index, with a consequent shift of the plasmonic resonance. At the same time,

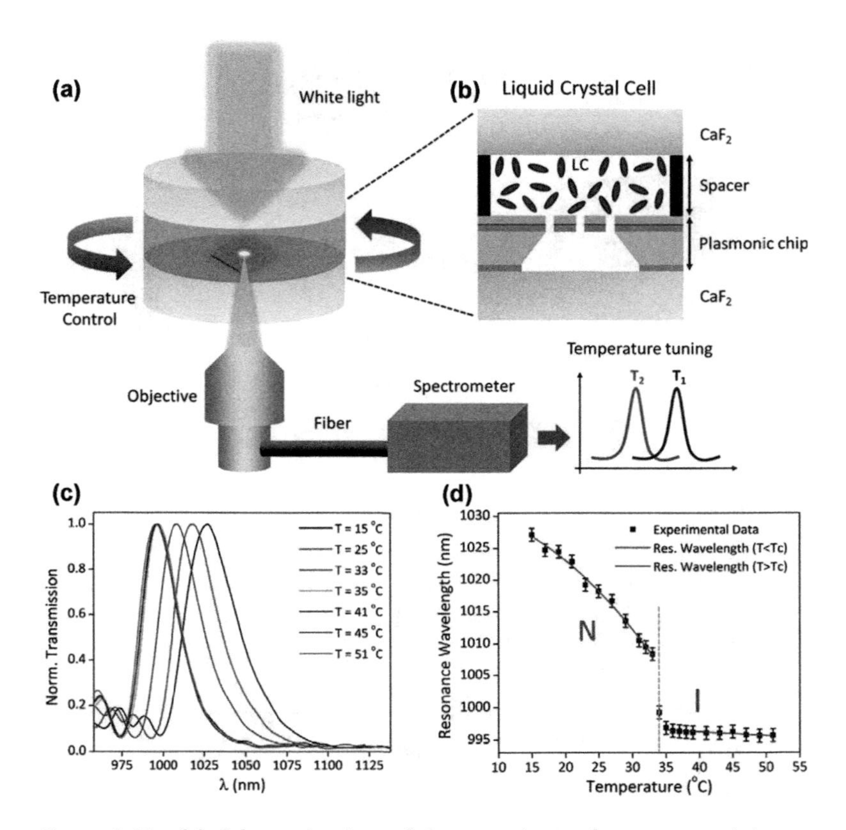

Figure 9.10 (a) Schematic view of the experimental setup containing a liquid crystal cell thermally controlled by a heat bath, white light incident source collected by an objective lens and a spectrometer collecting the plasmonic response for different temperature values; (b) zoomed schematic of the liquid crystal cell between the upper CaF_2 window and the plasmonic chip; (c) experimental transmission response of the nanohole array for different temperatures from 15°C to 51°C; and (d) spectral position of the transmission resonance for different temperatures. In (c), the red curve denotes the resonance points below the transition temperature T_c (showing the quadratic behavior) and the blue curve denotes the points above T_c (showing the linear behavior). The green dashed line represents the transition between nematic (N) and isotropic (I) phases. Reprinted from Ref. [105], with permission from John Wiley & Sons, Inc., Copyright 2013.

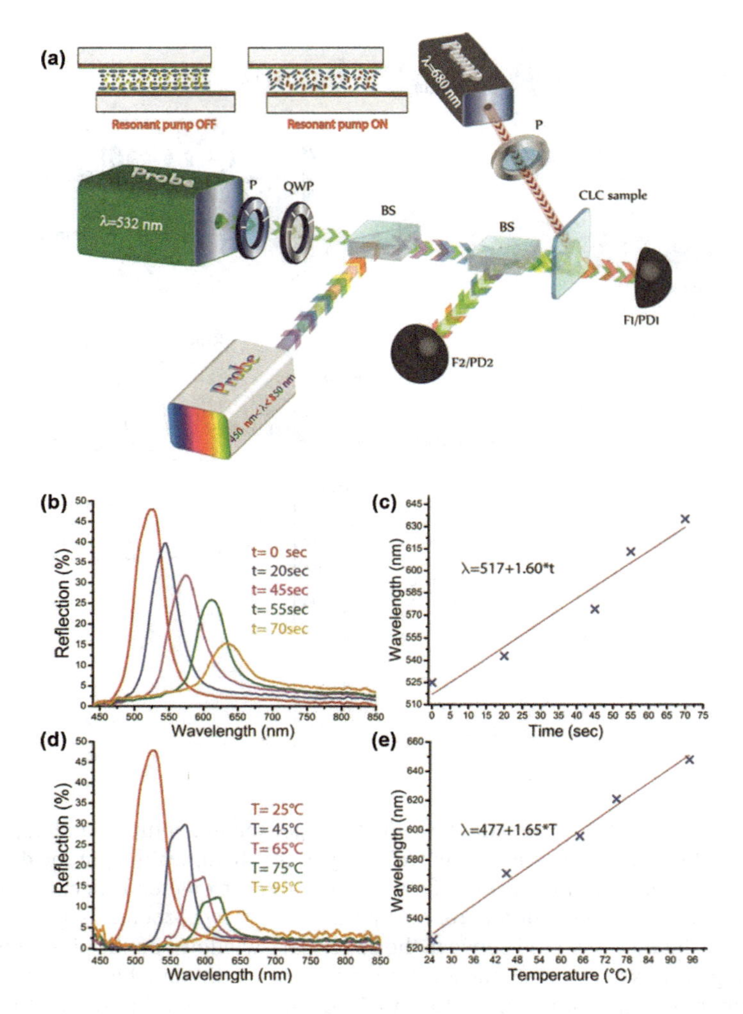

Figure 9.11 (a) All-optical setup for sample characterization. P: polarizer; QWP: quarter waveplate; BS: beam splitter; $F_{1,2}$: transmission and reflection fibers; $PD_{1,2}$: photodetectors. In the top-left, a sketch of the sample configuration with and without the action of the pump beam is shown. Reflection spectra of the sample for different values of illumination time (b) and temperature (d); linear fit of the position of the center of the reflection band versus illumination time (c) and temperature (e). Figure reprinted from Ref. [109], with permission from John Wiley & Sons, Inc., Copyright 2013.

the photoexcitation of gold nanorods induces the formation of a heated electron gas that subsequently cools rapidly by exchanging energy with the gold nanorod lattice; this process is followed by a phonon-photon interaction, where the gold nanorod lattice cools rapidly by exchanging energy with the surrounding medium [109]. This local heating induces a pitch elongation of the cholesteric liquid crystal (inset Fig. 9.11, resonant pump on) with a consequent linear red-shift of the reflection band, as shown in Fig. 9.11b,c. This underlying mechanism is also confirmed by a control experiment, wherein a linear red-shift of the reflection band was observed when the same sample was heated from 25°C up to 95°C (see Fig. 9.11d,e). The working mechanism can be exploited to control the position of the selective reflection exhibited by the cholesteric liquid crystal configuration. This synergy between plasmonics and photonics can be potentially used for building up a method for detecting the temperature around nanoparticles under optical illumination.

9.3 Outlook

In summary, we have reviewed the recent research progress and achievements in liquid-crystal-based plasmonics. Different tuning schemes of the liquid-crystal-based plasmonic devices have been categorized and discussed. The versatile driving methods of the liquid crystals provide enormous freedom in the design and implementation of the plasmonic devices. Although tremendous efforts have been focused on the development of liquid-crystal-based active plasmonic devices in recent years, many challenges still remain before they can be efficiently used in practice. In the following subsections, we provide a more specific description of these challenges as well as future demands.

9.3.1 *Current Challenges*

Currently, most liquid-crystal-based plasmonic devices are fabricated by mechanically assembling liquid crystals on passive plasmonic nanostructures on a substrate. However, large-area uniform liquid crystal alignment on the nanostructures is still a big challenge, given the fact that only the liquid crystal molecular layer that is near (generally <100 nm) to the plasmonic nanostructures affects the plasmonic signals. To achieve large-area uniform liquid crystal alignment in plasmonic devices, one possible way is to co-self-assemble liquid crystals and plasmonic nanoparticles [99, 110–113]. Based on the co-self-assembly method, uniform areas up to 100 μm^2 have been achieved, which is promising for device development.

9.3.2 *Future Demands*

9.3.2.1 Fast response liquid crystals

High speed is always demanding for active plasmonic devices, especially for future development of nanophotonic circuitry/chips with multiple functions assembled. Currently, most liquid-crystal-based plasmonic devices have a millisecond scale response using the electric-field-driven method. To have a faster speed, one could choose some special liquid crystals, such as ferroelectric liquid crystals, which in general have microsecond response time. By confining the liquid crystal in a polymer matrix, it will also help increase the response speed. For example, millisecond response time has been achieved in a PDLC system, in which microscale liquid crystal droplets are randomly confined in a polymer matrix [114, 115]. While in a holographic PDLC system, in which nanoscale liquid crystal droplets are periodically confined in a polymer matrix, microsecond response time has been demonstrated [116–118]. More recently, polymer network liquid crystal (PNLC) [119] and polymer-stabilized blue phase liquid crystal (BPLC) [120] have received much attention since both systems can achieve sub-millisecond response time.

9.3.2.2 Large birefringence liquid crystals

Currently, liquid crystals employed in most reported work have birefringence of 0.1–0.2. Although such a birefringence has already given detectable changes in terms of peak shift or intensity, there is much room for further improvement. To achieve more pronounced changes, liquid crystals possessing larger birefringence are highly desired. Thus far, liquid crystals with large birefringence of >0.4 [121–123] and even >0.7 [124] have been reported. We believe that much enhanced performance, such as large peak shift and intensity modulation, will be achieved by integrating the large birefringence liquid crystal into plasmonic nanostructures, hence beneficial to the development of active plasmonics.

9.3.2.3 Multi-function or multi-control integration

At the current stage, a single function of liquid-crystal-based plasmonic devices has been demonstrated in terms of proof-of-concept. Looking ahead, it will be great that a single device could do multiple functions. Liquid-crystal-enabled plasmonic devices look promising for achieving multiple functions. In addition, multi-mode controls of liquid crystals will also lend a convenient hand and allow us to choose their functions freely under certain circumstances. For example, azo-dye-doped liquid crystals or azobenzene liquid crystals can respond to not only electric fields but also light [125], which gives us more freedom to control them.

Acknowledgments

Liu, Y.J. and Leong, E.S.P. thank the funding support from Joint Council Office (JCO) of the Agency for Science, Technology and Research (A*STAR) under the grant No. 12302FG012.

References

1. Barnes, W. L., Dereux, A., and Ebbesen, T. W. (2003). Surface plasmon subwavelength optics, *Nature*, **424**, pp. 824–830.
2. Ebbesen, T. W., Lezec, H. J., Ghaemi, H. F., Thio, T., and Wolff, P. A. (1998). Extraordinary optical transmission through sub-wavelength hole arrays, *Nature*, **391**, pp. 667–669.
3. Bozhevolnyi, S. I. (2009). *Plasmonic Nanoguides and Circuits*, Pan Stanford Publishing, Singapore.
4. Degiron, A., and Smith, D. R. (2007). Numerical simulations of long-range plasmonic transmission lines, in *Surface Plasmon Nanophotonics*

(Brongersma, M. L., and Kik, P. G., eds), Springer, Netherlands, pp. 55–71.

5. Garcia-Vidal, F. J., Martin-Moreno, L., and Pendry, J. B. (2005). Surfaces with holes in them: New plasmonic metamaterials, *J. Opt. A Pure Appl. Opt.*, **7**, pp. S97–S101.
6. Cai, W., and Shalaev, V. (2010). *Optical Metamaterials*, Springer, Berlin, Germany.
7. Smith, S. J., and Purcell, E. M. (1953). Visible light from localized surface charges moving across a grating, *Phys. Rev.*, **92**, pp. 1069–1070.
8. Veselago, V. G. (1967). The electrodynamics of substances with simultaneous negative values of ε and μ, *Usp. Fiz. Nauk*, **92**, pp. 517–526.
9. Zayats, A. V., Smolyaninov, I. I., and Maradudin, A. A. (2005). Nano-optics of surface plasmon-polaritons, *Phys. Rep.*, **408**, pp. 131–314.
10. Smolyaninova, V. N., Smolyaninov, I. I., Kildishev, A. V., and Shalaev, V. M. (2010). Broadband transformation optics devices, *Materials*, **3**, pp. 4793–4810.
11. Dutta, N. Mirza, I. O. Shi, S. and Prather, D. W. (2010). Fabrication of large area fishnet optical metamaterial structures operational at near-IR wavelengths, *Materials*, **3**, pp. 5283–5292.
12. Alu, A., and Engheta, N. (2011). Emission enhancement in a plasmonic waveguide at cut-off, *Materials*, **4**, pp. 141–152.
13. Si, L., Jiang, T., Chang, K., Chen, T., Lv, X., Ran, L., and Xin, H. (2011). Active microwave metamaterials incorporating ideal gain devices, *Materials*, **4**, pp. 73–83.
14. Pendry, J. B. (2000). Negative refraction makes a perfect lens, *Phys. Rev. Lett.*, **85**, pp. 3966–3969.
15. Fang, N., Lee, H., Sun, C., and Zhang, X. (2005). Sub-diffraction-limited optical imaging with a silver superlens, *Science*, **308**, pp. 534–537.
16. Pendry, J. B., and Smith, D. R. (2006). The quest for the superlens, *Sci. Am.*, **295**, pp. 60–67.
17. Vincenti, M. A., D'Orazio, A., Cappeddu, M. G., Akozbek, N., Bloemer, M. J., and Scalora, M. (2009). Semiconductor-based superlens for subwavelength resolution below the diffraction limit at extreme ultraviolet frequencies, *J. Appl. Phys.*, **105**, pp. 103103.
18. Liu, Z., Durant, S., Lee, H., Pikus, Y., Fang, N., Xiong, Y., Sun, C., and Zhang, X. (2007). Far-field optical superlens, *Nano Lett.*, **7**, pp. 403–408.

19. Liu, Z., Durant, S., Lee, H., Pikus, Y., Xiong, Y., Sun, C., and Zhang, X. (2007). Experimental studies of far-field superlens for sub-diffractional optical imaging, *Opt. Express*, **15**, pp. 6947–6954.
20. Schurig, D., Mock, J. J., Justice, B. J., Cummer, S. A., Pendry, J. B., Starr, A. F., and Smith, D. R. (2006). Metamaterial electromagnetic cloak at microwave frequencies, *Science*, **314**, pp. 977–980.
21. Pendry, J. B., Schurig, D., and Smith, D. R. (2006). Controlling electromagnetic fields, *Science*, **312**, pp. 1780–1782.
22. Jacob, Z., and Narimanov, E. E. (2008). Semiclassical description of nonmagnetic cloaking, *Opt. Express*, **16**, pp. 4597–4604.
23. Smolyaninov, I. I., Smolyaninova, V. N., Kildishev, A. V., and Shalaev, V. M. (2009). Anisotropic metamaterials emulated by tapered waveguides: Application to optical cloaking, *Phys. Rev. Lett.*, **102**, pp. 213901.
24. Zhu, G., Lin, Z., Jing, Q., Bai, P., Pan, C., Yang, Y., Zhou, Y., and Wang, Z. L. (2013). Toward large-scale energy harvesting by a nanoparticle-enhanced triboelectric nanogenerator, *Nano Lett.*, **13**, pp. 847–853.
25. Walter, M. J., Borys, N. J., van Schooten, K. J., and Lupton, J. M. (2008). Light-harvesting action spectroscopy of single conjugated polymer nanowires, *Nano Lett.*, **8**, pp. 3330–3335.
26. Aubry, A., Lei, D. Y., Fernandez-Dominguez, A. I., Sonnefraud, Y., Maier, S. A., and Pendry, J. B. (2010). Plasmonic light-harvesting devices over the whole visible spectrum, *Nano Lett.*, **10**, pp. 2574–2579.
27. Andreussi, O. Biancardi, A. Corni, S. and Mennucci, B. (2013). Plasmon-controlled light-harvesting: Design rules for biohybrid devices via multiscale modeling, *Nano Lett.*, **13**, pp. 4475–4484.
28. Kim, I., Bender, S. L., Hranisavljevic, J., Utschig, L. M., Huang, L., Wiederrecht, G. P., and Tiede, D. M. (2011). Metal nanoparticle plasmon-enhanced light-harvesting in a photosystem I thin film, *Nano Lett.*, **11**, pp. 3091–3098.
29. Dang, X., Qi, J., Klug, M. T., Chen, P., Yun, D. S., Fang, N. X., Hammond, P. T., and Belcher, A. M. (2013). Tunable localized surface plasmon-enabled broadband light-harvesting enhancement for high-efficiency panchromatic dye-sensitized solar cells, *Nano Lett.*, **13**, pp. 637–642.
30. Si, G. Y., Zhao, Y. H., Lv, J., Lu, M., Wang, F., Liu, H., Xiang, N., Huang, T. J., Danner, A. J., Teng, J. H., and Liu, Y. J. (2013). Reflective plasmonic color filters based on lithographically patterned silver nanorod arrays, *Nanoscale*, **5**, pp. 6243–6248.

31. Huang, W., Qian, W., Jain, P. K., and El-Sayed, M. A. (2007). The effect of plasmon field on the coherent lattice phonon oscillation in electron-beam fabricated gold nanoparticle pairs, *Nano Lett.*, **7**, pp. 3227–3234.
32. Sun, S., and Leggett, G. J. (2004). Matching the resolution of electron beam lithography by scanning near-field photolithography, *Nano Lett.*, **4**, pp. 1381–1384.
33. Atay, T., Song, J., and Nurmikko, A. V. (2004). Strongly interacting plasmon nanoparticle pairs: From dipole–dipole interaction to conductively coupled regime, *Nano Lett.*, **4**, pp. 1627–1631.
34. Si, G. Y., Zhao, Y. H., Lv, J., Wang, F., Liu, H., Teng, J. H., and Liu, Y. J. (2013). Direct and accurate patterning of plasmonic nanostructures with ultrasmall gaps, *Nanoscale*, **5**, pp. 4309–4313.
35. Si, G. Y., Zhao, Y. H., Liu, H., Teo, S. L., Zhang, M. S., Huang, T. J., Danner A. J., and Teng, J. H. (2011). Annular aperture array based color filter, *Appl. Phys. Lett.*, **99**, pp. 033105.
36. Jiang, X., Gu, Q., Wang, F., Lv, J., Ma Z., and Si, G. (2013). Fabrication of coaxial plasmonic crystals by focused ion beam milling and electron-beam lithography, *Mater. Lett.*, **100**, pp. 192–194.
37. Si, G., Danner, A. J., Teo, S. L., Teo, E. J., Teng J., and Bettiol, A. A. (2011). Photonic crystal structures with ultrahigh aspect ratio in lithium niobate fabricated by focused ion beam milling, *J. Vac. Sci. Technol. B*, **29**, pp. 021205.
38. Si, G., Teo, E. J., Bettiol, A. A., Teng J., and Danner, A. J. (2010). Suspended slab and photonic crystal waveguides in lithium niobate, *J. Vac. Sci. Technol. B*, **28**, pp. 316–320.
39. Liu, Y. J., Si, G. Y., Leong, E. S. P., Xiang, N., Danner A. J., and Teng, J. H. (2012). Light-driven plasmonic color filters by overlaying photoresponsive liquid crystals on gold annular aperture arrays, *Adv. Mater.*, **24**, pp. OP131–OP135.
40. Liu, Y. J., Si, G. Y., Leong, E. S. P., Wang, B., Danner A. J., Yuan, X. C., and Teng, J. H. (2012). Optically tunable plasmonic color filters, *Appl. Phys. A*, **107**, pp. 49–54.
41. He, J., Huang, X., Li, Y., Liu, Y., Babu, T., Aronova, M. A., Wang, S., Lu, Z., Chen, X., and Nie, Z. (2013). Self-assembly of amphiphilic plasmonic micelle-like nanoparticles in selective solvents, *J. Am. Chem. Soc.*, **135**, pp. 7974–7984.
42. Gandra, N., Abbas, A., Tian, L., and Singamaneni, S. (2012). Plasmonic planet-satellite analogues: Hierarchical self-assembly of gold nanostructures, *Nano Lett.*, **12**, pp. 2645–2651.

43. Li, X., Cole, R. M., Milhano, C. A., Bartlett, P. N., Soares, B. F., Baumberg, J. J., and de Groot, C. H. (2009). The fabrication of plasmonic Au nanovoid trench arrays by guided self-assembly, *Nanotechnology*, **20**, pp. 285309.
44. Krasavin, A. V., and Zheludev, N. I. (2004). Active plasmonics: Controlling signals in Au/Ga waveguide using nanoscale structural transformations, *Appl. Phys. Lett.*, **84**, pp. 1416–1418.
45. Khoo, I. C. (2009). Nonlinear optics of liquid crystalline materials, *Phys. Rep.*, **471**, pp. 221–267.
46. Zhao, Y., Hao, Q., Ma, Y., Lu, M., Zhang, B., Lapsley, M., Khoo, I. C., and Huang, T. J. (2012). Light-driven tunable dual-band plasmonic absorber using liquid-crystal-coated asymmetric nanodisk array. *Appl. Phys. Lett.*, **100**, pp. 053119.
47. Smalley, J. S. T., Zhao, Y., Nawaz, A. A., Hao, Q., Ma, Y., Khoo, I. C., and Huang, T. J. (2011). High contrast modulation of plasmonic signals using nanoscale dual-frequency liquid crystals, *Opt. Express*, **19**, pp. 15265–15274.
48. Hao, Q., Zhao, Y., Juluri, B. K., Kiraly, B., Liou, J., Khoo, I. C., and Huang, T. J. (2011). Frequency-addressed tunable transmission in optically thin metallic nanohole arrays with dual-frequency liquid crystals, *J. Appl. Phys.*, **109**, pp. 084340.
49. Liu, Y. J., Leong, E. S. P., Wang, B., and Teng, J. H. (2011). Optical transmission enhancement and tuning by overlaying liquid crystals on a gold film with patterned nanoholes, *Plasmonics*, **6**, pp. 659–664.
50. Liu, Y. J., Ding, X., Lin, S.-C. S., Shi, J., Chiang, I., and Huang, T. J. (2011). Surface acoustic wave driven light shutters using polymer-dispersed liquid crystals, *Adv. Mater.*, **23**, pp. 1656–1659.
51. Liu, Y. J., Hao, Q., Smalley, J. S. T., Liou, J., Khoo, I. C., and Huang, T. J. (2010). A frequency-addressed plasmonic switch based on dual-frequency liquid crystal, *Appl. Phys. Lett.*, **97**, pp. 091101.
52. Liu, Y. J., Zheng, Y. B., Shi, J., Huang, H., Walker, T. R., and Huang, T. J. (2009). Optically switchable gratings based on azo-dye-doped, polymer-dispersed liquid crystals, *Opt. Lett.*, **34**, pp. 2351–2353.
53. Hsiao, V. K. S., Zheng, Y. B., Juluri, B. K., and Huang, T. J. (2008). Light-driven plasmonic switches based on Au nanodisk arrays and photoresponsive liquid crystals, *Adv. Mater.*, **20**, pp. 3528–3532.
54. Chu, K. C., Chao, C. Y., Chen, Y. F., Wu, Y. C., and Chen, C. C. (2006). Electrically controlled surface plasmon resonance frequency of gold nanorods, *Appl. Phys. Lett.*, **89**, pp. 103107.

55. Zografopoulos, D. C., and Beccherelli, R. (2013). Long-range plasmonic directional coupler switches controlled by nematic liquid crystals, *Opt. Express*, **21**, pp. 8240–8250.

56. Dickson, W., Wurtz, G. A., Evans, P. R., Pollard, R. J., and Zayats, A. V. (2008). Electronically controlled surface plasmon dispersion and optical transmission through metallic hole arrays using liquid crystal, *Nano Lett.*, **8**, pp. 281–286.

57. Kossyrev, P. A., Yin, A., Cloutier, S. G., Cardimona, D. A., Huang, D., Alsing, P. M., and Xu, J. M. (2005). Electric field tuning of plasmonic response of nanodot array in liquid crystal matrix, *Nano Lett.*, **5**, pp. 1978–1981.

58. De Sio, L., Cunningham, A., Verrina, V., Tone, C. M., Caputo, R., Burgi, T., and Umeton, C. (2012). Double active control of the plasmonic resonance of a gold nanoparticle array, *Nanoscale*, **4**, pp. 7619–7623.

59. Khatua, S., Chang, W. S., Swanglap, P., Olson, J., and Link, S. (2011). Active modulation of nanorod plasmons, *Nano Lett.*, **11**, pp. 3797–3802.

60. Vivekchand, S. R. C., Engel, C. J., Lubin, S. M., Blaber, M. G., Zhou, W., Suh, J. Y., Schatz, G. C., and Odom, T. W. (2012). Liquid plasmonics: Manipulating surface plasmon polaritons via phase transitions, *Nano Lett.*, **12**, pp. 4324–4328.

61. Zheng, Y., Yang, Y., Jensen, L., Fang, L., Juluri, B. K., Flood, A. H., Weiss, P. S., Stoddart J. F., and Huang, T. J. (2009). Active molecular plasmonics: Controlling plasmon resonances with molecular switches, *Nano Lett.*, **9**, pp. 819–825.

62. Zheng, Y. B., Hao, Q., Wang, Y., Kiraly, B., Chiang, I., and Huang, T. J. (2010). Light-driven artificial molucular machines, *J. Nanophoton.*, **4**, pp. 042501.

63. Geandier, G., Renault, P., Bourhis, E. L., Goudeau, P., Faurie, D., Bourlot, C., Djemia, P., Castelnau, O., and Cherif S. M. (2010). Elastic-strain distribution in metallic film-polymer substrate composites, *Appl. Phys. Lett.*, **96**, pp. 041905.

64. Yang, J., You, J., Chen, C., Hsu, W., Tan, H., Zhang, X. W., Hong, Z., and Yang, Y. (2011). Plasmonic polymer tandem solar cell, *ACS Nano*, **5**, pp. 6210–6217.

65. Chah, S., Noolandi, J., and Zare, R. N. (2005). Undulatory delamination of thin polymer films on gold surfaces, *J. Phys. Chem. B*, **109**, pp. 19416–19421.

66. Novo, C., Funston, A. M., and Mulvaney, P. (2008). Direct observation of chemical reactions on single gold nanocrystals using surface plasmon spectroscopy, *Nat. Nanotechnol.*, **3**, pp. 598–602.

67. Lioubimov, V., Kolomenskii, A., Mershin, A., Nanopoulos, D. V., and Schuessler, H. A. (2004). Effect of varying electric potential on surface-plasmon resonance sensing, *Appl. Opt.*, **43**, pp. 3426–3432.

68. Ung, T., Liz-Marzan, L. M., and Mulvaney, P. (1998). Controlled method for silica coating of silver colloids. Influence of coating on the rate of chemical reactions, *Langmuir*, **14**, pp. 3740–3748.

69. Yang, D. K., and Wu, S. T. (2006). *Fundamentals of Liquid Crystal Devices* John Wiley & Sons Inc, Hoboken, NJ, USA.

70. Khoo, I.-C. (2007). *Liquid Crystals*, 2nd ed. John Wiley & Sons Inc, Hoboken, NJ, USA.

71. Wu, S. T. (1986). Birefringence dispersions of liquid crystals, *Phys. Rev. A*, **33**, pp. 1270–1274.

72. Zografopoulos, D. C., and Beccherelli, R. (2013). Liquid-crystal-tunable metal–insulator–metal plasmonic waveguides and Bragg resonators, *J. Opt.*, **15**, pp. 055009.

73. Schadt, M. (1982). Dual-frequency addressing of field effects, *Mol. Cryst. Liq. Cryst.*, **89**, pp. 77–92.

74. Xianyu, H., Wu, S. T., and Lin, C. L. (2009). Dual frequency liquid crystals: A review, *Liq. Cryst.*, **36**, pp. 717–726.

75. Chang, W., Lassiter, J. B., Swanglap, P., Sobhani, H., Khatua, S., Nordlander, P., Halas, N. J., and Link, S. (2012). A plasmonic Fano switch, *Nano. Lett.*, **12**, pp. 4977–4982.

76. Liu, Y. J., Dai, H. T., and Sun, X. W. (2011). Holographic fabrication of azo-dye-functionalized photonic structures, *J. Mater. Chem.*, **21**, pp. 2982–2986.

77. Liu, Y. J., Zheng, Y. B., Shi, J., Huang, H., Walker, T. R., and Huang, T. J. (2009). Optically switchable gratings based on azo-dye-doped, polymer-dispersed liquid crystals, *Opt. Lett.*, **34**, pp. 2351–2353.

78. Liu, Y. J., Su, Y.-C., Hsu, Y.-J., and Hsiao, V. K. S. (2012). Light-induced spectral shifting generated from azo-dye doped holographic 2D gratings, *J. Mater. Chem.*, **22**, pp. 14191–14195.

79. De Sio, L., Tedesco, A., Tabirian, N., and Umeton, C. (2010). Optically controlled holographic beam splitter, *Appl. Phys. Lett.*, **97**, 183507.

80. De Sio, L., Serak, S., Tabirian, N., and Umeton, C. (2011). Mesogenic versus non-mesogenic azo dye confined in a soft-matter template for realization of optically switchable diffraction gratings, *J. Mater. Chem.*, **21**, pp. 6811–6814.

81. Liu, Y. J., Cai, Z. Y., Leong, E. S. P., Zhao, X. S., and Teng, J. H. (2012). Optically switchable photonic crystals based on inverse opals partially

infiltrated by photoresponsive liquid crystals, *J. Mater. Chem.*, **22**, pp. 7609–7613.

82. Liu, Y. J., Dai, H. T., Leong, E. S. P., Teng, J. H., and Sun, X. W. (2012). Azo-dye-doped absorbing photonic crystals with purely imaginary refractive index contrast and all-optically switchable diffraction properties, *Opt. Mater. Express*, **2**, pp. 55–61.
83. De Sio, L., Klein, G., Serak, S., Tabiryan, N., Cunningham, A., Tone, C. M., Ciuchi, F., Bürgi, T., Umeton, C., and Bunning, T. (2013). All-optical control of localized plasmonic resonance realized by photoalignment of liquid crystals, *J. Mater. Chem. C*, **1**, pp. 7483–7487.
84. Liu, Y. J., Zheng, Y. B., Liou, J., Chiang, I.-K., Khoo, I. C., and Huang, T. J. (2011). All-optical modulation of localized surface plasmon coupling in a hybrid system composed of photo-switchable gratings and Au nanodisk arrays, *J. Phys. Chem. C*, **115**, pp. 7717–7722.
85. Tamai, N., and Miyasaka, H. (2000). Ultrafast dynamics of photochromic systems, *Chem. Rev.*, **100**, pp. 1875–1890.
86. De Sio, L., Ricciardi, L., Serak, S., La Deda, M., Tabiryan, N., and Umeton, C. (2012). Photo-sensitive liquid crystals for optically controlled diffraction gratings, *J. Mater. Chem.*, **22**, pp. 6669–6673.
87. Hrozhyk, U. A., Serak, S. V., Tabiryan, N. V., Hoke, L., Steeves, D. M., and Kimball, B. R. (2010). Azobenzene liquid crystalline materials for efficient optical switching with pulsed and/or continuous wave laser beams, *Opt. Express*, **18**, pp. 8697–8704.
88. Hrozhyk, U. A., Serak, S. V., Tabiryan, N. V., Hoke, L., Steeves, D. M., Kimball, B., and Kedziora, G. (2008). Systematic study of absorption spectra of donor–acceptor azobenzene mesogenic structures, *Mol. Cryst. Liq. Cryst.*, **489**, pp. 257–272.
89. Faraday, M. (1846). Experimental researches in electricity, *Phil. Trans. R. Soc. Lond.*, **136**, pp. 1–20.
90. Kerr, J. (1877). On the rotation of the plane of polarization of light by reflection from the pole of a magnet, *Phil. Mag.*, **3**, pp. 321–343.
91. Kerr, J. (1878). On reflection of polarized light from the equatorial surface of a magnet, *Phil. Mag.*, **5**, pp. 161–177.
92. González-Díaz, J. B., García-Martín, A., Armelles, G., García-Martín, J. M., Clavero, C., Cebollada, A., Lukaszew, R. A., Skuza J. R., Kumah, D. P., and Clarke, R. (2007). Surface-magnetoplasmon nonreciprocity effects in noble-metal/ferromagnetic heterostructures, *Phys. Rev. B*, **76**, pp. 153402.

93. González-Díaz, J. B., García-Martín, A., García-Martín, J. M., Cebollada, A., Armelles, G., Sepúlveda, B., Alaverdyan, Y., and Käll, M. (2008). Plasmonic Au/Co/Au nanosandwiches with enhanced magneto-optical activity, *Small*, **4**, pp. 202–205.

94. Du, G. X., Mori, T., Suzuki, M., Saito, S., Fukuda, H., and Takahashi, M. (2010). Evidence of localized surface plasmon enhanced magneto-optical effect in nanodisk array, *Appl. Phys. Lett.*, **96**, pp. 081915.

95. Temnov, V. V., Armelles, G., Woggon, U., Guzatov, D., Cebollada, A., Garcia-Martin, A., Garcia-Martin, J. M., Thomay, T., Leitenstorfer, A., and Bratschitsch, R. (2010). Active magneto-plasmonics in hybrid metal–ferromagnet structures, *Nat. Photon.*, **4**, pp. 107–111.

96. Wang, L., Clavero, C., Huba, Z., Carroll, K. J., Carpenter, E. E., Gu, D., and Lukaszew, R. A. (2011). Plasmonics and enhanced magneto-optics in core–shell Co–Ag nanoparticles, *Nano Lett.*, **11**, pp. 1237–1240.

97. Jain, P. K., Xiao, Y., Walsworth, R., and Cohen, A. E. (2009). Surface plasmon resonance enhanced magneto-optics (SuPREMO): Faraday rotation enhancement in gold-coated iron oxide nanocrystals, *Nano Lett.*, **9**, pp. 1644–1650.

98. Belotelov, V. I., Akimov, I. A., Pohl, M., Kotov, V. A., Kasture, S., Vengurlekar, A. S., Gopal, A. V., Yakovlev, D. R., Zvezdin, A. K., and Bayer, M. (2011). Enhanced magneto-optical effects in magnetoplasmonic crystals, *Nat. Nanotech.*, **6**, pp. 370–376.

99. Liu, Q. K., Cui, Y. X., Gardner, D., Li, X., He, S. L., and Smalyukh, I. I. (2010). Self-alignment of plasmonic gold nanorods in reconfigurable anisotropic fluids for tunable bulk metamaterial applications, *Nano Lett.*, **10**, pp. 1347–1353.

100. Verluise, F., Laude, V., Cheng, Z., Spielmann, Ch., and Tournois, P. (2000). Amplitude and phase control of ultrashort pulses by use of an acousto-optic programmable dispersive filter: Pulse compression and shaping, *Opt. Lett.*, **25**, pp. 575–577.

101. Gérard, D., Laude, V., Sadani, B., Khelif, A., Van Labeke, D., and Guizal, B. (2007). Modulation of the extraordinary optical transmission by surface acoustic waves, *Phys. Rev. B*, **76**, 235427.

102. Li, J. Gauzia, S., and Wu, S. T. (2004). High temperature-gradient refractive index liquid crystals, *Opt. Express*, **12**, pp. 2002–2010.

103. Li, J., Wen, C. H., Gauza, S., Lu, R., and Wu, S. T. (2005). Refractive indices of liquid crystals for display applications, *J. Display Technol.*, **1**, pp. 51–61.

104. Li, J., Gauza, S., and Wu, S. T. (2004). Temperature effect on liquid crystal refractive indices, *J. Appl. Phys.*, **96**, pp. 19–24.

105. Cetin, A. E., Mertiri, A., Huang, M., Erramilli, S., and Altug, H. (2013). Thermal tuning of surface plasmon polaritons using liquid crystals, *Adv. Opt. Mater.*, **1**, pp. 915–920.

106. Bardhan, R., Lal, S., Joshi, A., and Halas, N. J. (2011). Theranostic nanoshells: From probe design to imaging and treatment of cancer, *Acc. Chem. Res.*, **44**, pp. 936–946.

107. Richardson, H. H., Hickman, Z. N., Gocorov, A. O., Thomas, A. C., Zhang, W., and Kordesch, M. E. (2006). Thermooptical properties of gold nanoparticles embedded in ice: Characterization of heat generation and melting, *Nano Lett.*, **6**, pp. 783–788.

108. Wilson, O. M., Hu, X., Cahill, D. G., and Braun, P. V. (2002). Colloidal metal particles as probes of nanoscale thermal transport in fluids, *Phys. Rev. B*, **66**, pp. 224301.

109. De Sio, L., Placido, T., Serak, S., Comparelli, R., Tamborra, M., Tabiryan, N., Curri, M. L., Bartolino, R., Umeton, C., and Bunning, T. (2013). Nano-localized heating source for photonics and plasmonics, *Adv. Opt. Mater.*, **1**, pp. 899–904.

110. Huang, X., Neretina, S., and El-Sayed, M. A. (2009). Gold nanorods: From synthesis and properties to biological and biomedical applications, *Adv. Mater.*, **21**, pp. 4880–4910.

111. Khatua, S., Manna, P., Chang, W.-S., Tcherniak, A., Friedlander, E., Zubarev, E. R., and Link, S. (2010). Plasmonic nanoparticles-liquid crystal composites, *J. Phys. Chem. C*, **114**, pp. 7251–7257.

112. Umadevi, S., Feng, X., and Hegmann, T. (2013). Large area self-assembly of nematic liquid-crystal-functionalized gold nanorods, *Adv. Funct. Mater.*, **23**, pp. 1393–1403.

113. Milette, J., Cowling, S. J., Toader, V., Lavigne, C., Saez, I. M., Bruce Lennox, R., Goodby, J. W., and Reven, L. (2012). Reversible long range network formation in gold nanoparticle-nematic liquid crystal composites, *Soft Matter*, **8**, pp. 173–179.

114. Wu, B.-G., Erdmann, J. H., and Doane, J. W. (1989). Response times and voltages for PDLC light shutters, *Liq. Cryst.*, **5**, pp. 1453–1465.

115. Liu, Y. J., and Sun, X. W. (2007). Electrically switchable computer-generated hologram recorded in polymer-dispersed liquid crystals, *Appl. Phys. Lett.*, **90**, pp. 191118.

116. Tondiglia, V. P., Natarajan, L. V., Sutherland, R. L., Bunning, T. J., and Adams, W. W. (1995). Volume holographic image storage and electro-

optical readout in a polymer-dispersed liquid-crystal film, *Opt. Lett.*, **20**, pp. 1325–1327.

117. Liu, Y. J., Sun, X. W., Liu, J. H., Dai, H. T., and Xu, K. S. (2005). A polarization insensitive 2 × 2 optical switch fabricated by liquid crystal-polymer composite, *Appl. Phys. Lett.*, **86**, pp. 041115.
118. Liu, Y. J., Sun, X. W., Dai, H. T., Liu, J. H., and Xu, K. S. (2005). Effect of surfactant on the electro-optical properties of holographic polymer dispersed liquid crystal Bragg gratings, *Opt. Mater.*, **27**, pp. 1451–1455.
119. Sun, J., and Wu, S. T. (2014). Recent advances in polymer network liquid crystal spatial light modulators, *J. Polym. Sci. Part B Polym. Phys.*, **52**, pp. 183–192.
120. Yan, J., Rao, L., Jiao, M., Li, Y., Cheng, H. C., and Wu, S. T. (2011). Polymer-stabilized optically isotropic liquid crystals for next-generation display and photonics applications, *J. Mater. Chem.*, **21**, pp. 7870–7877.
121. Gauza, S., Wang, H., Wen, C. H., Wu, S. T., Seed, A. J., and Dabrowski, R. (2003). High birefringence isothiocyanato tolane liquid crystals, *Jpn. J. Appl. Phys.*, **42**, pp. 3463–3466.
122. Dabrowski, R., Kula, P., and Herman, J. (2013). High birefringence liquid crystals, *Crystals*, **3**, pp. 443–482.
123. Arakawa, Y., Kang S., Nakajima, S., Sakajiri, K., Cho, Y., Kawauchi, S., Watanabe, J., and Konishi G. (2013). Diphenyltriacetylenes: Novel nematic liquid crystal materials and analysis of their nematic phase-transition and birefringence behaviours, *J. Mater. Chem. C*, **1**, pp. 8094–8102.
124. Gauza, S., Wen, C. H., Wu, S. T., Janarthanan, N., and Hsu, C. S. (2004). Super high birefringence isothiocyanato biphenyl-bistolane liquid crystals, *Jpn. J. Appl. Phys.*, **43**, pp. 7634–7638.
125. De Sio, L., and Umeton, C. (2010). Dual-mode control of light by two-dimensional periodic structures realized in liquid-crystalline composite materials, *Opt. Lett.*, **35**, pp. 2759–2761.

Chapter 10

Nonlinear Optical Enhancement with Plasmonic Core–Shell Nanowires

Rachel Grange
Institute for Quantum Electronics, Department of Physics, ETH Zurich, 8093 Zurich, Switzerland
grangera@ethz.ch

10.1 Introduction

Many innovations in medicine, optoelectronics, or computer sciences rely on the development of new materials. To overcome the limitations of the well-established fields of semiconductor (silicon) and dielectric (glass fiber) materials, metal plasmonics recently demonstrated high performances [1]. Indeed, since semiconductors are limited in speed due to electric interconnects and dielectrics are limited in size due to the diffraction limit, metallic nanostructures seems to be ideal to reach higher speed and keep the size small, even though they exhibit resistive losses due to the physical nature of metal [2]. Therefore, hybrid nanomaterials are anticipated to be part of new solutions for recently described paradigms in sciences, e.g. optical nanocircuits [2]. Adapted from Brongersma and Shalaev [1],

Active Plasmonic Nanomaterials
Edited by Luciano De Sio

ISBN 978-981-4613-00-2 (Hardcover), 978-981-4613-01-9 (eBook)
www.panstanford.com

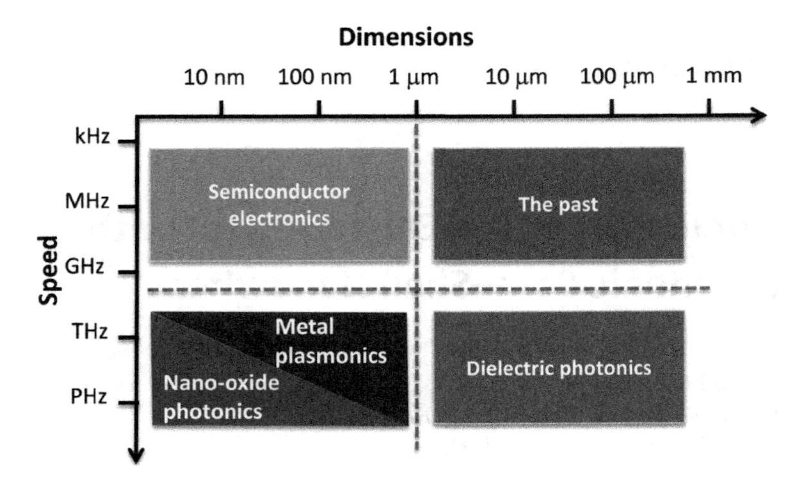

Figure 10.1 Widely studied materials, as semiconductor or dielectric, and their limitations in size and speed. Within the younger field of plasmonics emerges nano-oxide photonics based on hybrid nanostructures combining oxides and metallic nanomaterials. Reprinted from Ref. [1], with permission from AAAS.

Fig. 10.1 summarizes the advantages and limitations of the above-mentioned materials.

Recently we showed that combining a metallic shell with a nano-oxide core, either a $BaTiO_3$ nanosphere [3], or a $KNbO_3$ nanowire [4], can increase the functionality of the materials. Well known in bulk form or thin films, oxides such as $LiNbO_3$ with a special crystal structure are ferroelectric, piezoelectric and possess a nonlinear optical polarizability suitable for daily applications in telecommunication [5]. Compared to semiconductors' planar structures suited for the infrared spectral range and semiconductor quantum dots that need to be as small as 2–10 nm to sustain optical resonances in the visible range for fluorescence microscopy [6], oxide nanoparticles, either spherical or anisotropic, cover more easily a broad spectral range taking advantage of nonlinear physical processes such as second-harmonic generation (SHG), mostly regardless of feature sizes. The multiplexing effect that seems to be lost in the nonresonant process such as SHG can be in fact facilitated by other mechanisms such as polarization dependency. Since SHG is a coherent process, the orientation of the particles may be controlled and adjusted with polarization-sensitive illumination schemes [7]. The respective resonances can be engineered with metallic shells of different thicknesses taking advantage of plasmon-polaritonic effects. Moreover, oxide nanoparticles, particularly when thinly covered with a noble metal such as gold, seem to be much more biocompatible than metals used to fabricate quantum dots.

This chapter focuses on the nanowires hybrid nanostructures with an alkaline niobate core (Li-, K-, $NaNbO_3$) and a gold shell. Indeed, nanowires have interesting electronic and photonic properties suited for a wide range of applications such as nanolasers [8], solar absorbers [9, 10], generators [11], waveguides [12], or light-emitting diodes [13]. Biological applications are also of great interest and include nanowire-based devices such as nanoendoscopes or biosensors [14, 15]. We will present the state of the art of such hybrid nanostructures and their synthesis. Then we will show spectroscopic measurements and nonlinear optical experiments of single core–shell nanostructures to demonstrate the enhancement of the SHG using the localized surface plasmon resonance of the gold shell.

10.2 State of the Art

Bulk oxide materials such as lithium niobate ($LiNbO_3$), used in mobile telephones or optical modulators [5], still possess their optical properties at the nanoscale [16]. Furthermore, metals, first thought to be useless in optics except as mirrors, are now considered with great interest if the dimensions of their typical features are tens of nanometers, giving rise to resonant surface plasmon polaritons in the optical spectral range [17]. Therefore, a large choice of materials for nanophotonics seems available besides well-known semiconductors. In this section, we give an overview of the fields of nonlinear photonics, nano-oxide wires or spheres, and core–shell plasmonic structures that are currently available.

The field of nonlinear photonics has been at the cutting edge of optics and quantum electronics research since the first experiment on SHG in 1961 [18]. Many important advances are reported since then as the demonstration of stimulated Raman scattering

[19], soliton generation in optical fibers [20], and Bose–Einstein condensation [21]. Combining those well-established concepts with the recent progress and freedom of designs in nanotechnology is paving the way for the investigations of many exciting and unprecedented nanoscale systems.

The study of semiconductor nanomaterials is a well-established field yielding useful application-oriented research with 2D quantum wells [22], 1D nanowires [23, 24], or 0D quantum dots [25]. Nano-oxide syntheses of anisotropic alkaline materials are reported since 2005 using various bottom-up methods such as template-assisted pyrolysis resulting in regular arrays of tubes [26], solution-phase synthesis resulting in rod-like structures [27], or hydrothermal synthesis producing free-standing nanowires with high aspect ratio [28]. Up to now, the crystal properties of nanomaterials have been well characterized using standard material science methods such as X-ray diffraction, scanning electron microscopy (SEM), or transmission electron microscopy. However, nonlinear optical or electro-optical properties have been rarely studied and almost no further processing steps such as doping or coating have been applied to those nanostructures. Only a few applications have already used these types of nanowires by combining the various physical properties of perovskite alkaline materials and the anisotropic shape at the nanoscale level. A nanometric SHG light probe manipulated by optical tweezers and capable of guiding light has been already described [29, 30]. In 2009, we demonstrated localized SHG light sources in optofluidic environments [31].

The first experiment with metallic nanoshells considered an Au_2S dielectric core surrounded by a gold shell without independent control of the core size [32]. Later on, Oldenburg et al. used dielectric cores, such as silica nanospheres, and varied the thickness of the shell or of the core to engineer optical resonances [33]. However, most of the core materials are found to possess no additional interesting properties and the nanoshells are used only to interact with the surrounding medium, for instance by local heating [34]. In 2010, we showed that a $BaTiO_3$ core with nonlinear optical properties can greatly benefit from the presence of a metallic nanoshell [3]. Basically, the strong localization of the light at the nanoscale associated with the excitation of localized surface plasmon polaritons at the metallic nanostructures entails the possibility to locally enhance the electric field by several orders of magnitude, boosting nonlinear responses, such as SHG. The demonstration of this strong resonance may also contribute to plasmonic laser research using a thin metal planar layer and semiconductor nanowires [35]. There have been some attempts to coat ZnO nanowires with gold; however, the coating was achieved only via sputtering [36, 37]. No chemical synthesis enabling to surround the complete surface of a nanowire has been reported or modelled.

10.3 Fabrication

In this section, we focus on bottom-up fabrication methods of alkaline nanowires, and especially $KNbO_3$. However, a top-down approach was also demonstrated for $LiNbO_3$ nanowires based on an ion-beam-enhanced etching method that reduces chemical stability of $LiNbO_3$ after ion-beam irradiation [38].

10.3.1 *Bottom-Up Fabrication of the $KNbO_3$ Core*

Different chemistry-based approaches have been developed to fabricate crystalline alkaline niobate nanowires. Nanowires have been mainly produced by the sol-gel route [39], hydrothermal route [28, 40–44], and molten salt synthesis [45, 46].

For $KNbO_3$ nanowire used in the optical experiments of Sections 10.4 and 10.5, we used hydrothermal synthesis. This technique crystallizes substances under moderate temperatures (200–250°C) and high pressures. With an autoclave made of a thick-walled steel cylinder with a hermetic seal, anisotropic crystalline materials can be obtained in one step. Typical scanning electron microscopic images of bunch and isolated $KNbO_3$ nanowires can be seen in Fig. 10.2. This is a very convenient approach since a large amount of material can be fabricated in a one-step synthesis. Many synthesis parameters can be adjusted, such as temperature, time, pressure (by an external pressure or the degree of the autoclave filling), solid–liquid ratio, and additives, to control the properties of the end product. Thus, hydrothermal synthesis is powerful because

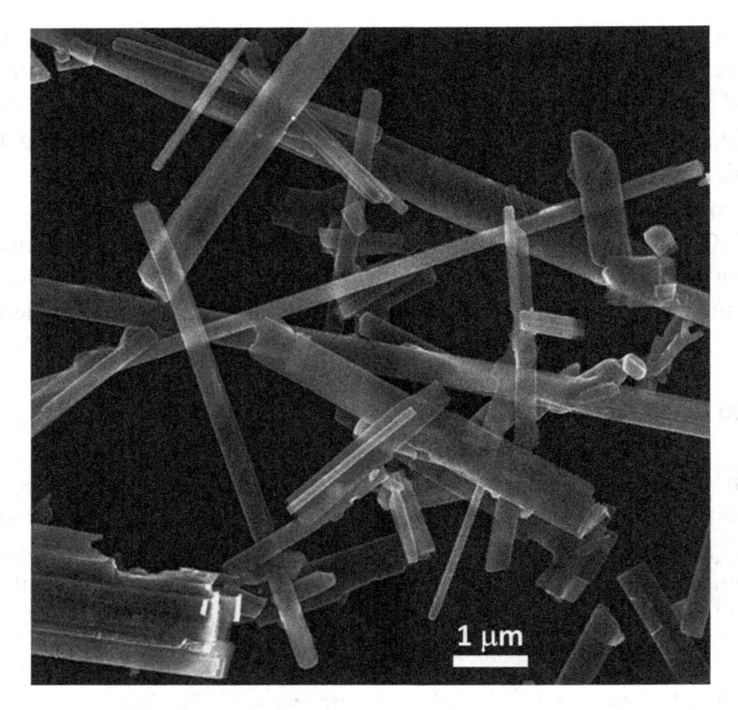

Figure 10.2 SEM images of typical $KNbO_3$ nanowires with aspect ratio up to 25.

many parameters can be modulated to control the particle size and morphology. Due to its simplicity, the hydrothermal technique has been widely studied and employed in inorganic synthesis for many years. A huge advantage of chemical synthesis in respect to chemical vapor deposition (CVD) or lithographic fabrication of nanowires is the ability to synthesize free-standing nanowires. Thus, no further step is needed to isolate or detach the nanowires, because no substrate is involved in the synthesis. If needed though, a substrate can be incorporated in the autoclave and nanoneedles have been demonstrated as well [47].

10.3.2 *Synthesis of the Gold Shell*

Since nonlinear optical processes such as SHG are generally inefficient at very small scale, we propose to take advantage of

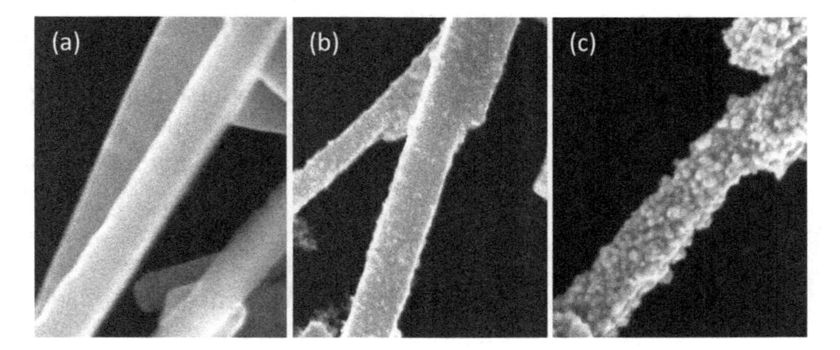

Figure 10.3 SEM image of $KNbO_3$ nanowires: (a) nanowires with polyelectrolyte coating, (b) nanowires seeded with gold nanoparticles, and (c) nanowire after the gold shell growth. The different nanowire diameter can be between 100 nm and 150 nm.

localized surface plasmon resonances from metallic nanostructures to strongly enhance nonlinear optical responses. We follow a typical silanization functionalization scheme [33] and demonstrate a $BaTiO_3$ core and gold shell structure [3]. Similarly, $KNbO_3$ nanowires are covered by a thin layer of gold to reach a near-infrared plasmonic resonance and enhance the SHG signal for optimized used as imaging probes or localized light source. The synthesis of the gold shell on a dielectric core is a process involving three stages: first, the functionalization of the core surface to make it positively charged (Fig. 10.3a); second, the adsorption of small gold particles negatively charged (seeds) onto the functionalized core (Fig. 10.3b); and third, the growth of the gold shell on the previous seeded structure (Fig. 10.3c). We follow Duff et al. for the synthesis of the gold seeds [48]. The seeds have a diameter of 1–5 nm, a negative surface charge, and a narrow size dispersion. Then the nanowire surface needs to be functionalized. Several schemes can be followed to make the nanowire positively charged, either amino silane layer as in [3] or by using a polymer (PDACMAC) to coat silica spheres [49]. Recently, we demonstrated the advantage of polymer concerning the functionalizing of $KNbO_3$ nanowires, since the seeding is much more uniform than with the amino silane process [4]. After the adsorption of the gold seeds, the nanowire is supposed to be uniformly covered by them. The final step is the shell growth around the seeds using an appropriate reducing agent, here hydroxylamine.

The resulting core–shell nanowire is not smooth (see Fig. 10.3c), but if the gold islands are touching each other, the collective oscillation of electrons can take place and consequently become a localized surface plasmon resonance.

10.4 Single Nanowire Spectrosocopy Measurements

The plasmonic resonances of the gold-coated $KNbO_3$ nanowires are studied by performing single particle spectroscopic measurements. Since the nanostructures are non-spherical and with large size distributions due to the chemical synthesis processes, ensemble measurements in a colloidal suspension are not possible. As the core–shell nanowires are produced in liquid environment, we drop them on a glass slide with a pipette. The glass substrates are patterned with numbers to recognize the position of the nanoparticles, either under an optical microscope or by SEM. The spectroscopic setup consists of an inverted microscope with a pinhole in the image plane to select the collected area with a spatial resolution of 1.6 μm. This enables precise measurements with a limited background signal due to the selective pinhole. Extinction measurements are done in transmission with bright field illumination of the single gold-covered $KNbO_3$ nanowire. Figure 10.4 shows the extinction spectrum of a single core–shell nanostructure with length and diameter of approximately 1.5 μm and 80 nm, respectively. We measured resonant peaks slightly below 700 nm and around 900 nm. According to the simulations (see next section), there are several core radii and thicknesses of the gold shell that exhibit plasmonic resonances close to such resonances. The geometry of the current wire corresponds to a shell thickness of 7.5 nm for a core radius of 35 nm to match with the resonance at 900 nm and to a shell thickness of 15 nm for the resonance at 700 nm. Those two different thicknesses are probably due to the nonhomogeneous gold shell along the nanowires. In nonlinear optical measurements, we therefore expect an enhancement of the excitation wavelength slightly below 700 nm and a strong enhancement at around 900 nm.

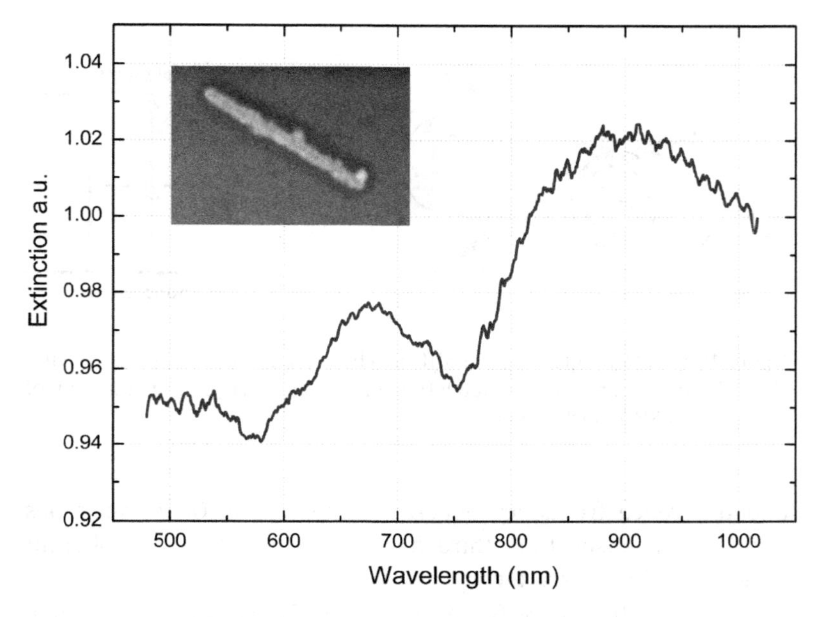

Figure 10.4 Single-wire spectroscopic measurements of the core–shell wire shown in the inset with length and diameter of approximately 1.5 μm and 80 nm, respectively. Reproduced from Ref. [4] by permission of The Royal Society of Chemistry.

10.5 Nonlinear Optical Measurements

The alkaline nanowires, such as $KNbO_3$, with non-centrosymmetric crystal structures possess second-order optical properties. Such nonlinear effect scales with the square of the electric field, and since this is a volume effect with the sixth power of the nanoparticle radius [50], contrary to weak surface SHG effect of centrosymmetric materials [51, 52]. The optical response is expressed by the polarization P as a power series of the electric field E as

$$\vec{P} = \varepsilon_0 \chi_1 \vec{E} + \varepsilon_0 \chi_2 \vec{E}^2 + \varepsilon_0 \chi_3 \vec{E}^3 + \ldots \qquad (10.1)$$

where ε_0 is the permittivity of free space and χ_i is the ith-order nonlinear optical susceptibility tensor [53]. Each χ_i represents a different optical effect that can be summarized as follows for a physical understanding of Eq. 10.1. χ_1, the linear susceptibility, is related to absorption and reflection of light. χ_2 encompasses sum

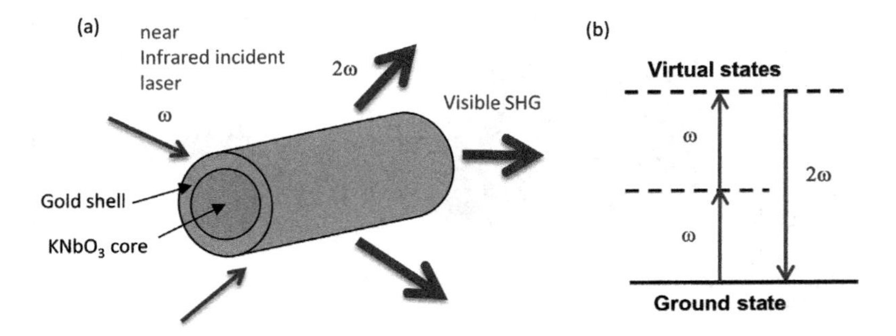

Figure 10.5 (a) Diagram of the SHG mechanism in a core–shell nanowire. The SHG arrows are bold to sketch the enhancement. (b) Energy diagram of the physical SHG mechanism.

and difference frequency generation such as SHG. χ_3 describes multiphoton absorption, third harmonic generation, or coherent anti-Stokes Raman scattering.

To illustrate the SHG effect, Fig. 10.5a shows a core–shell nanowire optically excited at a fundamental frequency ϖ; it emits the optical signal at the exact doubled frequency 2ϖ. The corresponding energy diagram of the SHG mechanism is displayed in Fig. 10.5b, where two photons are simultaneously reaching a virtual energy state prior to be recombined in a single photon of half the wavelength or doubled frequency.

The setup for the nonlinear optical characterization of the SHG signal is shown in Fig. 10.6. A near-infrared laser light is slightly focused onto a sample by a 10× objective, and the 100× objective collects the signal imaged through a 4f configuration. Filters are used to cut the fundamental frequency and detect only a narrow band around the SHG frequency onto an electron multiplying charges coupled device.

The measurement of the SHG signal of single core–shell nanowires uses the home-built transmission microscope setup described in Fig. 10.6 and with more details in Ref. [54]. The laser can be tuned from 690 nm up to 1040 nm wavelength. The averaged laser power incident on the wires ranges between 50 mW and 70 mW with a spot size of 4 μm, therefore illuminating the whole wire uniformly. We used the spectroscopically characterized core–

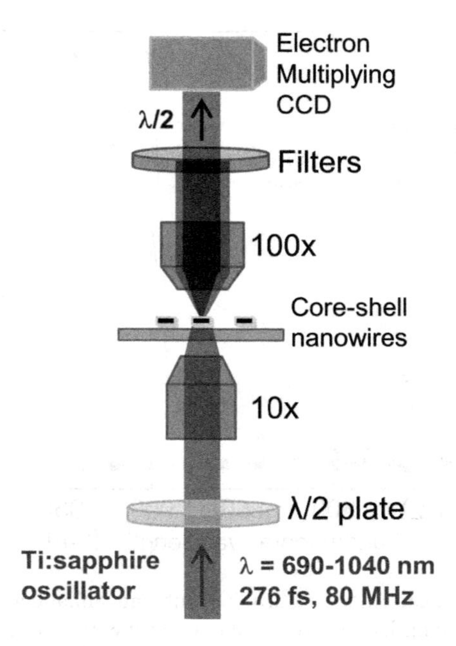

Figure 10.6 Setup for measuring SHG from individual nanowires dried on a microscope glass slide.

shell wire (Fig. 10.4) and a bare $KNbO_3$ nanowire, both lying on a glass substrate. The length and diameter of the gold-coated wire are approximately 1.5 μm and 80 nm, respectively, and that of the bare wire are 2 μm and 150 nm, respectively.

In Fig. 10.7, the results of the SHG measurements from the bare and the core–shell wire are displayed. The SHG responses of both wires are plotted from the excitation wavelength of 700 nm up to 1040 nm. The bare wire exhibits a nearly constant SHG output over the measured spectral range. But the core–shell nanowire shows an enhanced output at larger wavelengths with a peak position identified at 900 nm.

We normalized the SHG output of both wires on the respective wire volume and the pulse peak power to be able to calculate the enhancement between the bare and core–shell wire. The enhancement factor is plotted along the measurement curve of Fig. 10.7 and shows its highest value at an excitation wavelength of 960 nm reaching an enhancement of 250 times. Below 700 nm, it is

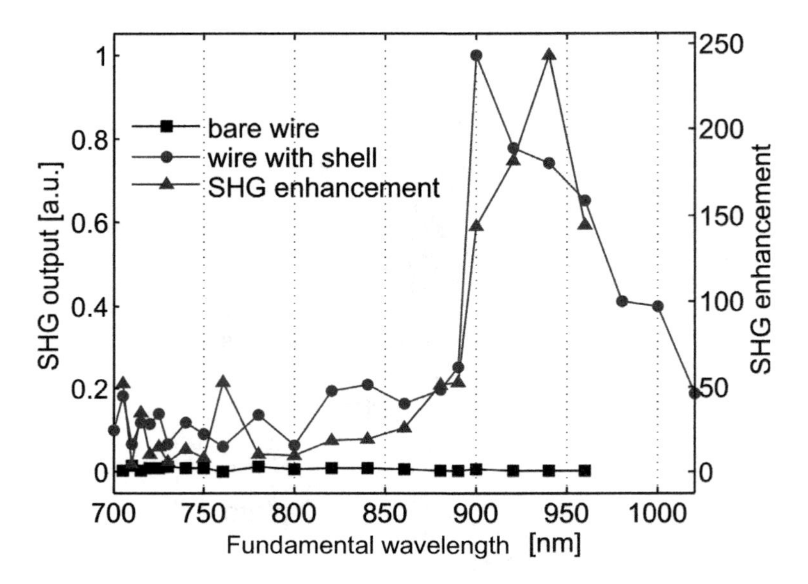

Figure 10.7 Measurement of SHG output and enhancement factor over excitation wavelength of a core–shell nanowire with a core radius of 35 nm and a shell thickness of 7.5 nm. Reproduced from Ref. [4] by permission of The Royal Society of Chemistry.

not possible to reliably measure the SHG due to the laser wavelength limitation.

To compare this experimental result with the expected value from the theory, we model a typical core–shell structure. With these simulations, we can analyze the optical measurements and determine the best geometrical parameters of gold shell and core radius.

The optical response to the excitation at the fundamental frequency and the emission of the generated second harmonic are calculated analytically by the expansion of the electric and magnetic fields into cylindrical harmonics [55–57]. The SHG is treated in the undepleted pump approximation, that is, the fundamental field in the core induces a nonlinear polarization at the second-harmonic frequency. The interaction between fundamental and second harmonic as well as the depletion of the fundamental wave are neglected. A more detailed description of the method can be found in Ref. [4]. The induced nonlinear polarization coefficients d_{ij}

within the core are given by [53, 58]

$$\begin{pmatrix} P_x^{NL}(2\omega) \\ P_y^{NL}(2\omega) \\ P_z^{NL}(2\omega) \end{pmatrix} = 2\varepsilon_0 \begin{pmatrix} 0 & 0 & 0 & 0 & d_{31} & 0 \\ 0 & 0 & 0 & d_{32} & 0 & 0 \\ d_{31} & d_{32} & d_{33} & 0 & 0 & 0 \end{pmatrix} \begin{pmatrix} E_x(\omega)^2 \\ E_y(\omega)^2 \\ E_z(\omega)^2 \\ 2E_y(\omega)E_z(\omega) \\ 2E_x(\omega)E_z(\omega) \\ 2E_x(\omega)E_y(\omega) \end{pmatrix} \tag{10.2}$$

Because the electric field at the fundamental frequency ω has only an E_x and an E_y component, the induced nonlinear polarization is z-polarized and second-harmonic emission is, therefore, TE (transverse electric) polarized.

We calculated the SHG (in analogy with two photon excitation) cross section σ_{SHG} both for bare $KNbO_3$ nanowires and for nanowires with an additional gold shell. The cross section is given by

$$\sigma_{SHG} = \frac{\hbar\omega}{2}\frac{P_{SHG}}{I_{in}^2} \tag{10.3}$$

where P_{SHG} is the emitted second-harmonic power per cylinder length, I_{in} is the incident intensity, and $\hbar\omega$ is the photon energy at

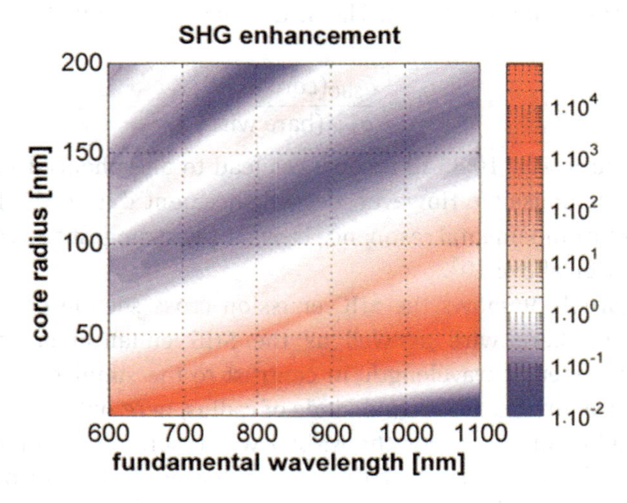

Figure 10.8 Simulation of the optical properties of $KNbO_3$ nanowires: resulting SHG emission enhancement. Reproduced from Ref. [4] by permission of The Royal Society of Chemistry.

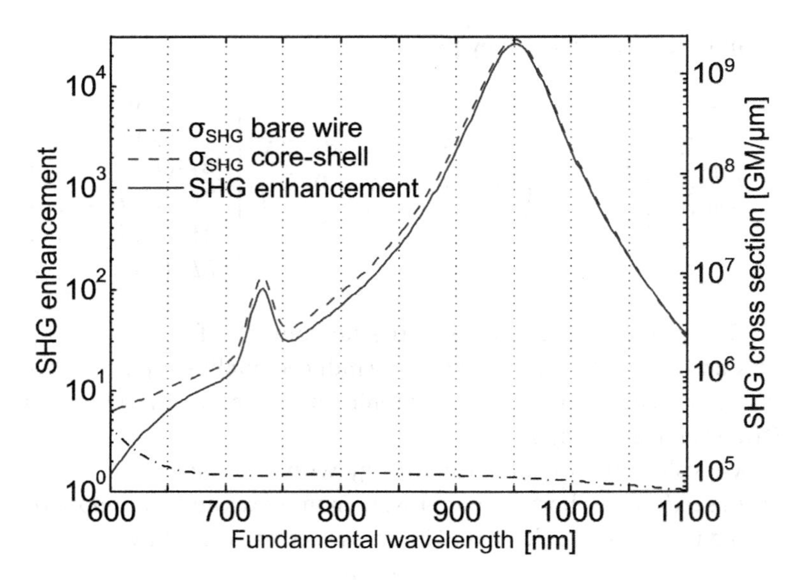

Figure 10.9 Simulated results for the SHG enhancement and SHG emission cross section of a core–shell nanowire with a core radius of 35 nm and a shell thickness of 7.5 nm. Reproduced from Ref. [4] by permission of The Royal Society of Chemistry.

the fundamental frequency. The SHG enhancement due to the gold shell

$$\Gamma = \frac{\sigma_{\text{SHG}}(\text{core–shell})}{\sigma_{\text{SHG}}(\text{bare wire})} \tag{10.4}$$

is shown in Fig. 10.8. All resonances lead to an enhanced second-harmonic emission. However, the enhancement due to the lowest energetic fundamental resonance is by far the strongest and reaches values well above 10^4.

Figure 10.9 shows the SHG emission cross section of a core–shell and bare wire as well as the SHG enhancement versus the fundamental wavelength. In contrast to the simulation results introduced in Fig. 10.8 for all kinds of radius, here only the specific core radius of 35 nm for the $KNbO_3$ was taken into account. The simulated enhancement of the second harmonic has its peak at 950 nm and reaches a value of 30,000, which is about two orders of magnitude larger than the measured value of 250. This discrepancy can be due to several factors. In the simulation, we assume a perfectly smooth gold shell and a plane wave excitation. But both conditions are not fully realized in the experiment. Indeed the shell is quite rough, and the nanowire is not perfectly cylindrical.

10.6 Conclusion

In this chapter, we showed how to enhance nonlinear optical signals as SHG with the growth of a plasmonic gold shell around $KNbO_3$ nanowires. We tested two types of functionalization to obtain the smoothest gold nanoshell. The gold shell with polyelectrolyte shows a more uniform coverage than silanized wires. We performed spectroscopic and SHG measurements on individual core–shell wires functionalized by the polyelectrolyte. Both measurements show a strong response slightly above 900 nm with an enhancement factor of the SHG signal of approximately 250 as compared to a bare wire. Such a strong signal enhancement is promising for applying the core–shell nanowire as a bright local light source in life sciences using the spectral filtering from nonlinear optical effects. Moreover, the chemically based synthesis is very simple to implement without any nanolithography processes involved.

Acknowledgments

The author thanks the Carl Zeiss Foundation, the DFG excellence program JSMC, and the Pro Chance Program of the Friedrich Schiller University for financial support. She is also grateful to Jessica Richter, Andrea Steinbrück, Anton Sergeyev, and Matthias Zilk for their support.

References

1. Brongersma ML and Shalaev VM (2010) The case for plasmonics. *Science* **328**, 440–441.
2. Engheta N (2007) Circuits with light at nanoscales: Optical nanocircuits inspired by metamaterials. *Science* **317**, 1698–1702.

3. Pu Y, Grange R, Hsieh C-L, and Psaltis D (2010) Nonlinear optical properties of core-shell nanocavities for enhanced second-harmonic generation. *Phys. Rev. Lett.* **104**, 207401–207405.
4. Richter J, Steinbrück A, Zilk M, Sergeyev A, Pertsch T, Tünnermann A, and Grange R (2014) Core-shell potassium niobate nanowires for enhanced nonlinear optical effects. *Nanoscale* **6**, 5200–5207
5. Wooten EL, Kissa KM, Yi-yan A, Murphy EJ, Member S, Lafaw DA, Hallemeier PF, Maack D, Attanasio DV, Fritz DJ, Mcbrien GJ, and Bossi DE (2000) A review of lithium niobate modulators for fiber-optic communications systems. *IEEE Sel. Top. Quantum Electron.* **6**, 69–82.
6. Michalet X, Pinaud FF, Bentolila LA, Tsay JM, Doose S, Li JJ, Sundaresan G, Wu AM, Gambhir SS, and Weiss S (2005) Quantum dots for live cells, in vivo imaging, and diagnostics. *Science* **307**, 538–544.
7. Hsieh C, Pu Y, Grange R, and Psaltis D (2010) Second-harmonic generation from nanocrystals under linearly and circularly polarized excitations. *Opt. Express.* **18**, 11917–11932.
8. Duan X, Huang Y, Agarwal R, Lieber CCM, and Fast CG (2003) Single-nanowire electrically driven lasers. *Nature* **421**, 241–245.
9. Colombo C, Heiß M, Grätzel M, and Fontcuberta i Morral A (2009) Gallium arsenide p-i-n radial structures for photovoltaic applications. *Appl. Phys. Lett.* **94**, 173108.
10. Cao L, Fan P, Vasudev AP, White JS, Yu Z, Cai W, Schuller J, Fan S, and Brongersma ML (2010) Semiconductor nanowire optical antenna solar absorbers. *Nano Lett.* **10**, 439–445.
11. Wang ZL and Song J (2012) Piezoelectric nanogenerators based on zinc oxide nanowire arrays. *Science* **312**, 242–246.
12. Voss T, Svacha GT, Mazur E, Mu S, Konjhodzic D, Marlow F, Müller S, and Ronning C (2007) High-order waveguide modes in ZnO nanowires. *Nano Lett.* **7**, 3675–3680.
13. Bao J, Zimmler MA, and Capasso F (2006) Broadband ZnO single-nanowire light-emitting diode. *Nano Lett.* **6**, 1719–1722.
14. Singhal R, Orynbayeva Z, Venkat R, Sundaram K, Niu JJ, Bhattacharyya S, Vitol EA, Schrlau MG, Papazoglou ES, Friedman G, and Gogotsi Y (2010) Multifunctional carbon-nanotube cellular endoscopes. *Nat. Nanotechnol.* **6**, 57–64.
15. Yan R, Park J, Choi Y, Heo C, Yang S, and Lee LP (2012) Nanowire-based single-cell endoscopy. *Nat. Nanotechnol.* **7**, 191–196.

16. Grange R, Dutto F, and Radenovic A (2011) Niobates nanowires: Synthesis, characterization and applications title, in *Nanowires/Book 2* (Hashim A, ed), Intech, pp. 509–524.

17. Atwater HA and Polman A (2010) Plasmonics for improved photovoltaic devices. *Nat. Mater.* **9**, 865.

18. Franken P, Hill A, Peters C, and Weinreich G (1961) Generation of optical harmonics. *Phys. Rev. Lett.* **7**, 118–119.

19. Eckhardt G, Hellwarth R, McClung F, Schwarz S, Weiner D, and Woodbury E (1962) Stimulated Raman scattering from organic liquids. *Phys. Rev. Lett.* **9**, 455–457.

20. Mollenauer LF, Stolen RH, and Gordon RH (1980) Experimental observation of picosecond pulse narrowing and solitons in optical fibers. *Phys. Rev. Lett.* **45**, 1095–1098.

21. Cornell E and Wieman C (2002) Nobel lecture: Bose–Einstein condensation in a dilute gas, the first 70 years and some recent experiments. *Rev. Mod. Phys.* **74**, 875.

22. Zory PS (1993) *Quantum Well Lasers*, Academic Press, San Diego.

23. Yazawa M, Koguchi M, Muto A, Ozawa M, and Hiruma K (1992) Effect of one monolayer of surface gold atoms on the epitaxial growth of InAs nanowhiskers. *Appl. Phys. Lett.* **61**, 2051–2053.

24. Yan RX, Gargas D, and Yang PD (2009) Nanowire photonics. *Nat. Photonics* **3**, 569–576.

25. Reed M, Randall J, Aggarwal R, Matyi R, Moore T, and Wetsel A (1988) Observation of discrete electronic states in a zero-dimensional semiconductor nanostructure. *Phys. Rev. Lett.* **60**, 535–537.

26. Zhao LL, Steinhart M, Yosef M, Lee SK, and Schlecht S (2005) Large-scale template-assisted growth of $LiNbO_3$ one-dimensional nanostructures for nano-sensors. *Sensors Actuators B* **109**, 86–90.

27. Wood BD, Mocanu V, and Gates BD (2008) Solution-phase synthesis of crystalline lithium niobate nanostructures. *Adv. Mater.* **20**, 4552–4556.

28. Magrez A, Vasco E, Seo JW, Dieker C, Setter N, and Forro L (2006) Growth of single-crystalline $KNbO_3$ nanostructures. *J. Phys. Chem. B* **110**, 58–61.

29. Nakayama Y, Pauzauskie PJ, Radenovic A, Onorato RM, Saykally RJ, Liphardt J, and Yang PD (2007) Tunable nanowire nonlinear optical probe. *Nature* **447**, 1098–1101.

30. Dutto F, Raillon C, Schenk K, and Radenovic A (2011) Nonlinear optical response in single alkaline niobate nanowires. *Nano Lett.* **11**, 2517–2521.

31. Grange R, Choi JW, Hsieh CL, Pu Y, Magrez A, Smajda R, Forró L, and Psaltis D (2009) Lithium niobate nanowires synthesis, optical properties, and manipulation. *Appl. Phys. Lett.* **95**, 143105–143105.
32. Zhou H, Honma I, Komiyama H, and Haus J (1994) Controlled synthesis and quantum-size effect in gold-coated nanoparticles. *Phys. Rev. B* **50**, 12052.
33. Oldenburg SJ, Averitt RD, Westcott SL, and Halas NJ (1998) Nanoengineering of optical resonances. *Chem. Phys. Lett.* **288**, 243–247.
34. Gobin M, Lee MH, Halas NJ, James WD, Drezek RA, West JL, and Gobin AM (2007) Near-infrared resonant nanoshells for combined optical imaging and photothermal cancer therapy. *Nano Lett.* **7**, 1929–1934.
35. Oulton RF, Sorger VJ, Zentgraf H, Ma R-M, Gladden C, Dai L, Bartal G, and Zhang X (2009) Plasmon lasers at deep subwavelength scale. *Nature* **461**(7264), 629–632.
36. Prokes SM, Glembocki OJ, Rendell RW, and Ancona MG (2007) Enhanced plasmon coupling in crossed dielectric/metal nanowire composite geometries and applications to surface-enhanced Raman spectroscopy. *Appl. Phys. Lett.* **90**, 093105–093105.
37. Sinha G, Depero LEL, and Alessandri I (2011) Recyclable SERS substrates based on Au-coated ZnO nanorods. *ACS Appl. Mater. Interfaces* **3**, 2557–2563.
38. Sergeyev A, Geiss R, Solntsev AS, Steinbrück A, Schrempel F, Kley E-B, Pertsch T, and Grange R (2013) Second-harmonic generation in lithium niobate nanowires for local fluorescence excitation. *Opt. Express* **21**, 19012–19021.
39. Pribosic I, Makovec D, and Drofenik M (2005) Formation of nanoneedles and nanoplatelets of KNbO3 perovskite during templated crystallization of the precursor gel. *Chem. Mater.* **17**, 2953–2958.
40. An CH, Tang KB, Wang CR, Shen GZ, Jin Y, and Qian YT (2002) Characterization of $LiNbO_3$ nanocrystals prepared via a convenient hydrothermal route. *Mater. Res. Bull.* **37**, 1791–1796.
41. Shi H, Li X, Wang D, Yuan Y, Zou Z, and Ye J (2009) $NaNbO_3$ nanostructures: Facile synthesis, characterization, and their photocatalytic properties. *Catal. Letters* **132**, 205–212.
42. Wang GZ, Selbach SM, Yu YD, Zhang XT, Grande T, and Einarsrud MA (2009) Hydrothermal synthesis and characterization of $KNbO_3$ nanorods. *CrystEngComm* **11**, 1958–1963.

43. Wang GZ, Yu YD, Grande T, and Einarsrud MA (2009) Synthesis of $KNbO_3$ nanorods by hydrothermal method. *J. Nanosci. Nanotechnol.* **9**, 1465–1469.

44. Wu SY, Liu XQ, and Chen XM (2010) Hydrothermal synthesis of $NaNbO_3$ with low NaOH concentration. *Ceram. Int.* **36**, 871–877.

45. Santulli AC, Zhou H, Berweger S, Raschke MB, Sutter E, and Wong SS (2010) Synthesis of single-crystalline one-dimensional $LiNbO_3$ nanowires. *CrystEngComm* **12**, 2675–2678.

46. Li L, Deng J, Chen J, Sun X, Yu R, Liu G, and Xing X (2009) Wire structure and morphology transformation of niobium oxide and niobates by molten salt synthesis. *Chem. Mater.* **21**, 1207–1213.

47. Wang Y, Chen Z, Ye Z, and Huang JY (2012) Synthesis and second-harmonic generation response of $KNbO_3$ nanoneedles. *J. Cryst. Growth* **341**, 42–45.

48. Duff DG, Baiker A, and Edwards PP (1993) A new hydrosol of gold clusters. 1. Formation and particle size variation. *Langmuir* **9**, 2301–2309.

49. Caruso F and Möhwald H (1999) Preparation and characterization of ordered nanoparticle and polymer composite multilayers on colloids. *Langmuir* **15**, 8276–8281.

50. Kim E, Steinbrück A, Buscaglia MT, Buscaglia V, Pertsch T, and Grange R (2013) Second-harmonic generation of single $BaTiO_3$ nanoparticles down to 22 nm diameter. *ACS Nano* **7**, 5343–5349.

51. Dadap JI, Shan J, Eisenthal KB, and Heinz TF (1999) Second-harmonic Rayleigh scattering from a sphere of centrosymmetric material. *Phys. Rev. Lett.* **83**, 4045–4048.

52. Dadap JI (2008) Optical second-harmonic scattering from cylindrical particles. *Phys. Rev. B* **78**, 205322.

53. Boyd RW (2003) *Nonlinear Optics*, 2nd edn, Academic Press, Amsterdam.

54. Grange R, Brönstrup G, Kiometzis M, Sergeyev A, Richter J, Leiterer C, Fritzsche W, Gutsche C, Lysov A, Prost W, Tegude F-J, Pertsch T, Tünnermann A, and Christiansen S (2012) Far-field imaging for direct visualization of light interferences in GaAs nanowires. *Nano Lett.* **12**, 5412–5417.

55. Bohren CF and Huffman DR (2004) *Absorption and Scattering of Light by Small Particles*, Wiley-VCH, Weinheim.

56. Lee S-C (1992) Scattering by closely-spaced radially-stratified parallel cylinders. *J. Quant. Spectrosc. Radiat. Transf.* **48**, 119–130.
57. Wu D, Liu X, and Li B (2011) Localized surface plasmon resonance properties of two-layered gold nanowire: Effects of geometry, incidence angle, and polarization. *J. Appl. Phys.* **109**, 083540.
58. Dmitriev VG, Gurzadyan GG, and Nikogosyan DN (1999) *Handbook of Nonlinear Optical Crystals*, 3rd edn, Springer, Berlin.

Chapter 11

Nanotechnology for Renewable Solar Energy

Xianhe Wei[a] and Ji Ma[b]
[a]*University of Michigan, 2300 Hayward St, Ann Arbor, MI 48109, USA*
[b]*Kent State University, 1425 University Esplande, Kent, OH 44242, USA*
jma2@kent.edu

Efficient and clean energy resources have been explored for energy consumptions. Renewable energy is becoming increasingly important since the limited fossil fuels are being depleted. In this chapter, renewable energy in photovoltaics (PVs) and thermoelectrics (TEs) will be discussed. Photovoltaic generators can absorb photons to create charge carriers and generate electricity, while thermoelectric generators can convert concentrated solar heat or other forms of heat source to electricity. Nanotechnologies have been developed to enhance the efficiencies of both types of renewable energy applications. Due to the unique nanoscale characteristics, nanomaterials, nanostructures, and nanoengineering can be used. The fundamentals, fabrication methods, and device applications related to nanotechnologies in photovoltaic and thermoelectric power generation are presented.

Active Plasmonic Nanomaterials
Edited by Luciano De Sio

ISBN 978-981-4613-00-2 (Hardcover), 978-981-4613-01-9 (eBook)
www.panstanford.com

11.1 Introduction

The global energy consumption is expected to grow at a steady rate. The increasing demand for energy has become a challenge considering the rapid depletion of fossil fuel energy resources. Not only are these resources limited, but also they produce environmental pollution [1]. So clean and renewable ways to generate power and highly efficient power usage are important. Among the many sustainable types of power generation, such as nuclear, tidal, biofuel, geoelectricity, hydropower, and wind, solar energy is a promising option due to the unlimited energy supply from the sun. Solar energy can be harvested by using either photovoltaics (PVs) or thermoelectrics (TEs). PVs can absorb photon energy to generate electricity, while TEs can convert concentrated solar heat to electricity. However, the efficiency of the device is low. Nanotechnologies have been introduced into these two fields. Through nanoscale engineering, significant enhancements of efficiency in both PVs and TEs have been made to realize practical devices toward broader applications.

11.2 Nanomaterials for Photovoltaics

The first practical photovolaic application, that is, solar cell, was used for powering satellites in 1958 [2]. Afterward, tremendous advancements have been made in this field, including solar-charged watches, solar-powered calculators, solar-driven heaters, and solar power generators. Compared with the environmental impact of burning fossil fuels, PV systems that convert the sun's solar capacity to generate electricity can minimize pollution. However, renewable energy from PVs and their applications are limited due to low efficiency and relatively high cost. The introduction of nanotechnology and its innovations will not only improve efficiencies, but also offer novel ways to reduce cost. In this section, we will focus on nanomaterials, such as nanoparticles, nanowires, and nanocrystal quantum dots, to enable new device architectures and improve efficiencies of PV devices.

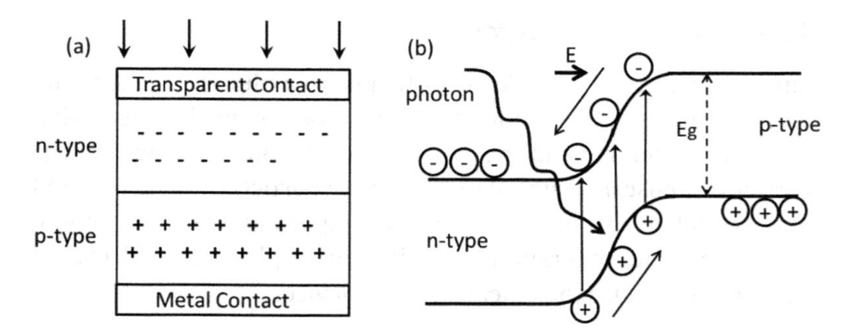

Figure 11.1 Schematics of a single-junction PV device: (a) device configuration and (b) photogeneration process in terms of energy levels.

11.2.1 *Basics of Photovoltaics*

A PV device works on the principle of converting solar energy to electricity by harvesting photo-excited electrons and holes, via a p-n junction or heterojunction [3]. A typical PV device configuration is shown in Fig. 11.1a, where a bilayer p-n junction is sandwiched between the top transparent conductive contact layer, such as indium tin oxide (ITO), and the bottom metal contact layer, such as Al. As shown in Fig. 11.1b, a built-in electric field (E) is formed across the junction due to carrier diffusion. When incident photons with energy ($h\gamma$) greater than band gap (E_g) reach the junction, pairs of electrons (negative charge carriers) and holes (positive charge carrier) are excited. The built-in field drives the photo-excited electrons toward the n-type section and holes toward the p-type section and thus allows the conductive layer to collect these charges to generate electricity.

The power conversion efficiency (η) of a PV device is

$$\eta = \frac{P_{\text{max}}}{P_{\text{in}}} \tag{11.1}$$

where P_{max} is the maximum power output and P_{in} is the incident illumination [4]. Theoretically, the maximum value of power conversion efficiency is the Schockley–Queisser limit, 34%, for a single-junction device [5].

11.2.2 *Nanomaterials for PVs*

Since the nature of a PV device is to absorb a photon and generate electric charge carriers, the efficiency depends on the absorption of incident photons and the collection of the photo-excited charge carriers. Nanoscale materials, such as nanoparticles, nanowires, and nanocrystal quantum dots, can play important roles on photon absorption and generation or collection of photo-excited charge carriers to enhance the efficiency of PV devices.

11.2.2.1 Nanoparticles

Metal nanoparticles represent an important class of materials to increase photon absorption in the PV devices because of the surface plasmon effect. The surface plasmons are the excitations of the conduction electrons at the metal–dielectric interface. The thickness of the absorption layer in the thin film solar cells needs to be thick enough for high absorption of incident light, but effective carrier collection requires the absorption layer thickness to be thin enough to transport. To overcome this conflict, metal nanoparticles can be adopted in the absorption layer in the PVs. The absorption of light can be enhanced by metal nanoparticles via light scattering, light concentration, and surface plasmon polaritons (SPPs). When located close to the interface of the two different dielectric materials, metal nanoparticles scatter the incident light preferentially toward the materials with higher dielectric constant, yielding scattered light with increased angular spread or longer corresponding optical path length in the dielectric material. The light-concentrating effect can be from the localized field of surface plasmons surrounding the particles, which enables higher photon absorption [6]. Adding metal nanoparticles to the interface between the bottom metal contact and the semiconductor section in the device structure in Fig. 11.1 can convert the incident light into SPPs, which are surface electromagnetic waves that travel along the interface. Close to the plasmon resonance frequency of the metal particles, the SSPs with wavelength greater than the dimensions of the particles can be confined near the interface, which effectively improve absorption [6]. Nanoparticles made from Au or Ag have been shown as effective

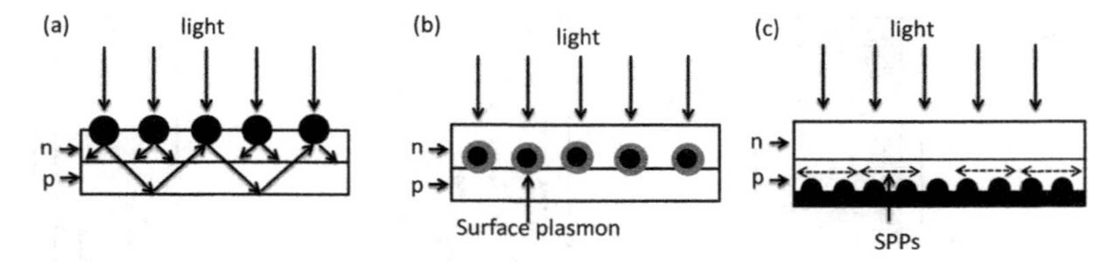

Figure 11.2 Example applications of metal nanoparticles in solar cells: (a) metal particles on the surface scatter the incident light, (b) imbedded particles concentrate the incident light, and (c) plasmonic particles on metal contact layer are for SPPs.

plasmonic absorption enhancers. For example, Ag nanoparticles can result in up to 16-fold absorption enhancement in Si-based solar cells [7]. The factors influencing the plasmonic enhancement of the nanoparticles include the metal itself, size, and geometry [8, 9].

Metal nanoparticles can be implemented in thin film solar cells to enhance photon energy absorption in different ways based on the mechanisms discussed above [6]. For example, as shown in Fig. 11.2a, the nanoparticles can be deposited on the surface of the solar cells to enhance the light scattering effect [7], where the absorption of light is increased by the multiple scattering paths. The nanoparticles can also be imbedded in the bulk to achieve increased light absorption due to the excited surface plasmons. As shown in Fig. 11.2b, the nanoparticles concentrate the energy from the incident light in the localized field of the surface plasmons, meaning more photon energy absorption in the surrounding semiconductor to induce extra photo-excited electron–hole pairs. Another implementation of the metal nanoparticles can be made by adding the particles on the bottom metal contact, as shown in Fig. 11.2c, to take advantage of the SPP effect [6].

11.2.2.2 Nanowires

The optical absorption in PVs can be enhanced by nanowires (arrays). Nanowires (arrays) can effectively reduce the optical loss due to the field concentration effect. In this case, the absorption cross section will be greater than the physical cross section of the nanowires [10]. Another enhancing effect of nanowire arrays has been achieved by using higher refractive index, which can

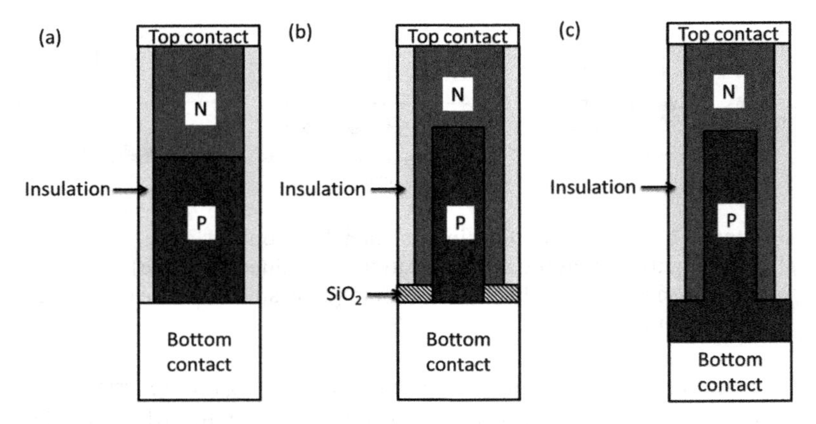

Figure 11.3 Schematics of a unit cell in the exemplary nanowire solar cell structures.

exhibit a cavity effect such that the light is trapped by the multiple reflections from the boundaries of wires to enhance efficiency [11]. The diameter and pitch of the nanowire arrays need to be on the same order of the incident light wavelength. And the dimension of the nanowires can be optimized. However, the effect would reach a plateau at an upper length limit for the cavity effect [12]. Effective nanowire arrays have been demonstrated using Si, InP, CdS, GaAs, ZnO, etc., offering many choices for different applications [13–17]. It is also possible to combine multiple nanostructures/nanomaterials into one design. For example, nanowires can be decorated with metal nanoparticles to be spectrally adjustable [18].

Nanowires can be implemented in the solar cells with many versatile designs. A nanowire with both a p-type section and an n-type section can form a unit cell, where the top n-type section utilizes the light-concentrating effect, and the dimension of the wire can be optimized to take advantage of the light cavity effect, as exemplified in the three structures in Fig. 11.3. The configuration of the n-type and p-type sections can be linear, as in Fig. 11.3a, or interlocking, as in Fig. 11.3b,c, depending on the process of fabrication [13, 14]. An array of such unit cells can be plainarized with a resin or spin-on glass (SOG). The tips of the wires can then be exposed by etching or chemical mechanical machining to allow contact with the transparent conducting oxide such as ITO [19].

11.2.2.3 Nanocrystal quantum dots

Beyond enhancing photoabsorption, nanomaterials have been utilized to boost carrier generation. The Schockley–Queisser limit (34%) was calculated assuming one single-excited electron–hole pair for each absorbed photon [20]. However, multiple exciton generation (MEG) has been demonstrated by nanocrystal quantum dots, making it possible to achieve power conversion efficiency even higher than 34% [21]. High energy photo-excited carries dissipate fast in bulk semiconductors because of carrier-phonon inelastic scattering resulting in fast dissipation, which can be slowed down by quantum dots. The discrete energy levels in quantum dots significantly lower the rate of the dissipation process, allowing the extra energy to generate a new pair of excitations, as illustrated in Fig. 11.4. By adjusting the size of semiconducting nanocrystals, quantum dots with discrete energy levels and altered band gap

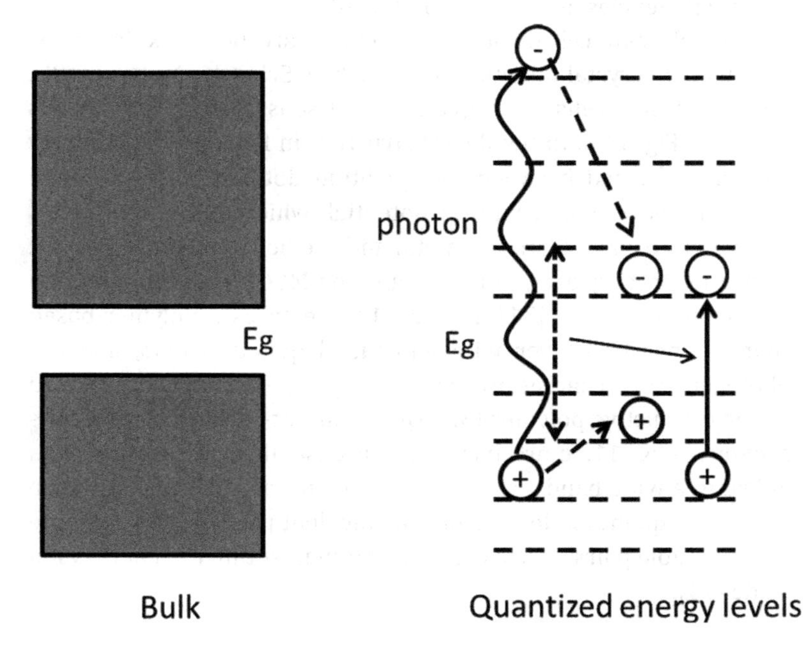

Figure 11.4 Quantized energy levels of quantum dots (right) as compared to the continuous energy levels on bulk semiconductor (left). The quantum confinement effects enable the generation of a second pair of carriers as the initial photoexcited carriers relax.

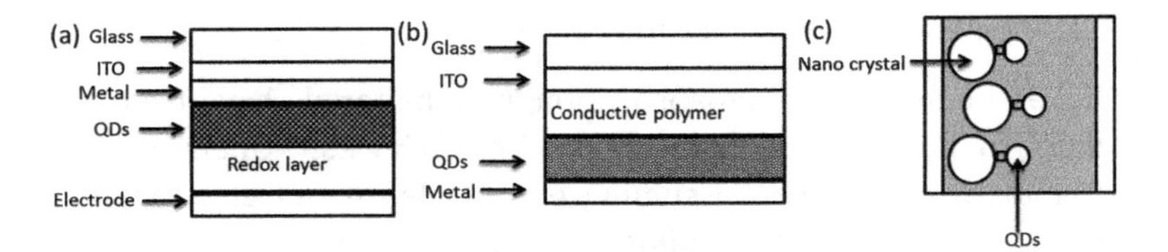

Figure 11.5 Three device structures of quantum dot solar cells: (a) metal/semiconductor junction cell, (b) polymer-based solar cell, and (c) quantum-dot-sensitized solar cells where quantum dots are linked to nanocrystals quantum dots.

can be achieved [22]. Such MEG effects have been demonstrated by quantum dots made from PbSe, PbS, PbTe, CdSe, Si, and InP, demonstrating the many material options to utilize this effect [23–28]. Complete extraction of the excess energy from the hot carriers could, in theory, break the Schockley–Queisser limit and enable device efficiencies up to 66% [21, 29, 30].

Several solar cell device architectures are being explored for various nanocrystal quantum dots, such as Schottky junction cells, polymer-based cells, and quantum-dot-sensitized cells [19], as shown in Fig. 11.5. In the device structure in Fig. 11.5a, a Schottky barrier is formed between the quantum dot film and the metal layer. The barrier is a built-in potential, which drives the photo-excited electrons toward the metal and the holes toward the redox hole transporting layer, thus enabling the MEGs from the quantum dots to be extracted [31]. Figure 11.5b features a polymer-based charge-generation layer with embedded quantum dots, and the photogenerated charges are extracted by the electric field provided by the conductive polymer [32]. Quantum-dot-sensitized solar cells shown in Fig. 11.5c are based on an ensemble of quantum dots linked to a wide band gap nanocrystal material (TiO_2 or ZnO) via a linker. The quantum dots absorb the incident photons and generate electron–hole pairs, which are then confined within the nanocrystal oxide [19].

11.2.3 *Fabrication of Nanomaterials for PVs*

Metallic nanoparticles can be fabricated by using a number of techniques, such as thermal evaporation and electrodeposition.

With thermal evaporation, a thin metal film in tens of nanometers can be formed on a substrate. Heating the substrate to a sufficiently high temperature induces agglomeration by surface tension, and an array of hemispherical nanoparticle with designed sizes suitable for plasmonic light trapping can be obtained. Techniques utilizing templates can allow more controls over the formation of the nanoparticles. For example, evaporation can be performed through a porous ceramic mask made by Al_2O_3 to give the control about the diameter and density of the particles. With this technique, Ag nanoparticles with 110 nm in diameter and 55 nm or 220 nm in height have been demonstrated [33]. The similar templated control has also been demonstrated with a lithographically defined mask followed by Ag evaporation and lift-off process [6, 34]. Electrodeposition of nanoparticles on a substrate suitable for PV applications can be achieved by using a pulsed current [35]. The particle size and density can be controlled by tuning current density and on/off time.

Nanowire arrays can be prepared by using either etching or growth approaches. For etching approaches, removing parts of an existing substrate by using either dry plasma or wet chemistry-based etching produces nanowires. Silicon is the most common materials in current commercial solar cells. It has the advantage that it is being used in the electronics industry extensively, where the dry etching method is a routine process in the industry. To obtain an ordered array of Si nanowires by dry etching, a template mask is usually required [36]. The mask can be generated by standard photolithography process or self-assembled particle arrays. For example, a monolayer of self-assembled silica beads can be formed on Si substrates with perfect two-dimensional hexagonal packing providing an ideal template for deep reactive ion etching to form Si nanowire arrays [13]. The shortcoming of reactive ion etching (RIE) is that the aspect ratio of the wires has limitation. Therefore, to obtain ordered Si nanowire arrays with high aspect ratios, electroless etching can be used by an ordered template made of precious metals such as Au, Ag, and Pt [37].

Contrary to the etching approach, the growth approach adds new nanowires to the underlying substrate. The catalytic growth method is a versatile technique that can enable nanowires of axial and

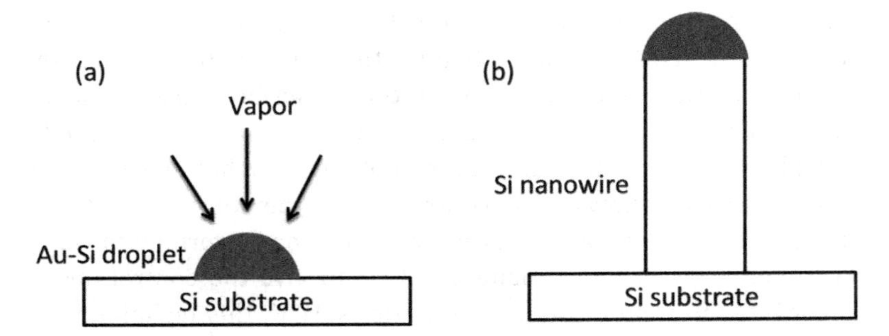

Figure 11.6 Schematic of growth of Si nanowires by VLS method: (a) starting with liquid droplet on Si substrate and (b) growing nanowire with the droplet on the top. Reprinted with permission from Ref. [38]. Copyright 2004, AIP Publishing LLC.

core–shell-type heterojunctions by changing the growth species. The most widely investigated growth approach is the vapor–liquid–solid (VLS) method, although other methods based on vapor–solid (VS) and electrochemical deposition techniques have been reported [39–42]. VLS has been demonstrated to grow both group IV semiconducting nanowires (Si, Ge) and III-V compound nanowires (GaAs, InAs, InP) [43]. In VLS methods, nanowires are formed on seed particles that collect the growth species in molecular beam epitaxy (MBE) or metal organic vapor phase epitaxy (MOVPE) systems. Several mechanisms are possibly involved in the growth process, including the direct impinging of growth species on the seed particles [43], the impingement of the species on the substrate area surrounding the seed particles followed by diffusion toward the seeds, and diffusion after impingement on the sidewalls after the initial formation of the wires [43]. All the mechanisms can contribute to the formation (Fig. 11.6a) and epitaxial growth (Fig. 11.6b) of the nanowires underneath the seed particles due to the supersaturation of the growth species [38, 44–48]. The options for the seed particle include Au droplets and Ga droplets. The Au droplets can be formed using evaporation, colloidal or aerosol deposition [45, 49], and the Ga droplets can be deposited via MBE on feature on native oxide [50–56]. Moreover, templates can be provided for VLS growth via various lithographic techniques to offer more control over the uniformity of the nanowires [45, 46].

For colloidal particles or semiconducting quantum dots, solution-based processing is commonly utilized for fabrication where the particles nucleate and grow in a controlled manner. The material systems typically require three components: chemical precursors of inorganic salts or organometallic compounds, organic surfactants, and solvents. The precursors chemically transform into monomers upon heating to a sufficiently high temperature and form the seeds of the nanocrystal quantum dots, which then grow under the influence of surfactants. This technique has been used to synthesize II-VI and I-VI semiconducting quantum dots by rapidly injecting the precursors into a heated and stirred solvent with organic surfactants. The resulting inorganic nanoparticles are inorganic in the core but carry an organic shell of surfactant ligands that prevent agglomeration in the solution [30]. Usually, these ligands are typically long-chain molecules with low conductivity. A ligand exchange can be performed to replace the long ligands with short and more conductive species to improve electro-transport properties [57]. The conductivity can be further improved by sintering the film at higher temperature to get higher device efficiencies [58]. The nanostructured materials in the PV devices as discussed above demonstrated the utility and versatility to enable new designs.

11.3 Nanostructures for Thermoelectrics

Thermoelectric power generators are another type of devices that can utilize (concentrated) solar heat to generate power. Concentrated solar power (CSP) is one of the potential candidates to harvest solar energy besides PVs. TE power generators using concentrated solar heat are a viable approach to realize CSP [59]. TE power generator from other heat forms can also increase efficiency of energy consumption. For fossil-fuel-based energy consumption, about 50% of the power is lost in waste heat accompanied by the emission of greenhouse gases [60]. TE devices convert heat to electricity without much pollution for waste heat recycling [61]. For example, recovering the exhaust heat from motor vehicles with a thermoelectric system with 10% efficiency can increase the mileage by up to 10% [62]. However, low device efficiency is hampering the application of TE devices. Recent research efforts have shown that reducing the dimension of semiconducting materials from bulk to nanoscale significantly renders superior thermoelectric properties, suggesting that implementing nanostructures in TE devices is a promising approach for higher efficiencies [60, 63, 64]. In this section, we will discuss the basics of TE power generation, the nanotechnologies for TEs, fabrication of nanostructures, and their applications in the devices.

11.3.1 *Basics of Thermoelectrics*

A representative configuration of a single-cell thermoelectric power generator [65] is shown in Fig. 11.7, where the p-type and n-type sections are joined via a conductor exposed to the heat source (T_h) and the cooler side (T_c). It can be configured to produce electricity to power an electric load R. Under the temperature difference (T_h–T_c), electrons in the n-type section and holes in the p-type section both move along the temperature gradient, meaning from T_h to T_c, which produces a voltage drop across the serially connected P-Conductor-N circuit.

The efficiency of TE materials (ZT) is characterized by a dimensionless figure of merit [66, 67], $ZT = S^2\sigma T/k$, where S is the thermoelectric power (Seebeck coefficient), defined as the thermoelectric voltage (V) per degree of temperature (T) difference ($S = dV/dT$); σ is the electrical conductivity; k is the thermal conductivity; and T is the temperature. The ZT of a thermoelectric material has to be above 3.0 to be competitive with present fossil-fuel-based technology.

Enhancing ZT entails maximizing S and σ while minimizing k, which is a challenging task because all these three parameters are typically interdependent. The interdependence of these parameters is best understood by examining their individual dependence on charge carriers [68]. Specifically, the Seebeck coefficient is related to carrier concentration (n) and carrier effect mass (m^*) as

$$S = \frac{8\pi^2 k_B^2}{3\varepsilon\hbar^2} m^* T \left(\frac{\pi}{3n}\right)^{2/3} \tag{11.2}$$

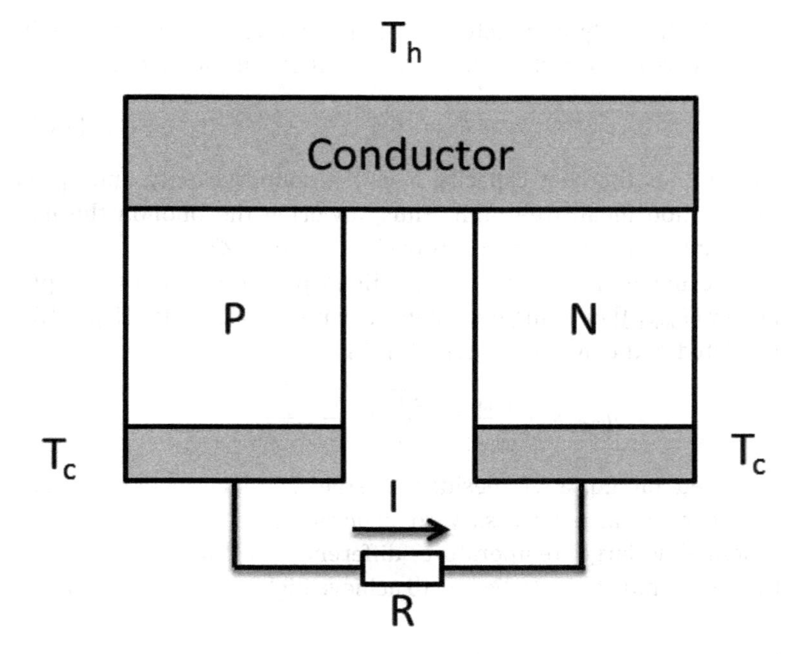

Figure 11.7 Schematic of a single p-n thermoelectric unit.

where k_B is the Boltzmann constant, e is the carrier charge, and h is the Plank constant. The electrical conductivity is related to charge carrier concentration by

$$\sigma = ne\mu \tag{11.3}$$

where μ is the carrier mobility, n is carrier concentration, and e is electron charge. Whereas thermal conductivity is the sum of the charge carrier thermal conductivity (k_e) and phonon thermal conductivity (k_l), that is,

$$k = k_l + k_e \tag{11.4}$$

where k_e is proportional to electrical conductivity according to

$$k_e = L\sigma T \tag{11.5}$$

where L is the Lorenz number. It can be seen from Eq. 11.2–11.5 that increasing electrical conductivity by increasing carrier concentration adversely affects thermopower and thermal conductivity.

Actually, the only parameter that is not influenced by changes in charge carrier characteristics is phonon thermal conductivity

$$k_l = \frac{C_v v_s \lambda_{ph}}{3} \tag{11.6}$$

where C_v is the heat capacity, v_s is the sound velocity, and λ_{ph} is the phonon mean free path. Thus, lowering the phonon thermal conductivity is an effective approach to improve *ZT*.

The maximum efficiency η_{max} [load power (P_{load}) over input power (P_{input})] of a single-element thermoelectric device (Fig. 11.7) is related to the figure of merit *ZT* [69] by

$$\eta_{\text{max}} = \frac{P_{\text{load}}}{P_{\text{input}}} = \frac{\Delta T}{T_h} \times \frac{\sqrt{1+ZT}-1}{\sqrt{1+ZT}+\frac{T_c}{T_h}} \tag{11.7}$$

Based on Eq. 11.7, besides a large *ZT*, a large temperature difference is also necessary to generate sufficient electricity. A sustainably large temperature difference in turn requires the thermal conductivity to be low to achieve higher efficiency for TEs.

11.3.2 *Nanotechnologies for TEs*

Nanostructures and nanomaterials have been reported to have significantly higher TE efficiency than their bulk counterparts [60, 61, 63, 70]. It can lower phonon thermal conductivity via increased phonon scattering when the dimension of the nanoscopic features in the material is less than phonon mean free path [71, 72]. Examples of nanotechnology in TEs include nanowire arrays, two-dimensional nanofeature arrays, and bulk nanocomposite materials with nanoscopic crystal grains [73].

Nanowires such as silicon nanowires have superior thermoelectric performance than their bulk counterparts, because of the low dimensions and large surface areas of the nanowires. Depending on the diameter of the nanowires, phonon scatterings on the surface and interface can greatly hinder the longitudinal transport of heat through the nanowires. As a result, a nanowire resists heat flow but remains electrically conductive, providing that the diameter is greater than the mean free paths of electrons. Based on this mechanism, it has been demonstrated that it is possible to increase the *ZT* value to about 1.0 at 200 K by using Si nanowires with cross-sectional area of 20 nm by 20 nm [63], which is two magnitudes higher than 0.01 for bulk Si. It is also possible to reach a *ZT* value of 0.6 at room temperature by using silicon nanowires with diameters of about 50 nm [60]. The preferential scatterings of phonons play a significant role when the nanowires have high aspect ratios because the phonons need to travel a longer distance from one end to the other end of the wires, which means more chances of phonon scattering to further reduce thermal conductivity and improve *ZT*.

Two-dimensional nanostructures can be made by using a thin film substrate, such as silicon-on-insulator (SOI) substrate with nanoscale features in the 10–20 nm ranges. Again, these dimensions are shorter than the typical mean free path of phonon in Si but longer than the electron mean free paths. Therefore, it offers a way to restrict phonon waves that carry heat without hindering the movement of electrons. For example, Si nanomesh with holes of less than 20 nm in diameter and a pitch of 34 nm on a 22 nm thick SOI has a thermal conductivity 100 times lower than a bulk Si, whereas its electrical conductivity is similar to that of a bulk material [64]. This reduction of thermal conductivity was attributed to phononic effects and the possible related modification of the bulk phonon dispersions [64], and it is consistent with recent atomistic simulations that demonstrated the effect of phonon scattering in nanopours silicon [74, 75]. Moreover, the preferential scattering of phonons over electrons in a periodic array of defects may also forbid phonons of certain of frequencies due to acoustic mismatch between the matrix and the defects, thus creating a phonon band gap [76, 77]. All these phononic effects are due to nanostructuring and increase the ZT values.

Beyond nanowires and two-dimensional nanofeatures, nanoscale engineering can be also incorporated into the bulk materials to reduce thermal conductivity. Bulk nanocomposite materials with nanoscale single crystal grains can reduce lattice thermal conductivity via increased grain boundary scattering of phonons, leaving electron transport less influenced because of the much smaller mean free paths of electrons than phonons [73]. This

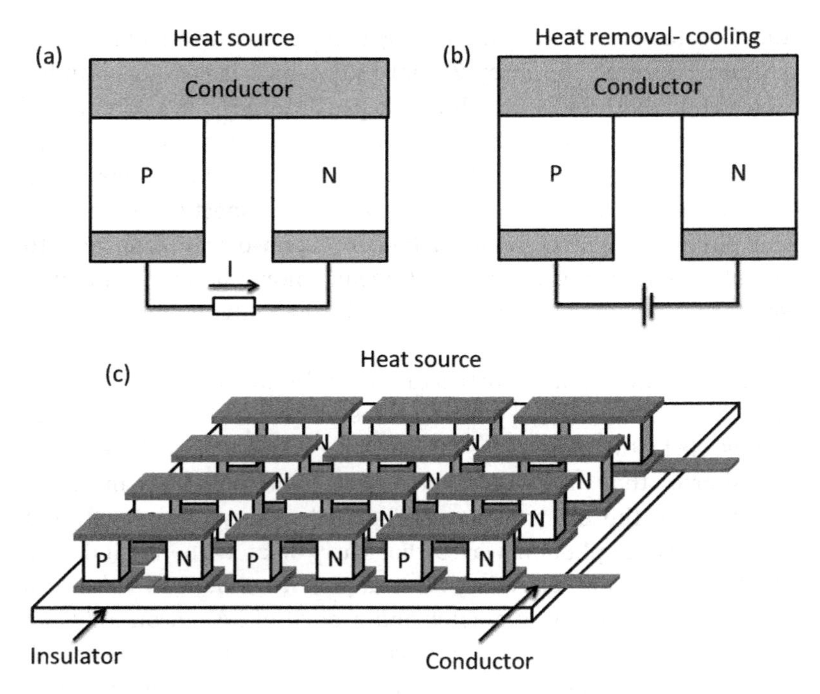

Figure 11.8 TE devices: (a) single unit for power generation mode, (b) single unit for cooling mode, and (c) the internal setup of unit array for practical TE power generator module.

approach has been applied to different compound polycrystalline materials based on Bi_2Te_3, PbTe, and SiGe [78]. Since these nanomaterials are produced in bulk form, they can be conveniently manufactured into TE devices, offering a cost advantage in the industry.

TE devices have relatively simple structural designs. A TE p-n junction unit can work for power generation mode or cooling mode, as illustrated in Fig. 11.8a,b. For power generation mode in Fig. 11.8a, recycling the heat from automobile engines, storing the energy in a battery, and converting the waste heats from nuclear engines to electricity in deep-space probes can be achieved [62]. For cooling mode in Fig. 11.8b, the TE device can effectively cool precision instruments without any moving parts. For example, TE

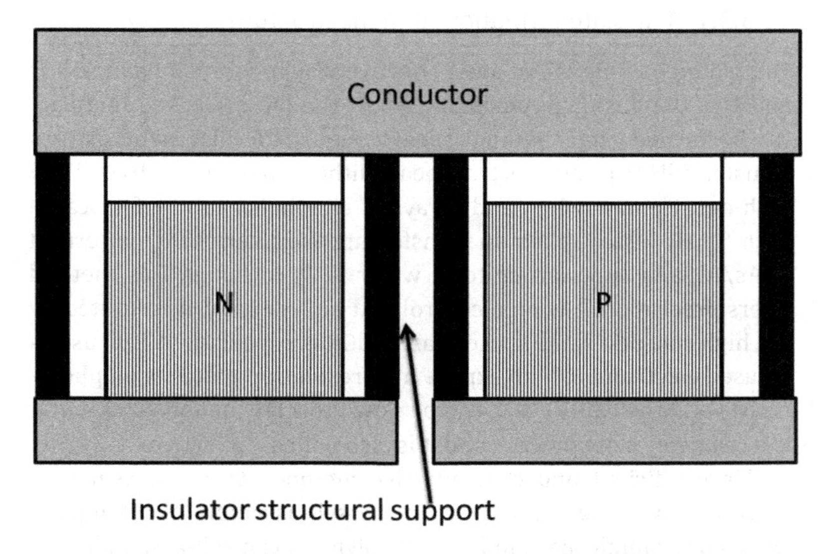

Figure 11.9 Schematic of a TE unit with n-p type nanowire arrays, and structural support is built into the system to deal with the mechanical stress.

coolers are used to produce a temperature difference of −80°C to cool down the infrared sensors in heat-tracking missiles and night vision imaging systems [62]. A thermoelectric module is typically a number of n-p junction unit arrays electrically connected in series but thermally parallel, as shown in Fig. 11.8c. This is an effective design to ensure that the thermoelectrically generated voltage in each unit cell adds up to the desired total voltage when functioning in power generation mode, while maintaining the same temperature difference across each unit.

All the different types of nanostructured TE materials can be implemented in a TE device with considerations about the mechanical reliability. It is especially simple for bulk nanocomposites since they can be easily shaped into pellets of the designed dimensions. Implementation of high-aspect-ratio nanowire arrays into a real device requires designs with spacers like insulator structural support as main load bearers when the device is under mechanical stress, as shown in Fig. 11.9.

11.3.3 *Fabrication of Nanostructure for TEs*

11.3.3.1 Template formation for nanostructure

Templated etching is the most effective way to fabricate nanowires and two-dimensional nanofeatures. At the lab scale, the template can be formed using methods such as superlattice nanowire pattern transfer (SNAP) and electron beam lithography (EBL) [63]. Both high-density nanowires and arrays of nanoholes can be fabricated with SNAP, which relies on transferring the alternating pattern of GaAs/AlGaAs superlattice to Si wafer [79]. Although this method offers precise dimensional control and high density, it is limited by the high cost and small size of GaAs/AlGaAs superlattice. EBL uses a focused electron beam to expose high resolution patterns on photoresists and the feature size can be down to a few nanometers, but it is a relatively slow process and it is also not cost effective.

More efficient and cost effective methods to produce nanostructures in a large scale rely on the self-assembly of a template layer. For example, certain block copolymers can self-assemble into nanoscale domains in a controlled environment and then be used to create templates for pattern transfer [80]. Because this template-generating process is solution based and only requires nanoscale film thickness, it is promising for large-scale fabrication. Templates based on block copolymer rely on the phase separation to form different morphologies. Diblock copolymers with blocks of polymer A and polymer B can form morphologies such as spheres of A in B (S), cylinders of A in B (C), or A-B alternating lamella (L). All of which can be transformed to achieve templates for nanopores, nanoholes, or nanowires [81]. While the detailed procedures may be dramatically different among the different methods to achieve these structures, they all rely on the sustainment or removal of one of the two blocks. As an example of template generation by removing the minor blocks that constitute either of the S, C, or L domains, the widely used PS-b-PMMA block copolymer with cylindrical domains of PMMA can be transformed to a PS template with hexagonal arrays of holes by soaking the thin film in acetic acid, which preferentially dissolves PMMA while keeping the PS matrix intact [82]. The PMMA domains can be also removed by exposure to UVC light followed by sonication in a good solvent such as toluene or benzene. UVC light has been shown to decompose the PMMA domain and, in the meanwhile, crosslinks the PS domain, creating a stronger PS template [83]. A nanoporous PS film generated with these methods can be used to transfer the hexagonal array to the underlying substrate by dry or wet etching methods. For some other copolymer systems, the minor domains can be retained and the major domain can be removed. One example of such systems is a sphere forming polystyrene-b-polydimethylsiloxane copolymer [80]. Due to the large Flory–Huggins interaction parameter between styrene and dimethylsiloxane, self-assembly easily happens at room temperature in a solvent environment to form spheres of PDMS in a matrix of PS. A short oxygen plasma treatment etches away the PS matrix and transforms the PDMS domains into a hexagonal array of silicon oxide spheres, thus retaining the minor spherical domains that can be used for pattern transfer [80].

11.3.3.2 Nanostructure formation for TEs

To achieve nanostructures suitable for TE devices, dry plasma and solution-based wet etching are commonly used to achieve nanowires or nanoholes/nanopores [60, 84]. A lithographic block copolymer template described above can be used to pattern the underlying substrate by a plasma etching process. This is especially useful for fabrication of thin film devices with highly dense nanoscopic features. However, the limitation is that most plasma etching recipes for etching thin films also attack the template, so the aspect ratio of the etched features is limited by the thickness of the template. The templates can be further processed to create metallic feature arrays that can be combined with catalytic wet etching to create features of high aspect ratio, such as Si nanowires. Etching of Si in solutions composed of HF and oxidants had been known to create porous silicon structures [85–88]. It was discovered later that such etching can be catalyzed by Al, in which the incubation time for the porous Si formation was greatly reduced [89]. Further investigation revealed that discontinuous film of noble metals (Au, Ag, Pt, or Au/Pd alloy) strongly catalyzes the etching of Si in solutions containing HF, H_2O_2, and EtOH, resulting in straight pores or columnar structures [84]. The commonly used method of metal-assisted chemical etching of Si involves submerging a substrate partially covered by noble metals in a solution of HF and H_2O_2. Due to the catalytic effects, Si underneath the metals is etched much faster than elsewhere so that the metal sinks into Si, leaving pores or wires behind with geometries matching the metallic catalyst.

11.3.3.3 Fabrication of bulk nanocomposite materials for TEs

To form bulk nanocomposites to be used in TE device, various approaches can be adopted. One method uses $Bi_{0.5}Sb_{1.5}Te_3$ nanopowers prepared by the ball-milling process, which were then hot-pressed to achieve crystal grains in a range of sizes as low as a few nanometers [78]. The same method has also been used to make SiGe-based nanocomposites, which have reduced thermal conductivity, increased thermopower, and slightly lowered electrical conductivity, yielding ZT as high as 1.3 at 1173 K [90]. Powders can be also prepared by melt-spinning an ingot to form thin ribbons followed by grinding. Spark plasma sintering of such powders resulted in an amorphous matrix with crystalline domains of nanocrystals ranging from 5 nm to 15 nm, which yielded a p-type BiSe-based bulk material with ZT as high as 1.56 at 300 K [91, 92]. More complex nanocomposites are possible by cooling slowly after melting to allow the formation of nanoscale precipitates in a single crystal matrix. Using this approach, a type of PbTe-based alloy with impurities of Ag and Sb has been fabricated with ZT as high as 2.2 at 800 K [93].

It should be mentioned that ZT of nanowires, two-dimensional nanofeatures, and bulk nanomaterials are still below 3, which is the necessary threshold to compete with fossil fuel technologies, despite the vast progresses have been achieved so far. Thus, the application of TE devices is far limited, where the cost is not a major concern but functionality and reliability. However, nanoscale engineering still holds promises to further reduce thermal conductivity to even below the amorphous limit by possibly altering phonon modes and phonon propagation [94]. So TE devices have the potential to be widely adopted in the near future with more efforts.

11.4 Conclusion

Based on our discussions in this chapter, it can be concluded that nanotechnologies offer many means to enhance the efficiencies of the PV and TE devices. For PVs, nanoparticles such as metal nanoparticles and quantum dots can enhance PV efficiency through enhanced absorption and multiple carrier generation, nanowires can enhance the photon absorption of PV devices through light concentration and cavity effect. For TEs, nanostructures (arrays) and bulk nanocomposite materials with nanoscale grains reduce thermal conductivity to enhance the efficiency of thermoelectric energy conversion. Adoption of these nanotechnologies is feasible with various designs in PV devices and TE devices. Although this chapter focuses on nanotechnologies in PVs and TEs, it is worth noting that nanoscale engineering also offers other opportunities for energy generation and storage, such as fuel cells, batteries, and bioapplications. With these promising characteristics, clean, green and highly efficient with the aid of nanotechnologies.

References

1. Snyder, G. J., and Toberer, E. S. (2008). Complex thermoelectric materials, *Nature Mater.*, **7**, pp. 105–114.
2. Zahl, H. A., and Ziegler, H. K. (1960). Power sources for satellites and space vehicles, *Sol. Energ.*, **4**, pp. 32–38.
3. Würfel, P. (2009). *Physics of Solar Cells: From Basic Principles to Advanced Concepts*, Wiley-VCH, Weinheim, Germany.
4. Nelson, J., and Ratner, M. (2004). The physics of solar cells, *Phys. Today*, **57**, pp. 71–71.
5. Henry, C. H. (1980). Limiting efficiencies of ideal single and multiple energy gap terrestrial solar cells, *J. Appl. Phys.*, **51**, pp. 4494–4500.
6. Atwater, H. A., and Polman, A. (2010). Plasmonics for improved photovoltaic devices, *Nature Mater.*, **9**, pp. 205–213.
7. Pillai, S., Catchpole, K. R., Trupke, T., and Green, M. A. (2007). Surface plasmon enhanced silicon solar cells, *J. Appl. Phys.*, **101**, pp. 093105–1.
8. Catchpole, K. R., and Polman, A. (2008). Design principles for particle plasmon enhanced solar cells, *Appl. Phys. Lett.*, **93**, pp. 191113.

9. Spinelli, P., Hebbink, M., van Lare, C., Verschuuren, M., de Waele, R., and Polman, A. (2010). Plasmonic anti-reflection coating for thin film solar cells, *Advanced Photonics and Renewable Energy*, OSA Technical Digest (Optical Society of America), paper PWE3.
10. Krogstrup, P., Jørgensen, H. I., Heiss, M., Demichel, O., Holm, J. V., Aagesen, M., Nygard, J., and Morral, A. F. (2013). Single-nanowire solar cells beyond the Shockley-Queisser limit, *Nature Photon.*, **7**, pp. 306–310.
11. Nobis, T., Kaidashev, E. M., Rahm, A., Lorenz, M., and Grundmann, M. (2004). Whispering gallery modes in nanosized dielectric resonators with hexagonal cross section, *Phys. Rev. Lett.*, **93**, pp. 103903.
12. Hu, Y., LaPierre, R. R., Li, M., Chen, K., and He, J. J. (2012). Optical characteristics of GaAs nanowire solar cells, *J. Appl. Phys.*, **112**, pp. 104311.
13. Garnett, E., and Yang, P. (2010). Light trapping in silicon nanowire solar cells, *Nano Lett.*, **10**, pp. 1082–1087.
14. Fan, Z., Razavi, H., Do, J., Moriwaki, A., Ergen, O., Chueh, Y.-L., Leu, P. W., Ho, J. C., Takahashi, T., Reichertz, L. A., Neale, S., Yu, K., Wu, M., Ager, J. W., and Javey, A. (2009). Three-dimensional nanopillar-array photovoltaics on low-cost and flexible substrates, *Nature Mater.*, **8**, pp. 648–653.
15. Cho, K., Ruebusch, D. J., Lee, M. H., Moon, J. H., Ford, A. C., Kapadia, R., Takei, K., Ergen, O., and Javey, A. (2011). Molecular monolayers for conformal, nanoscale doping of InP nanopillar photovoltaics, *Appl. Phys. Lett.*, **98**, pp. 203101.
16. Kelzenberg, M. D., Boettcher, S. W., Petykiewicz, J. A., Turner-Evans, D. B., Putnam, M. C., Warren, E. L., Spurgeon, J. M., Briggs, R. M., Lewis, N. S., and Atwater, H. A. (2010). Enhanced absorption and carrier collection in Si wire arrays for photovoltaic applications, *Nature Mater.*, **9**, pp. 239–244.
17. Colombo, C., Heiß, M., Gratzel, M., and Morral, A. F. (2009). Gallium arsenide p-i-n radial structures for photovoltaic applications, *Appl. Phys. Lett.*, **94**, pp. 173108.
18. Colombo, C., Krogstrup, P., Nygard, J., Brongersma, M. L., and Morral, A. F. (2011). Engineering light absorption in single-nanowire solar cells with metal nanoparticles, *New J. Phys.*, **13**, pp. 123026.
19. Kamat, P. V. (2008). Quantum dot solar cells. Semiconductor nanocrystals as light harvesters, *J. Phys. Chem. C*, **112**, pp. 18737–18753.
20. Shockley, W., and Queisser, H. J. (1961). Detailed balance limit of efficiency of p-n junction solar cells, *J. Appl. Phys.*, **32**, pp. 510–519.

21. Stolle, C. J., Harvey, T. B., and Korgel, B. A. (2013). Nanocrystal photovoltaics: A review of recent progress, *Curr. Opin. Chem. Eng.*, **2**, pp. 160–167.

22. Lin, Z., Franceschetti, A., and Lusk, M. T. (2011). Size dependence of the multiple exciton generation rate in CdSe quantum dots, *ACS Nano*, **5**, pp. 2503–2511.

23. Stubbs, S. K., Hardman, S. J. O., Graham, D. M., Spencer, B. F., Flavell, W. R., Glarvey, P., Masala, O., Pickett, N. L., and Binks, D. J. (2010). Efficient carrier multiplication in InP nanoparticles, *Phys. Rev. B*, **81**, pp. 081303-1–081303-4.

24. Schaller, R. D., and Klimov, V. I. (2004). High efficiency carrier multiplication in PbSe nanocrystals: Implications for solar energy conversion, *Phys. Rev. Lett.*, **92**, pp. 186601.

25. Ellingson, R. J., Beard, M. C., Johnson, J. C., Yu, P., Micic, O. I., Nozik, A. J., Shabaev, A., and Efros, A. L. (2005). Highly efficient multiple exciton generation in colloidal PbSe and PbS quantum dots, *Nano Lett.*, **5**, pp. 865–871.

26. Murphy, J. E., Beard, M. C., Norman, A. G., Ahrenkiel, S. P., Johnson, J. C., Yu, P., Micic, O. I., Ellingson, R. J., and Nozik, A. J. (2006). PbTe colloidal nanocrystals: Synthesis, characterization, and multiple exciton generation, *J. Am. Chem. Soc.*, **128**, pp. 3241–3247.

27. Beard, M. C., Knutsen, K. P., Yu, P., Luther, J. M., Song, Q., Metzger, W. K., Ellingson, R. J., and Nozik, A. J. (2007). Multiple exciton generation in colloidal silicon nanocrystals, *Nano Lett.*, **7**, pp. 2506–2512.

28. Schaller, R. D., Petruska, M. A., and Klimov, V. I. (2005). Effect of electronic structure on carrier multiplication efficiency: Comparative study of PbSe and CdSe nanocrystals, *Appl. Phys. Lett.*, **87**, pp. 253102.

29. Ross, R. T., and Nozik, A. J. (1982). Efficiency of hot-carrier solar energy converters, *J. Appl. Phys.*, **53**, pp. 3813–3818.

30. Klimov, V. I. (2003). Nanocrystal quantum dots, *Los Alamos Sci.*, **28**, pp. 214–220.

31. Luther, J. M., Law, M., Beard, M. C., Song, Q., Reese, M. O., Ellingson, R. J., and Nozik, A. J. (2008). Schottky solar cells based on colloidal nanocrystal films, *Nano Lett.*, **8**, pp. 3488–3492.

32. Kim, S. J., Kim, W. J., Cartwright, A. N., and Prasad, P. N. (2009). Self-passivating hybrid (organic/inorganic) tandem solar cell, *Sol. Energ. Mat. Sol. C.*, **93**, pp. 657–661.

33. Nakayama, K., Tanabe, K., and Atwater, H. A. (2008). Plasmonic nanoparticle enhanced light absorption in GaAs solar cells, *Appl. Phys. Lett.*, **93**, pp. 121904.
34. Beck, F. J., Polman, A., and Catchpole, K. R. (2009). Tunable light trapping for solar cells using localized surface plasmons, *J. Appl. Phys.*, **105**, pp. 114310-1–114310-7.
35. Kim, S.-S., Na, S.-I., Jo, J., Kim, D.-Y., and Nah, Y.-C. (2008). Plasmon enhanced performance of organic solar cells using electrodeposited Ag nanoparticles, *Appl. Phys. Lett.*, **93**, pp. 073307.
36. Li, J., Wang, D., and LaPierre, R. R. (2011). *Advances in III-V Semiconductor Nanowires and Nanodevices*, eBook by Bentham Science Publishers.
37. Peng, K., Lu, A., Zhang, R., and Lee, S. (2008). Motility of metal nanoparticles in silicon and induced anisotropic silicon etching, *Adv. Funct. Mater.*, **18**, pp. 3026–3035.
38. Wagner, R. S., and Ellis, W. C. (1964). Vapor-liquid-solid mechanism of single crystal growth, *Appl. Phys. Lett.*, **4**, pp. 89–90.
39. Fan, H. J., Bertram, F., Dadgar, A., Christen, J., Krost, A., and Zacharias, M. (2004). Self-assembly of ZnO nanowires and the spatial resolved characterization of their luminescence, *Nanotechnology*, **15**, pp. 1401–1404.
40. Wang, Z. L., Kong, X. Y., Ding, Y., Gao, P., Hughes, W. L., Yang, R., and Zhang, Y. (2004). Semiconducting and piezoelectric oxide nanostructures induced by polar surfaces, *Adv. Funct. Mater.*, **14**, pp. 943–956.
41. Martin, C. R. (1996). Membrane-based synthesis of nanomaterials, *Chem. Mater.*, **8**, pp. 1739–1746.
42. Xu, D., Xu, Y., Chen, D., Guo, G., Gui, L., and Tang, Y. (2000). Preparation of CdS single-crystal nanowires by electrochemically induced deposition, *Adv. Mater.*, **12**, pp. 520–522.
43. LaPierre, R. R., Chia, A. C. E., Gibson, S. J., Haapamaki, C. M., Boulanger, J., Yee, R., Kuyanov, P., Zhang, J., Tajik, N., and Jewell, N. (2013). III–V nanowire photovoltaics: Review of design for high efficiency, *Phys. Status Solidi Rapid Res. Lett.*, **7**, pp. 815–830.
44. Givargizov, E. I. (1975). Fundamental aspects of VLS growth, *J. Cryst. Growth*, **31**, pp. 20–30.
45. Martensson, T., Borgstrom, M., Seifert, W., Ohlsson, B. J., and Samuelson, L. (2003). Fabrication of individually seeded nanowire arrays by vapour–liquid–solid growth, *Nanotechnology*, **14**, pp. 1255–1258.
46. Martensson, T., Carlberg, P., Borgstrom, M., Montelius, L., Seifert, W., and Samuelson, L. (2004). Nanowire arrays defined by nanoimprint lithography, *Nano Lett.*, **4**, pp. 699–702.

47. Messing, M. E., Hillerich, K., Bolinsson, J., Storm, K., Johansson, J., Dick, K. A., and Deppert, K. (2010). A comparative study of the effect of gold seed particle preparation method on nanowire growth, *Nano Res.*, **3**, pp. 506–519.

48. Gustafsson, A., Hillerich, K., Messing, M. E., Storm, K., Dick, K. A., Deppert, K., and Bolinsson, J. (2012). A cathodoluminescence study of the influence of the seed particle preparation method on the optical properties of GaAs nanowires, *Nanotechnology*, **23**, pp. 265704.

49. Gudiksen, M. S., Wang, J., and Lieber, C. M. (2001). Synthetic control of the diameter and length of single crystal semiconductor nanowires, *J. Phys. Chem. B*, **105**, pp. 4062–4064.

50. Colombo, C., Spirkoska, D., Frimmer, M., Abstreiter, G., and Morral, A. F. (2008). Ga-assisted catalyst-free growth mechanism of GaAs nanowires by molecular beam epitaxy, *Phys. Rev. B*, **77**, pp. 155326.

51. Morral, A. F. (2011). Gold-free GaAs nanowire synthesis and optical properties, *IEEE J. Sel. Top. Quant. Electron.*, **17**, pp. 819–828.

52. Plissard, S., Dick, K. A., Larrieu, G., Godey, S., Addad, A., Wallart, X., and Caroff, P. (2010). Gold-free growth of GaAs nanowires on silicon: Arrays and polytypism, *Nanotechnology*, **21**, pp. 385602.

53. Mandl, B., Stangl, J., Hilner, E., Zakharov, A. A., Hillerich, K., Dey, A. W., Samuelson, L., Bauer, G., Deppert, K., and Mikkelsen, A. (2010). Growth mechanism of self-catalyzed group III–V nanowires, *Nano Lett.*, **10**, pp. 4443–4449.

54. Plissard, S., Larrieu, G., Wallart, X., and Caroff, P. (2011). High yield of self-catalyzed GaAs nanowire arrays grown on silicon via gallium droplet positioning, *Nanotechnology*, **22**, pp. 275602.

55. Morral, A. F., Colombo, C., Abstreiter, G., Arbiol, J., and Morante, J. R. (2008). Nucleation mechanism of gallium-assisted molecular beam epitaxy growth of gallium arsenide nanowires, *Appl. Phys. Lett.*, **92**, pp. 063112.

56. Bauer, B., Rudolph, A., Soda, M., Morral, A. F., Zweck, J., Schuh, D., and Reiger, E. (2010). Position controlled self-catalyzed growth of GaAs nanowires by molecular beam epitaxy, *Nanotechnology*, **21**, pp. 435601.

57. Semonin, O. E., Luther, J. M., Choi, S., Chen, H.-Y., Gao, J., Nozik, A. J., and Beard, M. C. (2011). Peak external photocurrent quantum efficiency exceeding 100% via MEG in a quantum dot solar cell, *Science*, **334**, pp. 1530–1533.

58. Guo, Q., Ford, G. M., Hillhouse, H. W., and Agrawal, R. (2009). Sulfide nanocrystal inks for dense Cu $(In_{1-x}Ga_x)(S_{1-y}Se_y)_2$ absorber films and their photovoltaic performance, *Nano Lett.*, **9**, pp. 3060–3065.

59. Tan, L., Singh, B., Date, A., and Akbarzadeh, A. (2012). Sustainable thermoelectric power system using concentrated solar energy and latent heat storage, *2012 IEEE International Conference on Power and Energy (PECon)*, pp. 105–109.
60. Hochbaum, A. I., Chen, R., Delgado, R. D., Liang, W., Garnett, E. C., Najarian, M., Majumdar, A., and Yang, P. (2008). Enhanced thermoelectric performance of rough silicon nanowires, *Nature*, **451**, pp. 163–167.
61. MacDonald, D. K. C. (2006). *Thermoelectricity: An Introduction to the Principles*, Dover Publications, Mineola, New York.
62. Bell, L. E. (2008). Cooling, heating, generating power, and recovering waste heat with thermoelectric systems, *Science*, **321**, pp. 1457–1461.
63. Boukai, A. I., Bunimovich, Y., Tahir-Kheli, J., Yu, J.-K., Goddard III, W. A., and Heath, J. R. (2008). Silicon nanowires as efficient thermoelectric materials, *Nature*, **451**, pp. 168–171.
64. Yu, J.-K., Mitrovic, S., Tham, D., Varghese, J., and Heath, J. R. (2010). Reduction of thermal conductivity in phononic nanomesh structures, *Nature Nanotech.*, **5**, pp. 718–721.
65. Nolas, G. S., Sharp, J., and Goldsmid, J. (2001). *Thermoelectrics: Basic Principles and New Materials Developments*, Springer, Berlin, New York.
66. Mahan, G., Sales, B., and Sharp, J. (1997). Thermoelectric materials: New approaches to an old problem, *Phys. Today*, **50**, pp. 42–47.
67. Chen, G., Dresselhaus, M. S., Dresselhaus, G., Fleurial, J.-P., and Caillat, T. (2003). Recent developments in thermoelectric materials, *Int. Mat. Rev.*, **48**, pp. 45–66.
68. Kaviany, M. (2008). *Heat Transfer Physics*, Cambridge University Press, Cambridge, UK.
69. Nemir, D., and Beck, J. (2010). On the significance of the thermoelectric figure of merit Z, *J. Electron. Mater.*, **39**, pp. 1897–1901.
70. Weber, L., and Gmelin, E. (1991). Transport properties of silicon, *Appl. Phys. A*, **53**, pp. 136–140.
71. Majumdar, A. (2004). Thermoelectricity in semiconductor nanostructures, *Science*, **303**, pp. 777–778.
72. Li, D., Wu, Y., Kim, P., Shi, L., Yang, P., and Majumdar, A. (2003). Thermal conductivity of individual silicon nanowires, *Appl. Phys. Lett.*, **83**, pp. 2934–2936.
73. Szczech, J. R., Higgins, J. M., and Jin, S. (2011). Enhancement of the thermoelectric properties in nanoscale and nanostructured materials, *J. Mater. Chem.*, **21**, pp. 4037–4055.

74. Lee, J.-H., Grossman, J. C., Reed, J., and Galli, G. (2007). Lattice thermal conductivity of nanoporous Si: Molecular dynamics study, *Appl. Phys. Lett.*, **91**, pp. 223110.

75. Lee, J.-H., Galli, G. A. and Grossman, J. C. (2008). Nanoporous Si as an efficient thermoelectric material, *Nano Lett*, **8**, pp. 3750–3754.

76. Kushwaha, M. S., Halevi, P., Dobrzynski, L. and Djafari-Rouhani, B. (1993). Acoustic band structure of periodic elastic composites, *Phys. Rev. Lett.*, **71**, pp. 2022–2025.

77. Vasseur, J. O., Deymier, P. A., Chenni, B., Djafari-Rouhani, B., Dobrzynski, L. and Prevost, D. (2001). Experimental and theoretical evidence for the existence of absolute acoustic band gaps in two-dimensional solid phononic crystals, *Phys. Rev. Lett.*, **86**, pp. 3012–3015.

78. Poudel, B., Hao, Q., Ma, Y., Lan, Y., Minnich, A., Yu, B., Yan, X., Wang, D., Muto, A. and Vashaee, D. (2008). High-thermoelectric performance of nanostructured bismuth antimony telluride bulk alloys, *Science*, **320**, pp. 634–638.

79. Melosh, N. A., Boukai, A., Diana, F., Gerardot, B., Badolato, A., Petroff, P. M. and Heath, J. R. (2003). Ultrahigh-density nanowire lattices and circuits, *Science*, **300**, pp. 112–115.

80. Jung, Y. S. and Ross, C. A. (2009). Well-ordered thin-film nanopore arrays formed using a block-copolymer template, *Small*, **5**, pp. 1654–1659.

81. Bates, F. S. and Fredrickson, G. H. (1999). Block copolymers-designer soft materials, *Phys. Today*, **52**, pp. 32–38.

82. Xu, T., Goldbach, J. T., Misner, M. J., Kim, S., Gibaud, A., Gang, O., Ocko, B., Guarini, K. W., Black, C. T., Hawker, C. J. and Russel, T. P. (2004). Scattering study on the selective solvent swelling induced surface reconstruction, *Macromolecules*, **37**, pp. 2972–2977.

83. Xiao, S., Yang, X., Edwards, E. W., La, Y.-H. and Nealey, P. F. (2005). Graphoepitaxy of cylinder-forming block copolymers for use as templates to pattern magnetic metal dot arrays, *Nanotechnology*, **16**, pp. S324–S329.

84. Li, X. and Bohn, P. W. (2000). Metal-assisted chemical etching in HF/H_2O_2 produces porous silicon, *Appl. Phys. Lett.*, **77**, pp. 2572–2574.

85. Uhlir, A. (1956). Electrolytic shaping of germanium and silicon, *Bell Syst. Tech. J.*, **35**, pp. 333–347.

86. Turner, D. R. (1958). Electropolishing silicon in hydrofluoric acid solutions, *J. Electrochem. Soc.*, **105**, pp. 402–408.

87. Watanabe, Y., Arita, Y., Yokoyama, T. and Igarashi, Y. (1975). Formation and properties of porous silicon and its application, *J. Electrochem. Soc.*, **122**, pp. 1351–1355.

88. Canham, L. (1990). Silicon quantum wire array fabrication by electrochemical and chemical dissolution of wafers, *Appl. Phys. Lett.*, **57**, pp. 1046–1048.

89. Dimova-Malinovska, D., Sendova-Vassileva, M., Tzenov, N. and Kamenova, M. (1997). Preparation of thin porous silicon layers by stain etching, *Thin Solid Films*, **297**, pp. 9–12.

90. Wang, X. W., Lee, H., Lan, Y. C., Zhu, G. H., Joshi, G., Wang, D. Z., Yang, J., Muto, A. J., Tang, M. Y., Klatsky, J., Song, S., Dresselhaus, M. S., Chen, G. and Ren, Z. F. (2008). Enhanced thermoelectric figure of merit in nanostructured n-type silicon germanium bulk alloy, *Appl. Phys. Lett.*, **93**, pp. 193121.

91. Xie, W., Tang, X., Yan, Y., Zhang, Q. and Tritt, T. M. (2009). High thermoelectric performance BiSbTe alloy with unique low-dimensional structure, *J. Appl. Phys.*, **105**, pp. 113713.

92. Xie, W., Tang, X., Yan, Y., Zhang, Q. and Tritt, T. M. (2009). Unique nanostructures and enhanced thermoelectric performance of melt-spun BiSbTe alloys, *Appl. Phys. Lett.*, **94**, pp. 102111.

93. Hsu, K. F., Loo, S., Guo, F., Chen, W., Dyck, J. S., Uher, C., Hogan, T., Polychroniadis, E. K. and Kanatzidis, M. G. (2004). Cubic $AgPb_mSbTe_{2+m}$: Bulk thermoelectric materials with high figure of merit, *Science*, **303**, pp. 818–821.

94. Chen, Z.-G., Han, G., Yang, L., Cheng, L. and Zou, J. (2012). Nanostructured thermoelectric materials: Current research and future challenge, *Prog. Nat Sci: Mater. Int.*, **22**, pp. 535–549.

Chapter 12

Plasmonic Coupling between Nanostructures: From Periodic and Rigid to Random and Flexible Systems

Ugo Cataldi,[a,b] Roberto Caputo,[a] Yuriy Kurylyak,[b] Gérard Klein,[a] Mahshid Chekini,[a] Cesare Umeton,[b] and Thomas Bürgi[a]

[a]*Department of Physical Chemistry, University of Geneva, 30 Quai Ernest-Ansermet, 1211, Geneva 4, Switzerland*
[b]*Department of Physics, Centre of Excellence for the Study of Innovative Functional Materials CEMIF-CAL, University of Calabria and LICRYL - IPCF (Liquid Crystals Laboratory, Institute for Chemical Physics Processes) CNR – UOS Cosenza, 87036 Arcavacata di Rende, Italy*
ugo.cataldi@unige.ch

In this chapter, the mechanism of plasmonic coupling of the near-fields that takes place when nanostructures are very close to each other is described. After a theoretical introduction illustrating the main physical principles governing plasmonic coupling, several experimental systems are considered. Interestingly, experimental results show that both periodic and rigid, random, and flexible systems of gold nanoparticles exhibit a universal scaling behavior and verify the plasmonic ruler equation.

Active Plasmonic Nanomaterials
Edited by Luciano De Sio

ISBN 978-981-4613-00-2 (Hardcover), 978-981-4613-01-9 (eBook)
www.panstanford.com

12.1 Introduction

Plasmonics is the study of interactions of electromagnetic fields with metals at the nanoscale. Peculiar properties mainly related to the size of the nanostructures, especially between zero and 100 nm, have attracted much interest for applications in sensing [1–5], design of metamaterials [6–8], for single molecule detection [9], and in spectroscopic techniques [10–12]. In fact, nanoparticles of such sizes, much smaller than the wavelength of light with which they interact, produce local nanometer-range electric fields because the electrons of the conduction band are subject to both a displacement force (due to the action of the impinging field) and a recalling force (due to the positive attraction of atoms nuclei). This phenomenon was initially studied by G. Mie [13, 14] who calculated the interaction cross section, in dipole approximation, under the hypothesis that the sizes of particles are much smaller than the wavelength of the exciting light. This extinction cross section can be written as the sum of the scattering and absorption cross section:

$$C_{\text{sca}} = \frac{k^4}{6\pi}|\alpha|^2 = \frac{8\pi}{3}k^4a^6\left|\frac{\varepsilon-\varepsilon_m}{\varepsilon+2\varepsilon_m}\right|^2$$

$$C_{\text{abs}} = k\text{Im}[\alpha] = 4\pi ka^3\text{Im}\left[\frac{\varepsilon-\varepsilon_m}{\varepsilon+2\varepsilon_m}\right] \tag{12.1}$$

$$C_{\text{ext}} = \sigma_{\text{ext}} = 9\frac{\omega}{c}\varepsilon_m^{3/2}V\frac{\varepsilon_2}{[\varepsilon_1+2\varepsilon_m]^2+\varepsilon_2^2} \tag{12.2}$$

The value of C_{ext} in Eq. 12.2 is maximum when the Fröhlich condition is fulfilled in the denominator:

$$\varepsilon_1(\omega) = -2\varepsilon_m \tag{12.3}$$

with ε_2 very small. From this formula, we can observe that a fundamental role is played by the permittivity of the medium surrounding the nanoparticle because a change in this one strongly influences the frequency of the localized surface plasmon resonance (LSPR) of the system.

It is worth noting that in the expression of the extinction cross section, there is no direct trace of variables linked to the shape of nanoparticles. According to Eq. 12.2, the nanoparticle shape does not influence the frequency of the LSPR resonance and the number of resonance frequencies. As such, the formula is valid in the case of spherical particles that only show one resonance peak. In case nanorods are considered, two resonance peaks are experimentally observed, one for longitudinal oscillations of electrons and another one for transversal oscillations. We can conclude that the number of resonant modes is linked to the symmetry of nanoparticles. It follows that the plasmonic response of nanostructures can be enhanced if the beam exciting probe beam is polarized in the direction of the axis of symmetry. At the same time, the light polarization can eventually point out an anisotropy in the nanostructure.

Interestingly indeed, nanoparticles with other shapes such as cubes, stars or triangles [15] show several resonance peaks that can be highlighted by a particular choice of the exciting beam polarization. In conclusion, the frequency of the LSPR resonance is then related to many intrinsic and extrinsic parameters of nanostructures.

When not a single nanostructure but a system of many of them is considered, their inter-distance becomes a fundamental parameter to study the near-field coupling. As it will be reported in the following, the near-field coupling in a system of many nanostructures can strongly influence the spectral position of the LSPR resonance. By properly playing with this inter-distance, it is then possible to realize systems where the coupling can be statically or dynamically or controlled. At this point, it is worth making a distinction between passive and active systems. We can define as a static passive system the one that always gives the same response when it undergoes in a given solicitation. This is the case of a well-ordered system of nanostructures deposited above a rigid substrate (e.g., array of nanodisks) [16, 17]. In this case, the plasmon coupling between nanostructures can be only realized through the fabrication of several samples with well-ordered nanostructures put at different inter-particle distances. On the other hand, a system can be considered active when the plasmon coupling can be induced and continuously tuned in a reversible way by external perturbations. This functionality can be exploited by utilizing a different substrate with properties of elasticity and transparency to the visible spectrum. Such a substrate can be coated with randomly distributed nanostructures and can be externally and reversibly solicited by mechanical strains [19–21]. As shown in the following, it is striking to observe that the universal scaling behavior of the plasmon coupling properties, initially discovered and applied to well-ordered array of nanostructures, can be applied equally well also to these amorphous systems made of a completely random distribution of nanostructures.

12.2 Coupling Between Near-Fields

The strong electric field [24] induced by the exciting light is confined to the surface of the nanoparticle, and in resonance condition, there is an enhancement of this near-field. In the dipolar limit, if E_0 represents the exciting field, the electric field on the surface of the nanoparticle can be written as:

$$E_{\text{surface}} = \frac{(1+k)\varepsilon_m}{(\varepsilon + k\varepsilon_m)} E_0 \tag{12.4}$$

In the previous formula, k is the shape factor that is related to the geometry of the nanoparticle surface ($k = 2$, in case of a sphere). In resonance conditions, $\text{Re}(\varepsilon) = -\text{k}\varepsilon_m$, and the ratio

$$\left(\frac{|E|}{|E_0|}\right)^2 \tag{12.5}$$

is maximum. As already written, if the typical size of the nanoparticle is much smaller than the wavelength of the exciting light, absorption effects mainly due to the electron–phonon interaction are prevalent. If the size increases, the scattering process starts being relevant. These two effects are responsible for the damping of the near-field with a radiative decay that increases with the size of the nanoparticle, because both effects are enhanced at the LSPR frequency. For a particle with radius r in an electrical field E_0, the near-field at a distance r along the direction parallel to the incident light polarization is given by the expansion in multipolar modes:

$$E_{\text{near field}} = \frac{2\alpha E_0}{4\pi\varepsilon_0 r^3} + \frac{3\beta \dot{E}_0}{4\pi\varepsilon_0 r^4} + \frac{4\gamma \ddot{E}_0}{4\pi\varepsilon_0 r^5} + \cdots \tag{12.6}$$

where $\alpha, \beta, \gamma, \ldots$ are, respectively, the dipole, quadrupole, octupole, $\ldots$ polarizability tensors of the particle. By getting closer to the particle, a reliable estimate of the near-field needs to take into account not only the dipolar contribution but also the quadrupolar and octupolar ones.

When two metal nanoparticles are close to each other, the electric field felt by each particle can be written as the sum of the incident field E_0 and the near-field E_{nf} [18, 23, 24]:

$$E = E_0 + E_{\text{nf}} \tag{12.7}$$

The near-field appearing on one particle can interact with that corresponding to the adjacent one, and this interaction between nano-entities gives rise to the coupling of plasmonic fields. This coupling modifies the resonant frequency of the near-field. In more detail, the macroscopic effect of a decrease in the inter-distance between nanoparticles is a strong red-shift of the resonance peak position (wavelength) and hence a lower energy is required to excite this coupled plasmon resonance.

12.3 Polarization and Inter-Distance Dependence of the Plasmon Coupling

Several attempts have been made to provide a model that can take into account plasmonic coupling [25–28]. While studying a system of elliptical gold nanoparticle pairs illuminated with an exciting beam parallel polarized to the particles' inter-axis, Su et al. [15] observed that the plasmon shift (scaled to the isolated particle plasmon wavelength) decays exponentially as a function of the inter-particle gap separation (scaled to the particle diameter). Successively, Jain et al., in similar experimental conditions, have experimentally and theoretically realized a systematic calibration and standardization procedure for the scaling behavior of the plasmon shift related to the coupling between ordered nanostructures. The result of this study was the finding of an empirical formula that gives information about the coupling in terms of the gap existing between nanostructures, calculated as fractions of the diameters of the involved nanostructures. The detailed derivation of this formula will be illustrated in Sections 12.4.2 and 12.4.3 [18, 24]. Interestingly, in both well-ordered and random systems of gold nanostructures, the polarization of the impinging electric field triggers the coupling between near-fields. In the following section, an analytical formulation is provided of the different behavior of the same system in two complementary polarization conditions: parallel and orthogonal to the inter-axis of a couple of nanostructures.

12.3.1 *Enhancement of Coupling Due to Polarization Effects of the Exciting Electromagnetic Field*

By considering the polarizability of a particle with a not well-defined shape:

$$\alpha = (1+k)\varepsilon_0 V \frac{(\varepsilon - \varepsilon_m)}{(\varepsilon + k\varepsilon_m)} \tag{12.8}$$

where $\varepsilon = \varepsilon_r + i\varepsilon_i$ is the dielectric function of the metal the nanoparticle is made of, ε_m is the permittivity of the surrounding medium, and ε_0 is the vacuum permittivity. Assuming that the imaginary part of the metal dielectric function is small or weakly dependent on the wavelength, in the following we make the approximation of $\varepsilon \approx \varepsilon_r$. The polarizability α is maximum when $\mathrm{Re}(\varepsilon) = -k\varepsilon_m$ and, considering the case of spheres, $k = 2$. In a quasi-static approximation, an isolated particle in an electric field E_0 holds an electric dipole moment μ given by:

$$\mu = \alpha\varepsilon_m E_0 \tag{12.9}$$

If we consider a dimer (made of two nanoparticles put at a distance r) in an electric field (Fig. 12.1), in the dipolar interaction approximation, the field is:

$$E = E_0 + \frac{\xi\mu'}{4\pi\varepsilon_m\varepsilon_0 r^3} \tag{12.10}$$

where ξ is the orientation factor that depends on the reciprocal alignment of the two single-particle dipoles (θ_{12}) and from the alignment of dipoles with respect to the wave field (θ_1, θ_2):

$$\xi = 3\cos(\theta_1)\cos(\theta_2) - \cos(\theta_{12}) \tag{12.11}$$

If the dipoles are parallel, $\theta_{12} = 0$ and the electric field is parallel to the dipoles' inter-axis ($\theta_1 = \theta_2 = 0$), it results $\xi = 2$; if the electric

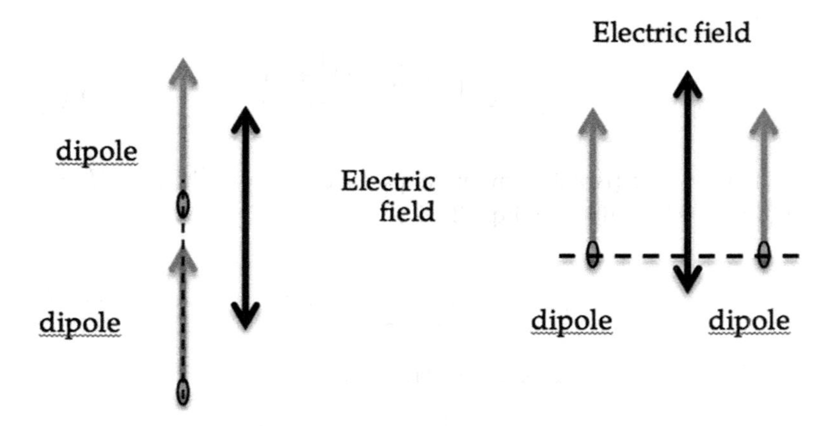

Figure 12.1 Two possible configurations of the electric field with respect to the dimer, parallel to the dimer inter-axis (left) and perpendicular to it (right).

field is orthogonal to the dipoles' inter-axis, the angle between the field and the dipoles is 90° for both and $\xi = -1$. The relative polarizability [17, 24] α' of the two-particle system is then given by:

$$\alpha' = \frac{\alpha}{1 - \frac{\xi\alpha}{4\pi\varepsilon_0 r^3}} \tag{12.12}$$

and from Eq. 12.8, for spheres with a volume $V = \pi D^3/6$, Eq. 12.12 becomes:

$$\alpha' = \frac{4\pi\varepsilon_0(\varepsilon_r - \varepsilon_m)D^3}{\varepsilon_r\left(8 - \xi\frac{D^3}{r^3}\right) + \varepsilon_m\left(16 + \xi\frac{D^3}{r^3}\right)} \tag{12.13}$$

which has a maximum for

$$\varepsilon_r = -\frac{\left(16 + \xi\frac{D^3}{r^3}\right)}{\left(8 - \xi\frac{D^3}{r^3}\right)}\varepsilon_m \tag{12.14}$$

If we introduce a gap s between the particles, the edge-to-edge distance defined as $s = r - D$, we have

$$\varepsilon_r = -\varepsilon_m\frac{\left[16\left(\frac{s}{D}+1\right)^3 + \xi\right]}{\left[8\left(\frac{s}{D}+1\right)^3 - \xi\right]} \tag{12.15}$$

with

$$k_{\text{system}} = \frac{\left[16\left(\frac{S}{D}+1\right)^3 + \xi\right]}{\left[8\left(\frac{S}{D}+1\right)^3 - \xi\right]} \tag{12.16}$$

For $S \to \infty$ (isolated spherical particle), we have $k_{\text{system}} = k = 2$ and $\varepsilon_r = -2\varepsilon_m$. If $\xi = 2$, Eq. 12.16 becomes:

$$\varepsilon_r\big|_{\xi=2} = -\varepsilon_m \frac{\left(8\left(\frac{S}{D}+1\right)^3 + 1\right)}{\left(4\left(\frac{S}{D}+1\right)^3 - 1\right)} \tag{12.17}$$

If $\xi = -1$, Eq. 12.16 can be written as:

$$\varepsilon_r\big|_{\xi=2} = -\varepsilon_m \frac{\left(16\left(\frac{S}{D}+1\right)^3 - 1\right)}{\left(8\left(\frac{S}{D}+1\right)^3 + 1\right)} \tag{12.18}$$

and the comparison between Eq. 12.17 and Eq. 12.18 gives:

$$\varepsilon_r\big|_{\xi=2} > \varepsilon_r\big|_{\xi=-1} \tag{12.19}$$

As such, the polarizability of the system (Eq. 12.13) for the parallel configuration $\xi = 2$ is larger than that for the orthogonal one $\xi = -1$.

12.4 Exponential Decay of Coupling for Well-Ordered Systems

12.4.1 *Two Interacting Gold Nanoparticles*

In 2003, Rechberger et al. [16] focused their attention on the near-field interaction of two identical gold nanoparticles. To study the inter-distance dependence of the plasmon coupling, they fabricated a bi-dimensional grating made of identical nanoparticles with the inter-distance between two successive columns gradually varied. Samples were fabricated by electron beam lithography. In Fig. 12.2, three sample structures are shown.

By exploiting this technique, only samples with small areas have been fabricated with a typical size of the gold nanoparticles of 150 nm (diameter) by 17 nm (height). The inter-distances (center to center) between each pair of nanostructures are, respectively,

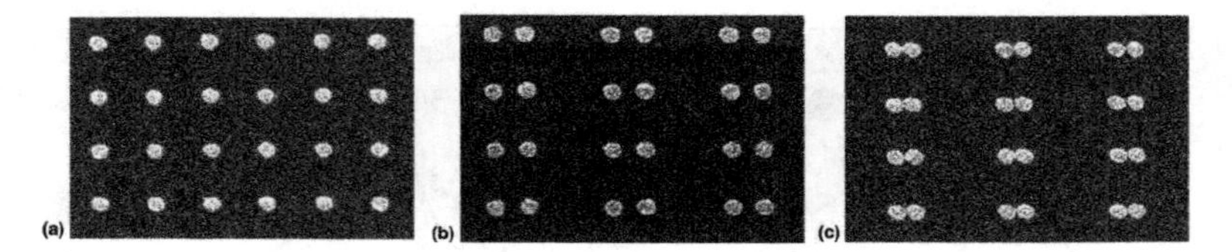

Figure 12.2 SEM images of nanoparticles arrays spaced (a) 450 nm, (b) 300 nm, and (c) 150 nm. The particle diameter is 150 nm with a thickness of 17 nm.

450 nm (Fig. 12.2a), 300 nm (Fig. 12.2b), and 150 nm (Fig. 12.2c). The UV-Vis spectra acquired from these samples are shown in Fig. 12.3. The exciting beam was polarized parallel to the long particle-pair axis and orthogonal to it. In the first case, the authors find a marked red-shift of the plasmon peak corresponding to a decrease in the inter-distance, while in the second case, they register only a limited blue-shift.

Summarizing these results in a graph where the plasmon shift is plotted as a function of the distance-diameter ratio, the authors find a different behavior for the two different polarizations (Fig. 12.3c). They explained these two different curves with a qualitative and simple dipole-dipole interaction model known in molecular systems. If the driven field is parallel to the long axis between the dipoles, the positive charge of one dipole faces the negative one.

The net effect is a weakening of the repulsive forces and an increase in the plasmon resonance frequency of the system. On the other hand, when the driven field is normal to the long axis between dipoles, charges of the same kind face each other resulting in an enhancement of the repulsive force and a decrease in the resonance wavelength. Figure 12.4 describes this qualitative model.

To quantitatively describe the phenomenon, the author used a retarded local field model, which considers the interaction of each dipole with all neighbor dipoles. The exponential decay of the red-shift with the increase in the inter-distance, highlighted by an s-polarized exciting light, is in agreement with other works. However, as reported in the following, a model explaining the inter-distance dependence of the plasmon coupling was successively proposed by Jain et al., who also considered well-ordered systems of nanostructures [18].

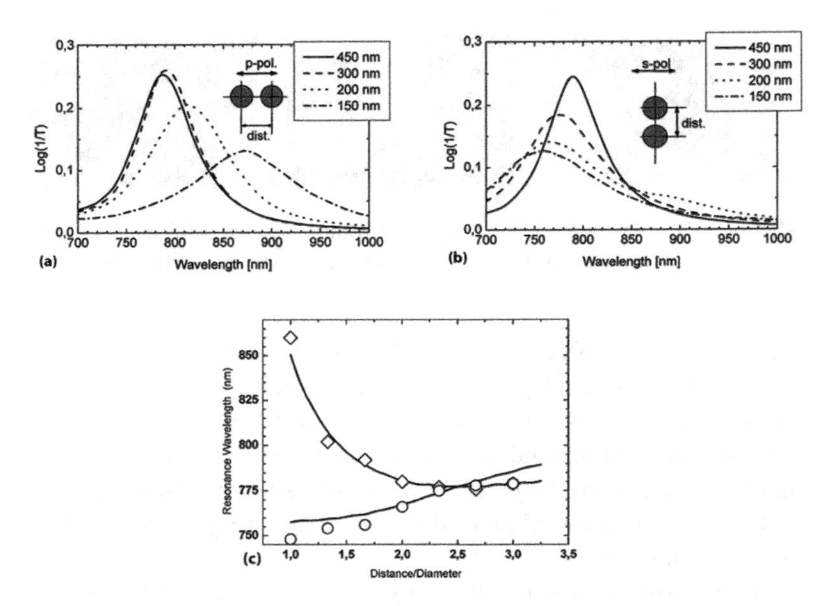

Figure 12.3 Extinction spectra of two-dimensional arrays of gold nanoparticles at different inter-particle spacing acquired with polarized beam probe (a) parallel to their inter-axes (p-pol) and (b) orthogonal to their inter-axes; (c) plasmon peak of extinction spectra versus inter-particle distance scaled with their diameters. Rhombus-shaped data are for parallel polarization, and circled ones are for orthogonal polarization.

12.4.2 *Pair of Ordered Nanodisc Array: Inter-Particle Plasmon Coupling and Universal Scaling Behavior*

To conduct this study [18], Jain et al. performed a complete optical characterization between pairs of gold nanodiscs, situated at different distances from each other. The nanostructures were fabricated on quartz slides by exploiting lithography techniques. Two-dimensional 80 μm × 80 μm arrays of 88 nm diameter gold nanodiscs with row spacing of 300 nm and 600 nm between particles within each row were fabricated. The edge-to-edge inter-distance between the nanodiscs is changed for every sample with values of 212, 27, 17, 12, 7, and 2 nm. Figure 12.5 reports the SEM image of the sample with a gap of 12 nm between the nanostructures.

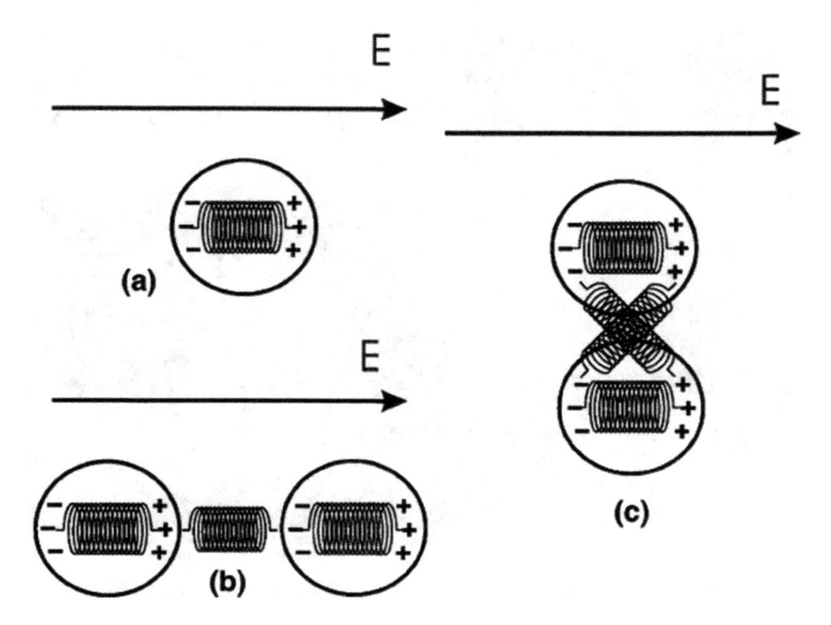

Figure 12.4 Illustration to give a qualitative explanation of interaction between the electromagnetic field and (a) single nanoparticle, (b) two nanoparticles and wave field parallel to the inter-axes, and (c) two nanoparticles and wave field orthogonal to the inter-axes between the nanostructures.

Also in this case, the exciting light utilized for the experiment has been polarized along the inter-axis between particles and orthogonal to it. The spectra acquired for the two polarizations (Fig. 12.6), with decreasing gap, are quite different: Under parallel polarization, there is a red-shift of the plasmon resonance wavelength, while under orthogonal polarization, the blue-shift is almost negligible. This behavior is very similar to the one observed in the previous section and can be interpreted on the basis of a dipole-dipole coupling model [29].

The dipole-dipole interaction is attractive for parallel polarization, while it is repulsive for the orthogonal one. The plot of the plasmon shift versus the inter-particle gap is reported in Fig. 12.6c. The data can be fit with an exponential decay function with a decay length of 15.5 ± 3.0 nm. Moreover, the authors realized a theoretical simulation of the plasmonic behavior of these nanodiscs with a

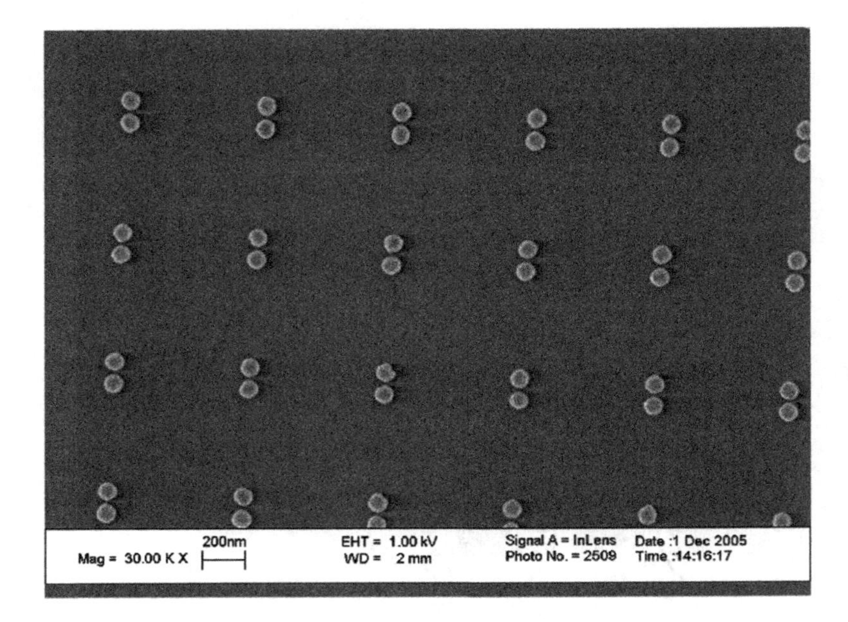

Figure 12.5 Array of nanodisc pair with 12 nm edge-to-edge gap of nanostructures. Each nanodisc has a diameter of 88 nm and a thickness of 25 nm.

DDA method. In the calculations, they considered a single couple of gold nanostructures because the distance between two neighboring dimers is much more than 2.5 diameters and then the dimers can be considered isolated. The calculated spectra for parallel and orthogonal polarization are shown in Fig. 12.7. The nanostructures were modelled as cylinders with a diameter of 86.5 nm and a height of 25.5 nm, with the cylinder axis normal to the surface. The simulation confirms a strong red-shift (while decreasing the gap between nanostructures), while a little blue-shift takes place and a small shoulder rises at 561 nm, as observed in the experimental spectra (Fig. 12.6).

Figure 12.7c shows the exponential decay of the shift that fits the simulated data. From this fit, the decay length was found to be 17.6 $\pm$ 2.5 nm, which is very close to the experimentally obtained decay length. In the simulation (Fig. 12.7a,b), the authors used the value 1.0 for the dielectric constant ε_m of the surrounding medium

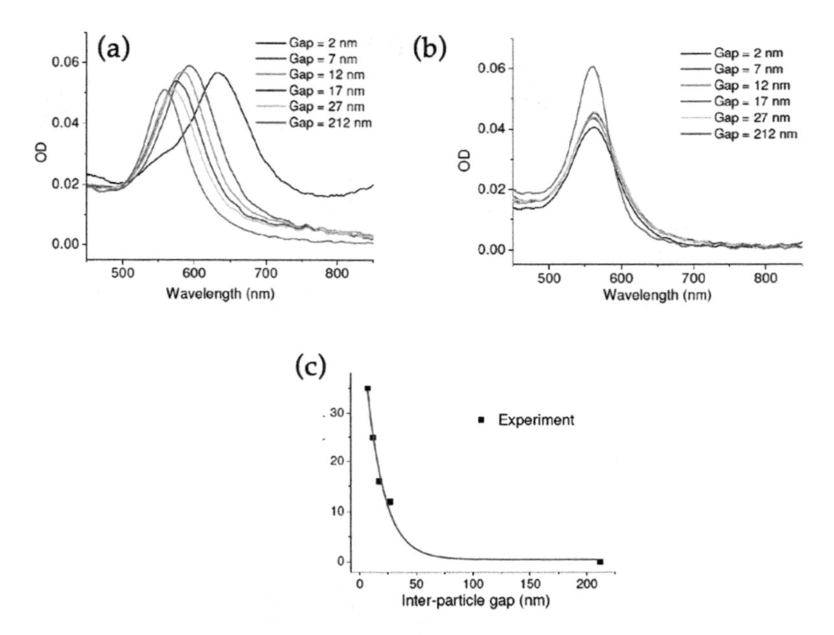

Figure 12.6 Experimental microabsorption spectra of different arrays with different gaps between nanodiscs: (a) spectra with parallel polarization with respect to the dimers' inter-axes, (b) spectra acquired with polarization orthogonal to the dimers' inter-axes, and (c) experimental plasmon shift plotted as a function of the inter-particle gap.

because they considered the nanostructures as surrounded by air. However, a more accurate treatment would require considering the dielectric behavior of the template, which is quartz covered with a very thin layer of chromium. Moreover, this theoretical approach assumes that the nanostructures are idealized as isotropic cylinders with homogeneous sizes and inter-particle gaps.

In the same study, again exploiting DDA simulations, the plasmon shift has been calculated as a function of the inter-particle distance (center to center) for discs of different diameters with a constant aspect ratio $h/D \approx 0.3$ and again a surrounding medium dielectric constant $\varepsilon_m = 1.0$. In the graph in Fig. 12.8, the plasmon shift versus the inter-particle distance for discs with sizes $D = 54$, 68, and 86 nm is shown. A double effect consisting of an increase in red-shift and

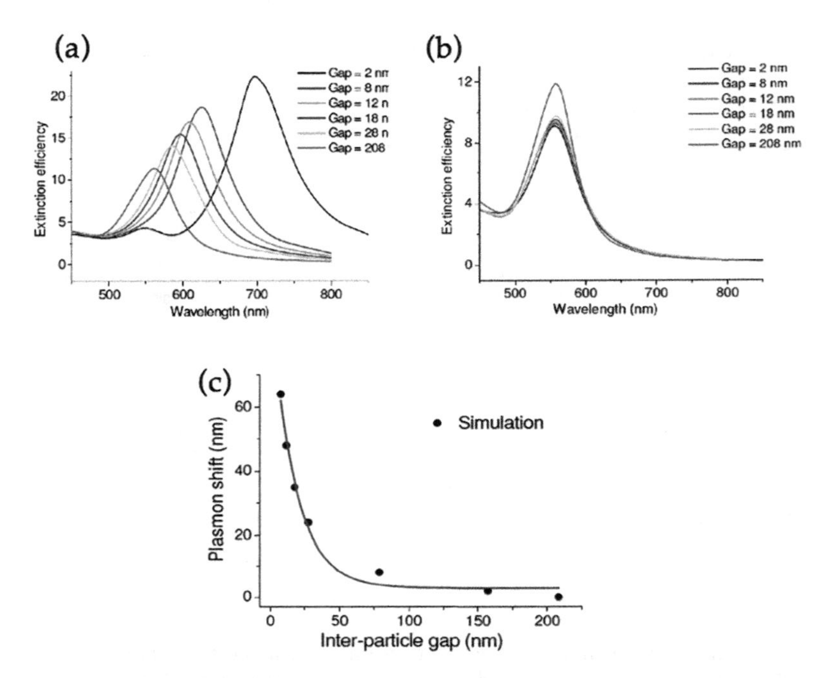

Figure 12.7 DDA-simulated extinction efficiency spectra of gold nanodisc pairs at different gaps between nanostructures with polarized beam probe (a) parallel and (b) orthogonal to the inter-axes between the nanostructures; (c) DDA-simulated plasmon shift versus the inter-particle gap for parallel polarization.

plasmon-coupling length can be observed only by decreasing the gap between nanostructures independently from their sizes (Fig. 12.8b).

Then, following the work of Su et al. [17], when both shift and gap (edge-to-edge distance) between these nanostructures are scaled, respectively, to the wavelength of un-coupled nanostructures and to the diameter, the decay becomes independent of the sizes of the nanostructures (Fig. 12.8b). Fitting these data with an exponential decay $y = a^{*}\exp(-x/\tau)$, they found a decay constant value $\tau = 0.23 \pm 0.03$ (Fig. 12.8b), which is well in agreement with the experimental value $\tau = 0.18 \pm 0.02$ (Fig. 12.8c). So independent of the sizes, the plasmon coupling decays over a length 0.2 times the diameter.

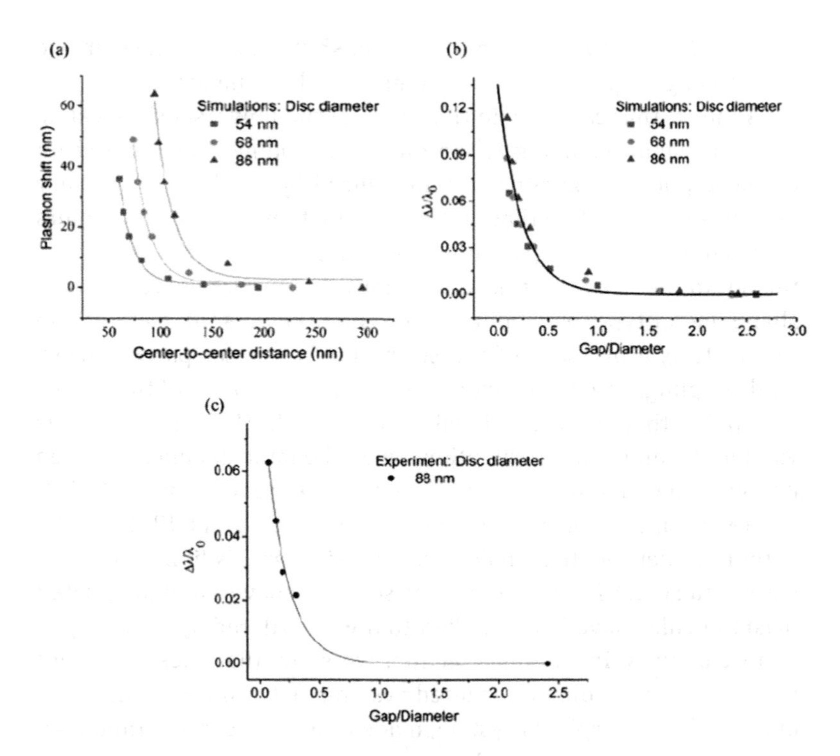

Figure 12.8 (a) Calculated plasmon shift for different sizes of nanodisc, (b) universal scaling behavior of these systems, (c) experimental scaling behavior for an 88 nm nanodisc.

12.4.3 *Plasmon Ruler Equation*

The fact that the same fitting curve verifies experimental data corresponding to particles of different size and shape has allowed the authors to introduce the concept of a universal scaling behavior of the plasmon coupling. Moreover, for this character of universality, the obtained empirical fitting equation has been called equation ruler. This equation is extremely useful because it allows to retrieve the distance between nanostructures from the experimental plasmon shift. The equation ruler can be used to fit experiments in which the exciting light has a polarization parallel to particles' inter-axes and is generally written in the exponential form:

$$\frac{\Delta\lambda}{\lambda_0} \approx 0.18 \exp\left(\frac{-(s/D)}{0.23}\right) \qquad (12.20)$$

where $\Delta\lambda/\lambda_0$ is the fractional plasmon shift, s is the inter-particle edge-to-edge separation, and D is the particle diameter.

A model that can explain the origin of the universal size-scaling of the plasmon coupling is the dipolar coupling model, based on dipole–dipole interactions and introduced by Kreibig and Vollmer [29]. It is worth noting that in the equation there are no parameters connected with their surrounding media. Thus the scaling behavior, related to the constant τ, is similar for different nanoparticle sizes, shape, metal type, or medium [18]. The relative shift $\Delta\lambda/\lambda_0$ is, in this picture, a measure of the strength of the inter-particle near-field coupling, and it is connected to the polarizability of the couple of dipoles that is proportional to (s/D) [3]. This description is valid in the dipole approximation, when the inter-distance between nanoparticles, and then the gap, is not smaller than the value $(0.5D)$ where the multipole interaction becomes important [23]. As the authors underline, this analysis is empirical and is limited to gaps larger than $0.1D$. In this section, the scaling behavior and the related plasmon ruler have been applied to a well-ordered system of gold nanostructures. In the next section, we show that these laws not necessarily apply only to ordered systems but can be verified also in case of an amorphous system, made of nanostructures randomly anchored on an elastic template.

12.5 Exponential Decay of Coupling for a Random Self-Assembled System of Gold Nanoparticles

12.5.1 *Introduction*

A simple method to control the coupling between plasmonic particles and the universal scaling behavior using a mechanical strain has been recently presented [30]. The large-scale samples have been prepared by first depositing and then further growing gold nanoparticles on a flexible PDMS tape. Upon stretching the tape, the particles move further apart in the direction of the stretching and closer together in the direction perpendicular to it. This leads to a drastic shift of the plasmon band and a color change of the sample. Furthermore, the plasmon shift is strongly polarization-dependent

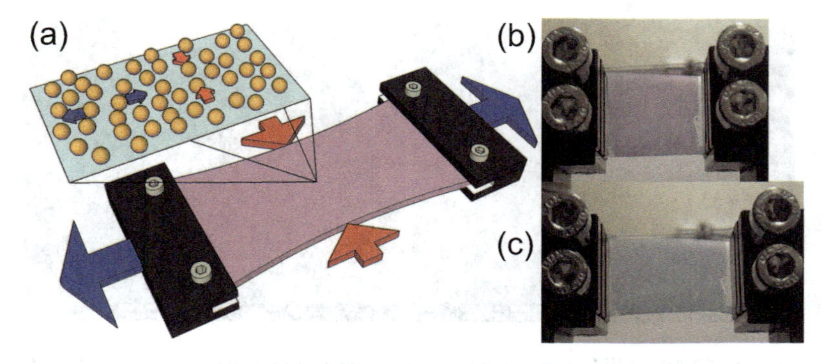

Figure 12.9 (a) Sketch of the experimental setup. Stretching of the sample is accompanied by a remarkable change of color from purple-red (b) to blue-violet (c).

for the amorphous and initially isotropic sample by stretching the sample by only a few percent. The observations are well described by a plasmon ruler equation that fits the scaling behavior of the system at rest and under stretching, indicating that a macroscopic mechanical strain with growing process of nanostructures allows one to control the coupling and, therefore, the electromagnetic field at the nanoscale.

12.5.2 *Mechanically Induced Plasmonic Coupling*

The idea at the basis of this work is sketched in Fig. 12.9a, whereas the real system is shown in Figs. 12.9b,c. As anticipated, upon stretching the sample, its color changes from magenta to blue-violet.

Samples have been prepared by using self-assembly techniques from nanochemistry. At the beginning, it functionalized a flexible polydimethylsiloxane (PDMS) substrate, so that gold nanoparticles are spontaneously adsorbed onto its surface in a charge-driven process. The resulting surface, Fig. 12.10a, is an amorphous arrangement of well-separated noninteracting particles (average diameter 23 nm) with an average nearest neighbor distance (center to center) of 39 nm. Afterward, this sample undergoes a chemically driven growth process for directly increasing the size of gold nanoparticles on-site [31–32]. The grown nanoparticles (Fig. 12.10b) reveal an average diameter of about 32 nm.

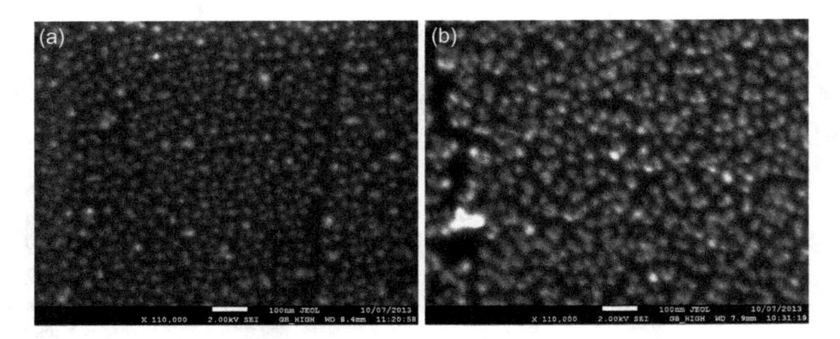

Figure 12.10 SEM micrographs of a gold-nanoparticle-coated PDMS substrate, taken (a) before and (b) after 12 growth cycles.

When solicited in elastic regime, the PDMS material has a Poisson ratio of about 0.5 (inset of Fig. 12.11a) [36]. As such, stretching the sample by 20% in one direction leads to a compression of 10% in the orthogonal direction. On a nanoscale, this should correspond to a slight reduction in the center-to-center distance between the particles: The 39 nm center-to-center distance drops down to about 3.9 nm, which corresponds to an important 56% change in the particle gap s, from 7 nm to 3.1 nm. The spectra acquired while stretching and then releasing the sample (Fig. 12.11a,b) show how this change in inter-particle distance induces a plasmonic coupling of the particles' near-fields. This coupling noticeably affects the position of the plasmon resonance peak of the system with a marked red-shift. In releasing the sample, it is possible to observe that the behavior is perfectly reversible (Fig. 12.11c).

To study the effect of the applied strain on the plasmonic behavior of the system, spectra after each growth cycle and at maximum stretching (Fig. 12.12a) have been compared with the corresponding spectra of the same samples at rest (Fig. 12.12b). Spectral positions of the maxima of the extinction band have been also reported for the same samples, both at rest and stretched (Fig. 12.12b, inset). In samples at rest, the growth of particles leads to a moderate shift of the localized surface plasmon resonance (triangles in the inset of Fig. 12.12b). When a strain is applied, from the ninth to the twelfth cycle, the stretching has a dramatic effect: An explosion of the plasmonic shift is observed.

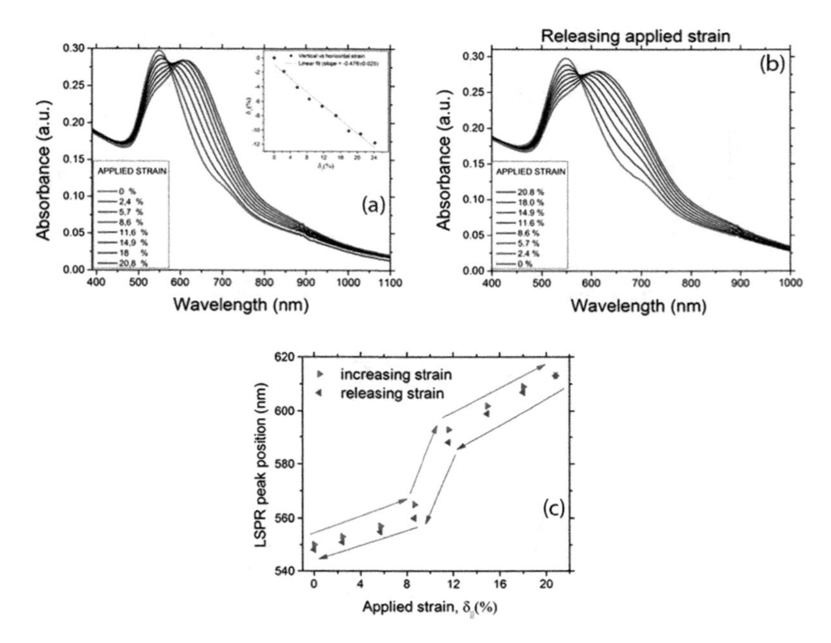

Figure 12.11 (a) Extinction spectra of the sample coated with gold nanoparticles (after 12 cycles of growth) acquired while increasing the applied strain from 0% to 20.8%. (b) Extinction spectra acquired by releasing the strain applied to the same sample from 20.8% to 0%. (c) Plasmon peaks as a function of applied strain during stretching and releasing: except a little hysteresis, the graph confirm the reversibility of process.

In Fig. 12.12c, the same analysis of spectral behavior but with polarization parallel to the applied strain at maximum stretching has been reported. Quite surprisingly, comparing Fig. 12.12a and Fig. 12.12c, these samples, which are amorphous and initially isotropic in the x–y plane, develop highly s-polarization-dependent optical properties after undergoing a stretching of about 20% only. For a more quantitative description of the phenomenon, it is convenient to evaluate the plasmonic shift normalized to the single-particle resonance wavelength ($\Delta\lambda/\lambda_0$) as a function of the normalized gap (s/D) between nanoparticles. To enable this calculation, a k-neighbors algorithm has been implemented, which exploits the SEM micrographs of samples before and after growing the particles

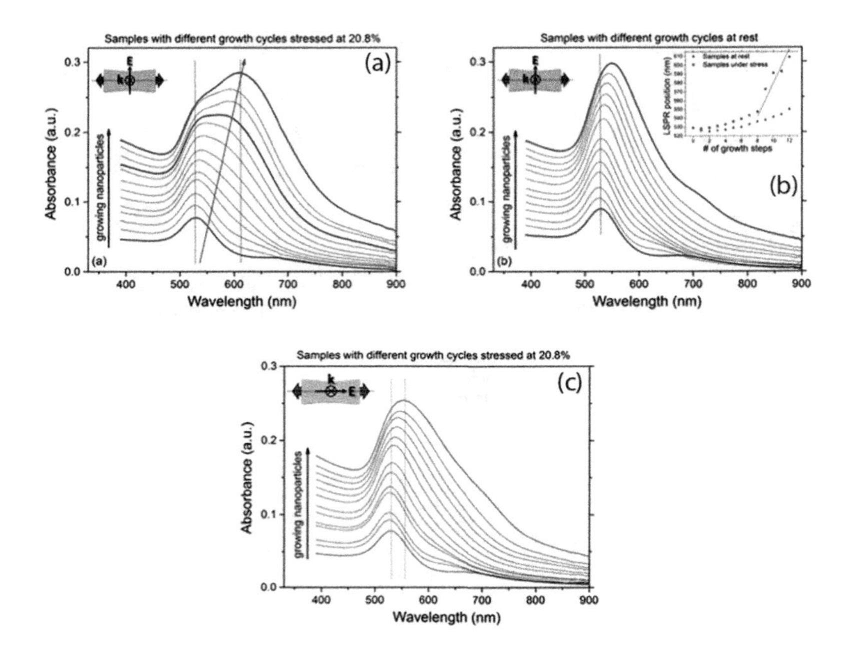

Figure 12.12 Extinction spectra of gold-nanoparticle-coated PDMS samples with different numbers of growth cycles of particles, in case that (a) an applied strain of 20.8% is applied and (b) samples are at rest. Samples are irradiated with s-polarized light. In the inset, the plasmon peak position is plotted as a function of the number of growth cycles, both for samples at rest (triangles) and under stress (stars). (c) Extinction spectra of sample irradiated with p-polarized light, at maximum stretching.

(Fig. 12.10a,b). The obtained experimental dependence of $\Delta\lambda/\lambda_0$ on s/D (Fig. 12.13b), for each step of growth and both for the sample at rest and under stretching, could be fit with the single exponential function (Eq. 12.21) (solid curve in Fig. 12.13b), with parameter values $k = 0.369 \pm 0.038$ and $\tau = 0.119 \pm 0.010$.

$$\frac{\Delta\lambda}{\lambda} = k^{*}\exp?\left(-\frac{s}{\tau D}\right) \qquad (12.21)$$

This proves that also this random and amorphous system verifies the conditions of applicability of scaling behavior and the plasmon ruler equation [17, 18, 23].

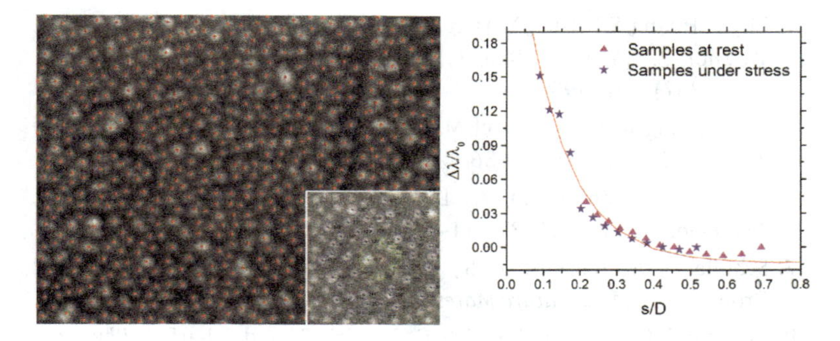

Figure 12.13 (a) Result of the image analysis performed on the SEM with MatLab code. (b) Universal scaling behavior of sample during the growth at rest (pink triangles) and under stretching (blue star) fitted with plasmon ruler equation.

12.6 Conclusion

In this chapter, we showed two kinds of plasmonic systems related to the different disposition of nanostructures (ordered and randomly distributed) above rigid and flexible templates. The authors, by studying the coupling between the near-fields of these nanostructures, obtained two main results: the first was a scaling behavior of coupling that confirmed the dependence of the phenomenon mainly from the gap between the nanostructures; and the second was the formulation of the empirical plasmon ruler equation that fits the scaling behavior and permits to calculate the inter-distance between nanostructures exploiting the decay of their coupling through the decay of the plasmon red-shift. The last considered work provides a further confirmation of the universality of the scaling behavior in conditions never considered before.

References

1. Mayer, K. M., and Hafner, J. H., *Chem. Rev.* 2011, **111**, 3828–3857.
2. Willets, K. A., and Van Duyne, R. P., *Ann. Rev. Phys. Chem.* 2007, **58**, 267–297.
3. Haes, A. J., Zou, S. L., Schatz, G. C., and Van Duyne, R. P., *J. Phys. Chem. B* 2004, **108**(1), 109–116.

4. Li, J. F., Huang, Y. F., Ding, Y., Yang, Z. L., Li, S. B., Zhou, X. S., Fan, F. R., Zhang, W., Zhou, Z. Y., Wu, D. Y., Ren, B., Wang, Z. L., and Tian, Z. Q., *Nature* 2010, **464**(7287), 392–395.
5. Liu, N., Tang, M. L., Hentschel, M., Giessen, H., and Alivisatos, A. P., *Nature Mater.* 2011, **10**(8), 631–636.
6. Mühlig, S., Cunningham, A., Dintinger, J., Scharf, T., and Bürgi, T., *Nanophotonics* 2013, **2**(3), 211–240.
7. Schuller, J. A., Barnard, E. S., Cai, W. S., Jun, Y. C., White, J. S., and Brongersma, M. L., *Nature Mater.* 2010, **9**(3), 193–204.
8. Rockstuhl, C., Lederer, F., Etrich, C., Pertsch, T., and Scharf, T., *Phys. Rev. Lett.* 2007, **99**(1), 017401.
9. Chen, M. W., Lang, X. Y., Qian, L. H., Guan, P. F., and Zi, J. A., *Appl. Phys. Lett.* 2011, **98**, 093701.
10. Pu, Y., Grange, R., Hsieh, C. L., and Psaltis, D., *Phys. Rev. Lett.* 2010, **104**, 207402.
11. Kennedy, D. C., Tay, L. L., Lyn, R. K., Rouleau, Y., Hulse, J., and Pezacki, J. P., *ACS Nano* 2009, **3**, 2329.
12. Bailo, E., and Deckert, V., *Chem. Soc. Rev.* 2008, **37**(5), 921–930.
13. Mie, G., *Ann. Phys.* 1908, **25**, 377.
14. Maier, S. A. 2007. *Plasmonics: Fundamentals and Applications*, Springer Science + Business Media LLC, ISBN 978-0387-33150-8.
15. Mock, J. J., Barbic, M., Smith, D. R., Schultz, D. A., and Schultz, S., *J. Chem. Phys.* 2002, **116**, 6755.
16. Rechberger, W., Hohenau, A., Leitner, A., Krenn, J. R., Lamprecht, B., and Aussenegg, F. R., *Opt. Commun.* 2003, **220**, 137–141.
17. Su, K. H., Wei, Q.-H., Zhang, X., Mock, J. J., Smith, D. R., and Schultz, S., *Nano Lett.* 2003, **3**, 1087–1090.
18. Jain, P. K., Huang, W., and El-Sayed, M. A., *Nano Lett.* 2007, **7**, 2080.
19. Huang, F., and Baumberg, J. J., *Nano Lett.*, 2010, **10**(5), 1787–1792.
20. Zhu, X. L., Xiao, S. S., Shi, L., Liu, X. H., Zi, J., Hansen, O., and Mortensen, N. A., *Opt. Exp.* 2012, **20**(5), 5237–5242.
21. Millyard, M. G., Huang, F. M., White, R., Spigone, E., Kivioja, J., and Baumberg, J. J., *Appl. Phys. Lett.* 2012, **100**(7).
22. Chiang, Y. L., Chen, C. W., Wang, C. H., Hsieh, C. Y., Chen, Y. T., Shih, H. Y., and Chen, Y. F., *Appl. Phys. Lett.* 2010, **96**(4).
23. Jain, P. K., and El-Sayed, M. A., *Chem. Phys. Lett.* 2010, **487**, 153–164.
24. Kinnan, M. K., Kachan, S. M., Simmons, C. K., and Chumanov, G., *J. Phys. Chem. C* 2009, **113**, 7079–7084.

25. Lee, K.-S., and El-Sayed, M. A., *J. Phys. Chem. B* 2006, **110**, 19220.
26. Reinhard, B. M., Siu, M., Agarwal, H., Alivisatos, A. P., and Liphardt, J., *Nano Lett.* 2005, **5**, 2246–2252.
27. Sih, B. C., and Wolf, M. O., *J. Phys. Chem. B* 2006, **110**, 22298–22301.
28. Malynych, S., and Chumanov, G., *J. Opt. A: Pure Appl. Opt.* 2006, 8, S144–S147.
29. Kreibig, U., and Volmer, V. 1995. *Optical Properties of Metallic Cluster*, Springer-Verlag, Berlin, Heidelberg.
30. Cataldi, U., Caputo, R., Kurylyak, Y., Klein, G., Chekini, M., Umeton, C., and Bürgi, T., *J. Mater. Chem. C* 2014, **2**, 7927–7933.
31. Brown, K. R., and Natan, M. J., *Langmuir* 1998, **14**(4), 726–728.
32. Brown, K. R., Lyon, L. A., Fox, A. P., Reiss, B. D., and Natan, M. J., *Chem. Mater.* 2000, **12**(2), 314–323.
33. Pritchard, R. H., Lava, P., Debruyne, D., and Terentjev, E. M., *Soft Matter* 2013, **9**(26), 6037–6045.

Index

9789814613002